Ethics and Clinical Neuroinnovation

Laura Weiss Roberts
Editor

Ethics and Clinical Neuroinnovation

Fundamentals, Stakeholders, Case Studies, and Emerging Issues

Springer

Editor
Laura Weiss Roberts
Department of Psychiatry and Behavioral Sciences
Stanford University School of Medicine
Palo Alto, CA, USA

ISBN 978-3-031-14338-0 ISBN 978-3-031-14339-7 (eBook)
https://doi.org/10.1007/978-3-031-14339-7

This Springer imprint is published by the registered company Springer Nature Switzerland AG
The registered company address is: Gewerbestrasse 11, 6330 Cham, Switzerland

Foreword

> At a gathering of brain scientists and philosophers, participants zeroed in on one portion of the world of worry about unbridled science called "neuroethics." It deals with the benefits and dangers of treating and manipulating our minds.
> William Safire (2002)[1]

Fascination with our minds, and ethical questions concerning them, can be traced back for millennia. The exploration of the physical brain as the source of the mind began in full force in seventeenth century England.[2] But "neuroethics," in its contemporary sense, was born in May 2002 at a conference in San Francisco sponsored by the Dana Foundation and co-hosted by Stanford University and the University of California at San Francisco.[3] At the same time, William Safire, former Nixon White House speechwriter and chairman of the board of the Dana Foundation popularized the use of the term "neuroethics" when he featured it in his New York Times column.

The past 20 years have seen great growth in the field of neuroethics, with the formation of an international scholarly society[4] in 2006, the subsequent creation of at least two scholarly journals[5] and the addition of neuroethics components to several national or regional projects or organizations.[6] They have also seen a rise in sustained grant funding for academic research into neuroethics. The BRAIN Initiative of the United States National Institutes of Health (NIH) began awarding

---

[1] William Safire, *The But-What-If Factor*, NY Times (May 16, 2002), available at https://www.nytimes.com/2002/05/16/opinion/the-but-what-if-factor.html.

[2] Carl Zimmer, Soul Made Flesh (Free Press, 2004)

[3] Neuroethics: Mapping the Field (ed. Steven J. Markus, The Dana Press, 2002)

[4] The International Neuroethics Society, https://www.neuroethicssociety.org.

[5] AJOB Neuroscience, https://www.tandfonline.com/toc/uabn20/12/4, and Neuroethics, https://www.springer.com/journal/12152.

[6] These include at least the Neuroethics Working Group of the NIH BRAIN® Initiative, https://braininitiative.nih.gov/about/neuroethics-working-group, the ethics components of the European Union's Human Brain Project, https://www.humanbrainproject.eu/en/social-ethical-reflective/about/, the International Brain Initiative, https://www.internationalbraininitiative.org, and the Global Neuroethics Summit, https://globalneuroethicssummit.com.

research grants for the study of neuroethics issues in 2016. This volume is one early result.

I generally view the role of a foreword in a book as akin to an appetizer at a meal, or, better, an amuse-bouche, a very small taste of what is to come. And so I will keep this foreword short and use it to do two things: First, I want to give you a sense of the book that lies in front of you. And second, I hope to try to place this book into its context in the field of neuroethics, both past and present.

The immediate source of this book was a 2017 NIH grant to Stanford University, entitled "Enabling Ethical Participation in Innovative Neuroscience on Mental Illness and Addiction: Towards a New Screening Tool Enhancing Informed Consent for Transformative Research on the Human Brain," with Professor Laura Weiss Roberts as the principal investigator.[7] But, in fact, as Chap. 12 makes clear, its genesis lies much deeper in the past, with work on informed consent done by co-investigator Laura Dunn in the early 2000s, joined by Laura Weiss Roberts in the early 2010s. Although the book covers much other useful and important ground, at its core is a synthesis of some of the fascinating work done under the BRAIN Initiative grant.

Ethics and Clinical Neuroinnovation comes in three parts. The first provides background information on mental illness, neuroscience, and neuroethics. The second looks in depth at several aspects of neuroethics and innovative neurotechnologies. And the third lays out the unprecedented work completed by Dr. Roberts' team under the BRAIN Initiative grant to understand better what the stakeholders in the innovative neurotechnologies—patients, neuroscience researchers, ethicists, and others—think about these issues.

The initial part contains six chapters. The first lays out, in painful numbers, the vast amount of human suffering created by psychiatric, addiction-related, co-occurring, and behavioral disorders. If you do not know of a friend, family member, or loved one whose life has been blighted by one (or more) of these conditions, just wait—you will. The second chapter is the longest in the book but one of the best, as it lays out the history and current status of neuroscience, neuroimaging, and other forms of neuroinnovation. It does so with impressive panache and some humor—"Transcranial Electrical Stimulation, which … directly stimulates cortical tissue with high voltage electric shocks to the scalp (it's as painful as it sounds)." The third chapter looks at the basic approaches of neuroethics and how they may apply to machine learning algorithms and brain–machine interfaces.

Chapter 4 looks at changes in the context of innovation, from the ubiquity of digital data and its problems, to blurred lines between clinical care, research, and commerce, and the growing impact of both patient and patient advocacy organizations and other non-scientist communities on innovation. Chapter 5 analyzes what makes innovation in the brain different from general medical innovation, with a focus on the brain's special role as the source of consciousness—people would be much less concerned about, say, gall bladder innovation. And the last chapter in

[7] The grant is described at https://reporter.nih.gov/project-details/9419223.

section one reviews the NIH's "Brain Research Through Advancing Innovative Neurotechnologies" or BRAIN® (yes, the acronym is trademarked) Initiative, its neuroethics component, and the grant from NIH that resulted in the project described in this book.

The second part of the book comprises five chapters that dive into neuroethics issues in particular settings. Chapter 7 looks at the ethics of neurostimulation via neurosurgery as a way to treat intractable, dangerous obesity. The eighth chapter focuses on the fascinating question of "covert consciousness": how neurotechnologies have been and may be used to detect consciousness in unresponsive patients and what we should worry about in those efforts. Chapter 9 examines the ethics of human studies with psychedelic drugs, their substantial promise, and their equally large challenges. The tenth chapter analyzes the criminal justice system's uses of neuroscience technologies, especially in three ways: looking back at the time of the crime, support for a clinical diagnosis and evidence to bolster a claim of diminished capacity, while the third looks at the present for immediate issues like a witness's truthfulness, the validity of eyewitness identification, and implicit biases. Chapter 11, the last chapter of this section, emphasizes how innovation is skipping over academic labs and happening directly in firms, and the implications of that shift.

Part three is the core of this book. It describes many of the results of the empirical research projects undertaken by its chapters' authors as part of their BRAIN Initiative neuroethics grant. Chapter 12, its initial chapter, describes the genesis and development of the project and its two main components. The first component, aim 1 of the grant, uses semi-structured interviews with stakeholders to identify what distinctive ethical questions are raised by innovative neuroscience research in mental illness and addiction. The second, aim 2, uses a large survey of possible research participants to seek to understand what affects decisions whether or not to participate in such research. The survey seeks to test and refine the Roberts Valence Model for Ethical Engagement in Research, a tool that members of the group have been building over several years.

The remaining chapters of part three further describe this work. The next four cover the semi-structured interviews, beginning with Chap. 13, which details who was asked what, and how (and, importantly, how the answers were coded for analysis). The three chapters that follow, Chaps. 14, 15, and 16, analyze the interviews with 44 professional stakeholders—neuroscience researchers, IRB members, and ethicists. They probe the stakeholders' views on, respectively, ethical considerations in innovative neuroscience research involving human participants; the contexts in which research occurs and the special effects those contexts have on psychiatry and neuroscience research; and clinical innovation in psychiatry and neuroscience. As far as I know, these chapters and the work behind them make up a unique resource for understanding how they are engaged in, or overseeing, neuroscience research and what they are doing. They will provide valuable insights to inform this kind of research going forward.

The final chapter deals with the survey aspect of the project. Chapter 17 is an interesting and enlightening look at Mechanical Turk (widely known as "MTurk"), the Amazon survey tool that, because of its ease and low cost, has become widespread in research, both academic and otherwise. As someone who has read much research using MTurk, I was delighted finally to understand how it works—and particularly taken by the ethical questions the chapter raises about MTurk itself. Chapter 17 is followed by an Appendix that sets out some of the survey results. These are not results from the full 1000-person survey planned, but from one pilot survey of 151 people. Although pilot studies only, they provide some novel and interesting findings, and leave me eager to read the results of the full survey.

This is a useful and interesting book, but how does it fit into today's neuroethics? And, at least as importantly, just what *is* neuroethics today?

It may be useful to look back two decades to William Safire's op-ed. In it he raises many possibilities: drugs to enhance memory or alertness, technical manipulation of memories, neuroscientific lie detection, combining our heads' "wetware" with computers, and "a kind of Botox for the brain to smooth out wrinkled temperaments." Neuroethics analyzed, and argued about, all of these issues and more for years, convinced that if they were not already reality, they soon would be. I wrote about most of them myself. But 20 years later, they remain hypotheticals—still intriguing and still unreal, or, at least, unrealized. Astounded by rapid neuroscience progress, particularly using functional magnetic resonance imaging (fMRI), we were too optimistic—or, from some perspectives, pessimistic—about what the future would bring, and how soon. (Interestingly, at the same time, two other high profile bioscience fields, genomics and stem cell research, created similarly inflated hopes and fears.)

All of Safire's issues may well yet come to pass, but it turns out we did not need FDA regulation of fMRI-based lie detection in 2005 in spite of an article I wrote that year urging it.[8] The tools we had were astounding and excellent at giving us a much better understanding of "the human brain" than ever before. But usually that understanding was of group averages, not of individual brains, and did not provide the detail needed to understand *your* brain or *mine*. In some ways, the big lesson of the last 20 years in neuroscience is that human brains are even more complicated than we imagined.

So, until the next, and better, generation of tools—the creation of which is the main goal of the BRAIN Initiative—neuroethics is more usefully deployed in questioning the tools that are closer to hand, and the research being done to improve them. ETHICS AND CLINICAL NEUROINNOVATION does just that. This kind of neuroethics is less likely to show up in headlines, or in nightmares, but it is, for now, much more useful—useful as one part of the morally compelling effort to relieve the vast human suffering caused by diseases of the brain…very much a "neuroethical" goal.

Henry T. Greely
Stanford Law School

[8] Henry T. Greely, *Premarket Approval Regulation for Lie Detection: An Idea Whose Time May Be Coming*, AM. J. BIOETHICS, 5(2):50–52 (March-April 2005)

Preface

Necessity is the mother of invention. Necessity inspires creativity and novel approaches to consequential challenges. Necessity, unfortunately, is also the mother of failure (when solutions do not exist), expedience and compromise (when resources are costly, out of reach, or insufficient), and of neglect and stigma (when needs are simply too overwhelming).

The needs experienced by people living with mental illness and addiction throughout history have been immense and, for the most part, unmet. Failure, expedience, compromise, neglect, and stigma have been common themes. In recent decades, however, these needs have been increasingly recognized by society and have become an inspiration for pioneers—pioneers in the neurosciences, clinical medicine, and health professions—policy makers, and scholars. Invention, creativity, and novel approaches related to brain disorders and brain health have brought along their companions, promise, hope, and compassion.

This book covers this rich array of issues, broadly conceived under the notion of neuroethics in relation to innovation for the purpose of advancing the health and well-being of people living with mental illnesses, including addiction. The book embraces existing scholarship and, more importantly, qualitative and early quantitative data drawn from stakeholders with vastly different experiences. The book embraces this complexity, with areas of commonality and diversity, congruence, and contradiction, in an effort to help illuminate ethically salient dimensions of neuroinnovation in society at this moment.

This moment is exceptional in that we are living in a time of technological advance, scientific brilliance, and accelerated impact of entrepreneurialism in society. We are living in a time of pandemic and heightened realization of the connections amongst all people, past, present, and future. And we are living in a time of dynamic societal attitudes that are rapidly consolidating based on a variety of influences, in which skepticism in science seems equal to the greater need for belief in science as a path toward health and a better future. As we note in Chapter 1, this book is intended to bring forward a variety of perspectives for deeper consideration. Many impressions shared in this text may be corrected and many new findings may emerge in the coming years that serve to reinforce or to revise the ideas presented here. Through this book, we intend to strengthen the foundations of neuroethics during a time of immense change.

I thank people with lived experience of illness for sharing their invaluable and often neglected insights to this book, I thank the research professionals and IRB members who spoke with us for their perspectives and expertise, and I thank my wonderful colleagues for their great work and partnership. My sincere thanks to the National Institute of Mental Health for funding this project; to our Program Officer James Churchill; to my colleagues at Springer, Richard Lansing, Diane Lamsback, and Anila Vijayan, for seeing the value in our proposal and publishing this unique book; to our intrepid contributors who wrote and revised chapters of this book even in the midst of a novel global pandemic; to Hank Greely for the foreword; to our stakeholders for sharing their words with us; and to my team, including Max Kasun, Gabriel Termuehlen, and especially Jodi Paik, MFA, who helped shepherd this project.

Palo Alto, CA, USA Laura Weiss Roberts

Contents

Part I
Foundations of Ethics in Clinical Neuroinnovation

Chapter 1
The Case for Neuroinnovation: Health Burdens Associated with Psychiatric, Addiction-Related, and Co-occurring Disorders

Laura Weiss Roberts and Katie Ryan

The Global Burden of Mental Illness and Addiction

Mental, neurological, and substance use disorders are a source of great personal suffering for hundreds of millions of individuals across the globe. These disorders—the causes of which are often a combination of genetic, environmental, biological, and societal factors—have historically been stigmatized, underfunded, and undertreated. As an increasingly globalized world has allowed for unprecedented connectivity and insights, the devastating consequences of mental illness and addiction on both personal and socioeconomic scales have become fully apparent.

People in every nation, community, and family are affected by the direct and indirect burdens of mental illness. One in five American adults experience some form of mental illness in any given year, while one in every 20 lives with a serious mental illness [1]. This pattern holds true for populations across the globe—it is estimated that one in four individuals globally will experience mental illness in their lifetime [2]. Over 12 billion working days are lost to mental illness every year, and mental illness is estimated to cost the world $16 trillion by 2030 [3].

The mental health repercussions of the SARS-CoV-2 pandemic are incalculable and far-reaching, with anticipated impact for generations [4]. Psychosocial dimensions of the pandemic include isolation, loneliness, grief, family disruption, and poor coping, including use of addictive substances [5]. People living with mental disorders experienced disproportionate burden of infection and diminished access to appropriate physical and mental health services [6–8]. The full spectrum of neuropsychiatric sequelae of viral infections of the brain is just beginning to be recognized, with heightened risk for mortality and enduring morbidity [9]. The

L. W. Roberts · K. Ryan (✉)
Department of Psychiatry and Behavioral Sciences, Stanford University School of Medicine, Stanford, CA, USA
e-mail: robertsl@stanford.edu; kryan2@stanford.edu

L. W. Roberts (ed.), *Ethics and Clinical Neuroinnovation*,
https://doi.org/10.1007/978-3-031-14339-7_1

superimposed effects of the pandemic atop the opioid and substance epidemics felt in multiple nations are immense [10].

Although the global disease burden for mental illness is often cited as accounting for around 23% of years lived with disability (YLDs) [11] and 7.4% of disability-adjusted life years (DALYs) [12], Vigo et al. [13] argue that these are vast underestimates. Using published data, they estimate that mental illness accounts for 32.4%—nearly one-third—of YLDs across the globe, and 13.0% of DALYs. Comprehensive pandemic-related data assessing mental health consequences have yet to be gathered. Given these updated estimates, mental illness is the resounding leading cause of global burden of disease in terms of YLDs and historically has been as debilitating as cardiovascular and circulatory disease when it comes to DALYs.

Unlike many physical illnesses which primarily affect older adults, the burden of mental illness and substance use disorders is unfortunately shared amongst individuals across the lifespan. Mental illness and substance use disorders account for 25% of all YLDs in children and youth and are the leading cause of disability in children and youth globally [14]. Furthermore, mental illness and addiction are responsible for 5.7% of DALYs amongst children and are the sixth leading cause of DALYs amongst children [14].

While economically established nations have seen improvements in the identification, prevention, and eradication of many communicable and non-communicable diseases, mental illness and addiction remain common and undertreated. Despite the prevalence of mental illness and addiction and its documented impact on overall health and quality of life, as many as two-thirds of people who live with a mental illness in the United States may not receive any form of treatment [15].

The Individual Burden of Mental Illness and Addiction

While the socioeconomic burden of mental illness and addiction is immense, the impact of these disorders at the individual level is equally overwhelming. Living with mental illness or addiction can impact nearly all aspects of one's life, from personal and familial relationships, to career prospects and opportunities, to physical well-being and health outcomes. In the United States, individuals who live with a mental illness or addiction are significantly more likely than the general population to experience homelessness at some point in their lifetime [16]. Unemployment is also more common among those with a mental illness or addiction [17]. People with severe mental illness are also more likely to suffer from a range of physical illnesses when compared to the general population [18] and experience excess mortality two to three times greater than the general population, leading to a shortened life expectancy by 10–25 years [19]. During the pandemic, though initially under-recognized, people with mental disorders including addiction were disproportionately affected by infections and experienced greater mortality than other individuals [20, 21].

Adding to this tragedy, many living with mental disorders or addiction are victimized for their conditions and become targets of stigma and discrimination. Approximately 60% of people express an unwillingness to work closely with a person with a severe mental illness or addiction, and a similar percentage of people believe that those with mental illness or addiction are violent [22]. In reality, individuals with severe mental illness are more likely to be the victims of violence than other community members [23]. This stigmatization and victimization can lead individuals with mental illness and addiction to self-stigmatize and can ultimately discourage them from seeking appropriate care and treatment [24]. The negative impact of isolation during the pandemic was felt greatly by people living with mental disorders, in part because their social networks and resources are intrinsically more fragile [25–28].

Beyond the facts and figures, the nature of these types of disorders qualitatively impacts the day-to-day life of individuals in ways that are very difficult, if not impossible, to measure quantitatively. By definition, mental disorders involve a decline in one's capacity to function well and with fulfillment and joy. It is thus unsurprising that mental, neurological, and substance use disorders are serious risk factors for premature death. Of people who commit suicide, 45% have a known mental health condition [29].

Progress and the Ongoing Burden

Over the past decades, efforts to reduce the burden of mental illness and addiction have been initiated at local, national, and international levels. In 2015, the United Nations General Assembly adopted the World Health Organization's Sustainable Development Goals (SDGs), which addressed global health targets for the upcoming 15 years. This adoption of the SDGs was the first time that world leaders promoted mental health as a health priority within the global development agenda and declared it an integral piece of sustainable development [30]. Beyond promoting the reduction of premature mortality through increased mental health care and treatment, the SDGs also targeted strengthening the prevention and treatment of substance abuse disorder.

At a national level, the United States' National Institute of Health launched the BRAIN Initiative in 2013, with the goal of developing innovative tools and technologies necessary to better understanding the structure and functioning of the brain. As of the start of 2020, the BRAIN Initiative has awarded over $1.3 billion to over 700 investigators. Similar initiatives have emerged across the globe, and in 2017, a Declaration of Intent for an International Brain Initiative was announced, with representation from Japan, Korea, Europe, the United States, Australia, Canada, and China [31].

At state and local levels, awareness of mental health issues has increased through community outreach programs, marketing campaigns, and the use of social media. In 2013, the state of California launched a large-scale social marketing campaign that was intended to reduce stigma surrounding mental health issues. Preliminary findings show that individuals who were mentally ill who were exposed to the campaign were more likely to receive treatment for their illness, and it was estimated that if all Californians with a mental illness had been exposed to the campaign, the number of those seeking treatment may have increased by one-third [32].

Although this progress is extremely promising for the future of treating mental illness and addiction, the known number of individuals living with serious mental illness continues to increase [2]. Due to the complexity of mental illness, treatments that are effective for some do not provide any benefit to others, and access to effective care and treatment often remains limited to those who do not have adequate resources or support. In the context of the pandemic, which led to nearly 5.5 million deaths world-wide as of December 2021, many health resources were redirected to respond to the overwhelming crisis of infection with the SARS-CoV-2 virus. In addition, many health care providers tragically died as a result of the pandemic, and the workforce was further diminished by physical and psychological risk, burnout, compassion fatigue, and exhaustion associated with prolonged and unrelenting effort and exposures in health care activities. This shift and reduction in resources have been felt greatly by people with chronic and co-occurring conditions, including many with mental disorders. While scientific progress in all medical fields is a slow, concerted effort, the complexity of the brain and the difficulty involved in accessing it create additional challenges that can further impede advancement in the fields of mental health, neuroscience, and psychiatry.

Relief Through Innovation

As researchers, doctors, governments, and individuals continue to gain a more nuanced understanding about how mental illness and substance abuse impact individuals, families, and communities, we turn toward increasingly innovative and novel research on these conditions in hopes that progress toward a healthier and less-burdened world is possible. Recent advancements in technology, computing power, and public understanding of mental illness and addiction have set the stage for major developments toward the understanding of the human brain and the treatment of various of major mental illnesses.

These advancements have occurred, and continue to occur, across all levels of psychiatry and neuroscience. For example, since its discovery in 2005, the field of optogenetics has flourished, leading to unprecedented discoveries about how clusters of individual neurons communicate [33] and how the brain changes after a stroke [34], in addition to allowing for more precise mapping of the brain [35]. Advancements in cloud computing and internet speeds have allowed for the development of open-source data-sharing databases such as OpenfMRI, which permit researchers from across the globe to share neuroscience data, with the goal of increasing data validity and replication in order to better address questions regarding human brain structure and function, and ultimately to better treat mental illnesses [36–38].

Advances in basic science and technology have additionally moved beyond the laboratory and into the lives of patients who suffer from mental illness and addiction. Deep brain stimulation (DBS), a neurosurgical procedure where an implantable pulse generator is placed directly against relevant structures in the brain, is approved as a treatment for advanced Parkinson's disease and dystonia and is currently being studied as a therapeutic intervention for obesity and obsessive-compulsive disorder [39]. Within the past decades, developments in the non-invasive procedure of transcranial magnetic stimulation (TMS) have allowed 30–40% of patients with treatment-resistant depression to experience remission of depressive symptoms, with fewer side effects than antidepressant medications [40, 41]. Certain specific types of TMS, administered in novel ways, have led to dramatic recovery in even very treatment-resistant individuals [42, 43]. The FDA approval of intranasal ketamine in 2019 has provided a similar cohort of patients with an opportunity to ameliorate their symptoms [44]. Innovation in telehealth and the use of algorithms and precision psychiatry strategies to identify individuals who would most benefit from intervention have led to scalable opportunities that are unprecedented in the field of mental health [45].

Neuroethics and the Foundation for this Book

There is great hope that, through continued innovation in neuroscience, the global burden of mental illness and addiction can be relieved. With this hope and advancement, however, it is important to recognize the unique and important circumstances of the people and populations affected by brain disorders with mental health, addiction, behavioral, and psychosocial dimensions.

Mental illness in particular "affects aspects of life that we define as fundamental to being human," and the treatment of mental illness "involves techniques that require exploration of intimate aspects of patients' lives and interventions that in some cases may limit the freedoms of patients" [46, p. 3–4]. These distinctive aspects of mental illness, and brain disorders more broadly, paired with the misunderstanding, isolation, and stigmatization that often come hand-in-hand, form a population of individuals who may be exceptionally vulnerable in research and medical contexts.

As research involving populations with mental illness and addiction continues to progress into more innovative and hopefully more beneficial realms, it is important to keep in mind concerns related to the nature of these illnesses, stigma, lack of resources, and public trust in research institutions and researchers. Investigation of the place of neuroinnovation and clinical neuroscience in society, including the ethical dimensions of these domains and safeguards that undergird public trust, is imperative.

Table 1.1 Examples of funded grant proposals related to research ethics led by Dr. Laura Weiss Roberts (Principal Investigator), 1997-present

Years	Grant proposal title	Funder
2018–2020	*Enabling ethical participation in innovative neuroscience on Alzheimer's Disease and Related Dementias* (administrative supplement to R01 MH114856)	National Institutes of Health
2017–2021	*Enabling ethical participation in innovative neuroscience on mental illness and addiction: towards a new screening tool enhancing informed consent for transformative research on the brain* (R01 MH114856)	National Institutes of Health
2014–2015	*Ethical implications of excluding the mentally ill from medical treatment research*[a]	Greenwall Foundation
2008–2010	*Research for a healthier tomorrow—program development fund*	A component of the advancing a healthier Wisconsin endowment at the Medical College of Wisconsin
2006–2012	*Ethics and safeguards in genetics research* (R01 MH074080)	National Institute of Mental Health and National Human Genome Research Institute
2004–2007	*Genetics and ethics: worker perspectives* (DE-FG02-04ER63772)	U.S. Department of Energy
2002–2004	*Barriers to care for rural runaway youth* (administrative supplement to DA013139)	National Institute on Drug Abuse
2000–2002	*Healthy, ill, and working individuals' perspectives on ethical, legal, and social implications in complex genetic disorders* (ER63018–2387)	U.S. Department of Energy
1999–2004	*Stigma and rurality: drug abuse, HIV/STD and mental illness* (R01 DA013139)	National Institute on Drug Abuse
1999–2004	*The ethics of psychiatric research: Science and safeguards* (K02 MH001918)	National Institute of Mental Health
1999–2002	*Informed consent and surrogate decision-making in schizophrenia: perspectives of patients and their families*	National Alliance for research on schizophrenia and depression
1997–2000	*Vulnerability and informed consent in clinical research* (R01 MH058102)	National Institute of Mental Health and National Institute on Drug Abuse

[a]Laura Weiss Roberts served as Co-Principal Investigator, Keith Humphreys served as Principal Investigator, Philip Lavori served as Co-Investigator

This central concern is the impetus for the research led by one of us (LWR) over decades (see Table 1.1) and represents the fundamental premise of this book on neuroinnovation and ethics. **By anticipating, eliciting, and addressing the ethical issues that may emerge alongside innovative research on conditions originating in or affecting the brain, public trust in clinical neuroscience and psychiatry can be strengthened. By being rigorous, honest, self-observing, and deeply connected to the ecology of neuroscience and psychiatry, we can work**

collaboratively with stakeholders across society to ensure that the greatest benefits possible can be reaped from scientific advancement and at the same time do our best to ensure that the greatest harms and risks are identified and avoided.

The intention of this book is thus to further understanding of the developing field of neuroethics, specifically in the context of innovation and scientific inquiry related to clinical neurosciences. This first section, *Foundations of Ethics in Clinical Neuroinnovation*, lays the groundwork for further discussion by exploring the historical, ethical, and contextual roots of the subject. Specific use cases of neuroinnovation, and the ethical issues they may reveal, are discussed in section two, *Special Topics in Clinical Neuroinnovation.* In section three, *Neuroethics and Innovation: Inquiry informed by the Roberts Valence Model,* we document our team's research into better understanding the ethically salient perspectives of various stakeholders involved in neuroinnovative projects.

The scope of this book is limited to foundational and special topics in clinical neuroinnovation and the framework and qualitative phase of our project on neuroethics funded by the National Institutes of Health BRAIN Initiative. We have also included an introduction to the quantitative work associated with the pilot portion of our project in an Appendix (see Appendix 1). The full quantitative findings of our project and our related competitive supplement project on Alzheimer's disease and innovation are beyond what is possible to cover in this book.

The editor (LWR), authors, and research team who have developed this book may not agree with everything that appears in the chapters that follow. And many impressions may be corrected and many facts may emerge in the coming years. This book documents a spectrum of views and findings. We consider this work to be a "snapshot" that captures many different viewpoints, including, very importantly, perspectives of people living with mental health concerns and addiction, investigators, ethicists, scholars, policymakers, and thought leaders, at this time. The content of this book is, by its nature, complex and newly emerging. Shared understanding, principles, and societal congruence regarding neuroethics does not yet exist but we hope that work, such as recorded here, will help create this new foundation. These chapters, and the varied perspectives and the data proffered, will help define an ethical framework for clinical neuroinnovation. Further elucidation of this framework is critical if the benefits of highly innovative neuroscience are to be realized.

Key Points

1. Mental disorders account for nearly one-third of years lived with disability (YLD) across the globe, and 13.0% of disability-adjusted life years (DALY), making them the resounding leading cause of global burden of disease in terms of YLDs and level with cardiovascular and circulatory disease when it comes to DALYs.
2. Especially in light of the mental health and neuropsychiatric impact, including regarding addiction, of the COVID-19 pandemic, these estimates are low and insufficient to capture the true impact of mental illness.

3. The need to better understand, prevent, and treat mental illness has gained traction politically, with specific commitments toward focus and funding from the United Nations, the World Health Organization, and, nationally, the U.S. Government's BRAIN Initiative.
4. Highly innovative neuroscience has great transformative potential in reducing the burden of mental illness and related disorders.
5. Ethical frameworks that specifically address innovative research in the context of neuroscience are fundamental requirements to fully realize the potential of clinical neuroinnovation.

Questions to Consider

1. How has globalization influenced our understanding of mental illness and its effects?
2. How does innovation in the realm of neuroscience compare to innovation in other sectors (technology, medicine, etc.)? Where are the ethical concerns similar and where might they diverge?
3. What distinctive characteristics of mental illness affect our understanding of ethics in research?

References

1. National Alliance on Mental Illness. Individuals with mental illness [Internet]. [Cited 2021 Aug 23]. https://www.nami.org/Your-Journey/Individuals-with-Mental-Illness.
2. United Nations Department of Economic and Social Affairs. Mental Health and Development. [Cited 2021 Aug 23]. https://www.un.org/development/desa/disabilities/issues/mental-health-and-development.html.
3. Patel V, Saxena S, Lund C, Thornicroft G, Baingana F, Bolton P, et al. The *lancet* commission on global mental health and sustainable development. Lancet. 2018;392(10157):1553–98.
4. Sanchez-Ramirez DC, Normand K, Zhaoyun Y, Torres-Castro R. Long-term impact of COVID-19: a systematic review of the literature and meta-analysis. Biomedicine. 2021;9(8):900.
5. Dubey S, Biswas P, Ghosh R, Chatterjee S, Dubey MJ, Chatterjee S, et al. Psychosocial impact of COVID-19. Diabetes Metab Syndr Clin Res Rev. 2020;14(5):779–88.
6. De Hert M, Mazereel V, Detraux J, Van Assche K. Prioritizing COVID-19 vaccination for people with severe mental illness. World Psychiatry. 2021;20(1):54.
7. Melamed OC, Hahn MK, Agarwal SM, Taylor VH, Mulsant BH, Selby P. Physical health among people with serious mental illness in the face of COVID-19: concerns and mitigation strategies. Gen Hosp Psychiatry. 2020;66:30–3.
8. Warren N, Kisely S, Siskind D. Maximizing the uptake of a COVID-19 vaccine in people with severe mental illness: a public health priority. JAMA Psychiat. 2021;78(6):589–90.
9. Yelin D, Wirtheim E, Vetter P, Kalil AC, Bruchfeld J, Runold M, et al. Long-term consequences of COVID-19: research needs. Lancet Infect Dis. 2020;20(10):1115–7.
10. Alexander GC, Stoller KB, Haffajee RL, Saloner B. An epidemic in the midst of a pandemic: opioid use disorder and COVID-19. Ann Intern Med. 2020;173(1):57–8.
11. Vos T, Flaxman AD, Naghavi M, Lozano R, Michaud C, Ezzati M, et al. Years lived with disability (YLDs) for 1160 sequelae of 289 diseases and injuries 1990-2010: a systematic analysis for the global burden of disease study 2010. Lancet. 2012;380(9859):2163–96.

12. Murray CJL, Vos T, Lozano R, Naghavi M, Flaxman AD, Michaud C, et al. Disability-adjusted life years (DALYs) for 291 diseases and injuries in: a systematic analysis for the global burden of disease study 2010. Lancet. 2012;380(9859):2197–223.
13. Vigo D, Thornicroft G, Atun R. Estimating the true global burden of mental illness. Lancet Psychiatry. 2016;3(2):171–8.
14. Erskine HE, Moffitt TE, Copeland WE, Costello EJ, Ferrari AJ, Patton G, et al. A heavy burden on young minds: the global burden of mental and substance use disorders in children and youth. Psychol Med. 2015;45(7):1551–63.
15. Kessler RC, Demler O, Frank RG, Olfson M, Pincus A, Walters EE, et al. Prevalence and treatment of mental disorders, 1990 to 2003. N Engl J Med. 2005;352(24):2515–23.
16. National Coalition for the Homeless. Mental illness and homelessness [internet]. 2009. [Cited 2021 Aug 27]. https://www.nationalhomeless.org/factsheets/Mental_Illness.pdf.
17. National Alliance on Mental Illness. Mental illness by the numbers [internet]. 2021. [Cited 2021 Aug 30]. https://nami.org/mhstats.
18. De Hert M, Correll CU, Bobes J, Cetkovich-Bakmas M, Cohen D, et al. Physical illness in patients with severe mental disorders. I. Prevalence, impact of medications and disparities in health care. World Psychiatry. 2011;10(1):52–77.
19. World Health Organization. Information sheet: premature death among people with severe mental disorders [internet]. [Cited 2021 Aug 30]. https://www.who.int/mental_health/management/info_sheet.pdf.
20. Mazereel V, Van Assche K, Detraux J, De Hert M. COVID-19 vaccination for people with severe mental illness: why, what, and how? Lancet Psychiatry. 2021;8(5):444–50.
21. Marsden J, Darke S, Hall W, Hickman M, Holmes J, Humphreys K, et al. Mitigating and learning from the impact of COVID-19 infection on addictive disorders. Addiction. 2020;115(6):1007–10.
22. Pescosolido BA, Martin JK, Long JS, Medina TR, Phelan JC, Link BG. "A disease like any other"? A decade of change in public reactions to schizophrenia, depression, and alcohol dependence. Am J Psychiatry. 2010;167(11):1321–30.
23. Latalova K, Kamaradova D, Prasko J. Violent victimization of adult patients with severe mental illness: a systematic review. Neuropsychiatr Dis Treat. 2014;10:1925–39.
24. Corrigan PW, Druss BG, Perlick DA. The impact of mental illness stigma on seeking and participating in mental health care. Psychol Sci Public Interest. 2014;15(2):37–70.
25. Druss BG. Addressing the COVID-19 pandemic in populations with serious mental illness. JAMA Psychiat. 2020;77(9):891–2.
26. Liebrenz M, Bhugra D, Buadze A, Schleifer R. Caring for persons in detention suffering with mental illness during the Covid-19 outbreak. Forensic Sci Int Mind Law. 2020;1:100013.
27. Hamada K, Fan X. The impact of COVID-19 on individuals living with serious mental illness. Schizophr Res. 2020;222:3.
28. Marroquín B, Vine V, Morgan R. Mental health during the COVID-19 pandemic: effects of stay-at-home policies, social distancing behavior, and social resources. Psychiatry Res. 2020;293:113419.
29. National Alliance on Mental Illness. Risk of suicide [internet]. 2019. [Cited 2021 Aug 30]. https://www.nami.org/About-Mental-Illness/Common-with-Mental-Illness/Risk-of-Suicide.
30. United Nations General Assembly. Resolution adopted by the General Assembly on 25 September 2015 [internet]. 2015. [Cited 2021 Aug 30]. https://www.un.org/en/development/desa/population/migration/generalassembly/docs/globalcompact/A_RES_70_1_E.pdf.
31. International Brain Initiative [Internet]. [Cited 2021 Aug 30]. https://www.internationalbrain-initiative.org/.
32. RAND Corporation. Mental health stigma reduction campaign increased use of services, study finds. *Medical Xpress*. 2019. [Cited 2021 Aug 30]. https://medicalxpress.com/news/2019-06-mental-health-stigma-reduction-campaign.html.
33. Lim DH, LeDue J, Mohajerani MH, Vanni MP, Murphy TH. Optogenetic approaches for functional mouse brain mapping. Front Neurosci. 2013;7:54.

34. Lim DH, LeDue J, Mohajerani MH, Murphy TH. Optogenetic mapping after stroke reveals network-wide scaling of functional connections and heterogeneous recovery of the peri-infarct. J Neurosci. 2014;34(49):16455–66.
35. Deisseroth K. Optogenetics: 10 years of microbial opsins in neuroscience. Nat Neurosci. 2015;18(9):1213–25.
36. Gorgolewski KJ, Wheeler K, Halchenko YO, Poline JB, Poldrack RA. The impact of shared data in neuroimaging: the case of OpenfMRI.org. Presented at the 21st Annual meeting of the Organization for Human Brain Mapping. 2015. [Cited 2021 Aug 30]. https://doi.org/10.7490/f1000research.1110040.1.
37. Ra P, Barch DM, Mitchell JP, Wager TD, Wagner AD. Toward open sharing of task-based fMRI data: the OpenfMRI project. Front Neuroinform. 2013;7:1–12.
38. Poldrack RA, Gorgolewski KJ. Making big data open: data sharing in neuroimaging. Nat Neurosci. 2014;17(11):1510–7.
39. Formolo DA, Gaspar JM, Melo HM, Eichwald T, Zepeda RJ, et al. Deep brain stimulation for obesity: a review and future directions. Front Neurosci. 2019;13:323.
40. Rizvi S, Khan AM. Use of Transcranial magnetic stimulation for depression. Cureus. 2019;11(5):e4736.
41. Cole EJ, Phillips AL, Bentzley BS, Stimpson KH, Nejad R, Barmak F, et al. Stanford neuromodulation therapy (SNT): a double-blind randomized controlled trial. American Journal of Psychiatry. 2022;179(2):132–41.
42. Li H, Cui L, Li J, Liu Y, Chen Y. Comparative efficacy and acceptability of neuromodulation procedures in the treatment of treatment-resistant depression: a network meta-analysis of randomized controlled trials. J Affect Disord. 2021;287:115–24.
43. Cosmo C, Zandvakili A, Petrosino NJ, Berlow YA, Philip NS. Repetitive Transcranial magnetic stimulation for treatment-resistant depression: recent critical advances in patient care. Curr Treat Options Psychiatry. 2021;8(2):47–63.
44. Lapidus KAB, Levitch CF, Perez AM, Brallier JW, Parides MK, et al. A randomized controlled trial of intranasal ketamine in major depressive disorder. Biol Psychiatry. 2014;76(12):970–6.
45. Williams LM, Ball TM, Kircos CL. Precision psychiatry. In: Roberts LW, editor. Textbook of psychiatry. 7th ed. Washington DC: American Psychiatric Association Publishing; 2019. p. 1016–42.
46. Roberts L. A clinical guide to psychiatric ethics. Arlington, VA: American Psychiatric Association Publishing; 2016.

Further Reading

Roberts L. A clinical guide to psychiatric ethics. Arlington, VA: American Psychiatric Association Publishing; 2016.

Chapter 2
Neuroinnovation in Medicine: History and Future

Octavio Choi

Neuroscience is currently in a golden age [1] made possible by the ever-accelerating pace of new tool development. On the one hand, advances in neuroimaging techniques such as diffusion tensor imaging (DTI) have enabled researchers to elucidate high-resolution wiring blueprints of the human brain [2]. On the other hand, the development of fundamental interventional tools such as optogenetics [3], deep brain stimulation (DBS) [4], and transcranial magnetic stimulation (TMS) [5] have allowed researchers to probe and modulate brain circuits with unprecedented precision. Increasingly, insights derived from basic research are being translated into clinical therapeutics. We are entering an era in which neuroinnovation-driven advances in knowledge of the brain are sophisticated enough to allow for development of effective, rationally designed treatments for a large and increasing number of psychiatric conditions (such as major depressive disorder (MDD) and obsessive-compulsive disorder (OCD)), giving rise to the new field of interventional psychiatry [6]. This has not always been the case.

A Historical Perspective

For most of history, the origin and causes of mental illnesses were unknown, and descriptions of mental illnesses were based on behavioral observations and subjective reports. A limited understanding of the neurobiological basis of mental disorders resulted in many individuals subjected to questionable treatments such as surgical frontal lobotomy [7]. Psychiatry lacked a neuroscientific

O. Choi (✉)
Department of Psychiatry and Behavioral Sciences, Stanford University School of Medicine, Stanford, CA, USA
e-mail: ochoi@stanford.edu

L. W. Roberts (ed.), *Ethics and Clinical Neuroinnovation*,
https://doi.org/10.1007/978-3-031-14339-7_2

foundation on which to appropriately diagnose and treat patients due to limited knowledge and insufficient tools to visualize, probe, and manipulate brain activity.

Things began to change in the twentieth century when innovations in neuroscience provided a framework for characterizing and treating mental illnesses. The development of the microscope led to pioneering work by Camillo Golgi and Santiago Ramon y Cajal, leading to the elucidation of the neuron as the fundamental unit of the nervous system [8]. Advances in biochemistry and electrophysiology helped characterize the chemical and electrical properties of neurons, establishing the molecular basis of neurotransmission. This in turn gave rise to the field of psychopharmacology and the development of modern psychiatric drugs. To this day, the vast majority of psychiatric treatments involve medications, such as selective serotonin reuptake inhibitors (SSRIs), whose fundamental mechanism of action appears to be modulation of neurotransmission at the synapse, although other theories have been proposed [9, 10].

As advanced and useful as psychotropic medications may be, one persistent problem has been the nonspecific distribution and action of such medications throughout the entire brain, leading to side effects. For example, most antipsychotic medications are thought to exert antipsychotic effects by blocking dopamine-2 (D2) receptors in areas of the brain responsible for cognition and perception but may also cause motor side effects (so-called extrapyramidal symptoms) by blockage of the same D2 receptors in the basal ganglia [11]. Another problem is treatment resistance; up to 30% of patients with major depressive disorder fail to remit with standard pharmaceutical interventions [12], indicating the need to develop alternative modalities of treatment.

Developing more precise and effective brain treatments required an increasing understanding of the neural basis of disease and the development of interventional tools to safely modulate brain activity. Prior to the advent of modern neuroimaging, establishing correlations between brain and behavior was slow, painstaking work. Neuroanatomists had long observed relationships between localized brain lesions and distinctive psychological and behavioral abnormalities (for example, Broca and Wernicke's work in the mid-nineteenth century [13]), but progress was slow due to the invasive nature of then-available analysis tools of autopsy and gross examination of the post-mortem brain.

The advent of noninvasive neuroimaging, first detailing brain structure, then elucidating brain activity, vastly accelerated the knowledge of human brain–behavior relationships, and with it our understanding of the neural basis of psychiatric and neurologic illness, setting the stage for the subsequent development of neuroinnovative treatments.

A Brief History of Neuroimaging

For much of the twentieth century, medicine actively sought search a noninvasive, high-resolution method to image the living human brain.

The discovery of X-rays by Wilhelm Roentgen in 1895 revolutionized medical imaging but unfortunately did little to shed light on brain structure, which remained a largely invisible "dark continent" [14]; X-ray technology at the time could not distinguish between different soft tissues. In 1918, the neurosurgeon Walter Dandy hit upon the idea of introducing contrast materials such as air into the ventricles (*air ventriculography*) of his patients, allowing for crude X-ray visualization of the ventricular system. Later, in 1927, the Portuguese neurologist Egaz Moniz pioneered and subsequently refined the technique of *cerebral angiography*, which allowed indirect visualization of brain structures via the introduction of contrast medium into the cerebral vasculature [14].

The development of computerized axial tomography (CAT) by Godfrey Hounsfield in the 1960s revolutionized brain imaging. Hounsfield's insight, based on principles developed by Alan Cormack, was that X-ray images taken from numerous angles ("axial") could be reconstructed by computer algorithms ("computed tomography") to generate three-dimensional images that could distinguish between various types of soft tissues. In 1968, he produced the first picture of a human brain (encased in lucite) that could distinguish gray matter from white matter [14]. Because of it's obvious potential, the British Medical Research Council helped fund the rapid development of a prototype that could scan a living human head. The first scan of a living patient was conducted on October 1, 1971 at Atkinson Morley's Hospital in London. Although the resulting brain image was crude by today's standards (the image was only 80 by 80 pixels), it was good enough to visualize a frontal brain tumor in the patient, which was promptly resected. Within 5 years, 475 CT scanners were in use in US hospitals, and by 1981 CT scanners were installed in 46% of all large hospitals in the US [14].

As impressive as CT brain scans were at the time, they could only visualize brain structure, not brain activity. Researchers soon realized, however, that principles of computed tomography could be applied to visualize the distribution of radioactive tracers injected into the brain's blood supply, and the positron emission tomography (PET) scan was born [15]. PET scans are based on the principle that radionuclide tracers injected into the bloodstream concentrate in areas of increased neural activity. Radionuclides, which are unstable, emit positrons as they spontaneously decay. These positrons travel an average distance of 2–3 mm before eventually colliding with an electron, resulting in mutual annihilation and the generation of a pair of gamma rays which are detected by an array of gamma ray detectors arranged around the head. By applying principles of computed tomography, a 3-dimensional image reflecting the spatial distribution of radionuclides can be reconstructed [16]. Depending on the radionuclide tracer used, different aspects of brain function can be measured and localized, such as oxygen consumption (using $^{15}O_2$), glucose utilization (using ^{18}F-deoxyglucose), and blood flow (using $H_2{}^{15}O$). Indeed, one of the great strengths of PET imaging is the large variety of radioactive tracers available which can quantitatively measure a large array of brain functions [17].

PET scans, however, suffer from several significant limitations. The spatial resolution of PET imaging is relatively poor due to the fact that emitted positrons travel an average of 2–3 mm from their source before colliding with an electron (the event which generates the gamma rays used for localization), thus limiting spatial resolution to typically 6–8 mm^3 voxels [16]. *Voxels* are three-dimensional pixels which comprise the basic "building blocks" of three-dimensional images; smaller voxels result in higher resolution images. In addition, the expense of PET scan machines and the need to have particle accelerators nearby to generate radionuclides with short half-lives limit the number of PET studies possible. Finally, while PET scans are noninvasive, they do involve the injection of radioactive materials, raising safety concerns for participants.

Functional Magnetic Resonance Imaging (fMRI)

The development of functional MRI largely circumvented the limitations of PET scanning, thus becoming the functional imaging modality of choice in the modern era. Machines capable of acquiring fMRI scans are widely available, as they are captured using the same machines that perform structural MRIs. Further, MRI scans do not involve the use of radioactive tracers, use magnetic fields which are considered safe, and routinely achieve spatial resolutions down to less than 1 mm^3 [18]. Depending on the strength of the main magnetic coil (stronger magnets produce higher resolution images), resolutions as fine as 0.1 mm may be theoretically achieved [19]. Structural MRIs are based on the principle that many nuclei, such as hydrogen ions, possess magnetic properties (*angular momentum*) which vary depending on their surrounding chemical environment. These magnetic properties can be probed by the application of strong magnetic fields and radiofrequency pulses, forming the basis of identification of chemical compounds by *nuclear magnetic resonance* (NMR) spectroscopy. In 1973, Paul Lauterbur and Peter Mansfield hit upon the idea of applying graded magnetic fields to localize NMR signals in space, forming the basis of *magnetic resonance imaging* [14]. The resulting MRI images could differentiate different types of biological matter (for example, cerebrospinal fluid, white matter, and gray matter) based on their differing magnetic properties [18].

Early attempts to measure brain activity with MRI focused on techniques to measure cerebral blood flow, taking advantage of the fact (established in earlier PET studies [20]) that regional blood flow and regional brain activity are highly correlated. The exact mechanism of this *cerebral autoregulation* is still unclear, but from a functional perspective, it appears to be based on the fact that neurons are entirely dependent on glucose as an energy source. Since the brain contains very limited glucose reserves, increased neural activity must be supported by an increased rate of delivery of glucose, which is accomplished by increased blood flow [21].

Initially, researchers injected magnetic contrast agents such as gadolinium into the bloodstream, which, by virtue of its sequestration in the intravascular space, could be imaged to measure localized cerebral blood volumes [22]. Using this technique, Belliveau and colleagues were able to map out human visual cortex using MRI by visualizing areas of increased blood flow in response to a flickering stimulus known to strongly drive activity in visual cortical neurons [23]. It was Ogawa and colleagues, however, who revolutionized functional MRI (fMRI) with the discovery of blood oxygenation level-dependent (BOLD) contrast [24]. In essence,

Ogawa and colleagues rediscovered Linus Pauling's original 1936 finding [25] that hemoglobin (the principal oxygen-carrying molecule in red blood cells) has slightly differing magnetic qualities when bound or unbound to oxygen. Ogawa serendipitously found that these differences could be visualized by MRI, enabling the creation of real-time maps of blood oxygenation levels in the brain without the need for contrast agents. Relative blood oxygenation levels (the basis of the BOLD signal) could then be used to infer regional brain activity (regions of the brain that "work hard" recruit more blood flow, raising regional blood oxygen levels). Soon after Ogawa's discovery, a slew of studies demonstrated the use of the BOLD signal to detect regional increase of neural activity, and the fMRI was born [26–29].

Distributed Processing and Functional Specialization

The advent of functional brain imaging laid to rest a long-running debate about the nature of brain computing, characterized at one extreme by *localists* such as Franz Joseph Gall, and on the other by *holists* such as Pierre Flourens. Gall first proposed in the early 1800s his theory that the mind arose from operations of the brain, with each mental faculty localizing in a 1:1 manner to a specific brain area. He identified at least 27 distinct regions which were purported to correspond to a wide range of behaviors and mental states such as generosity, secretiveness, and religiosity [30]. Gall's ideas led to the development of the (now) much-maligned field of *phrenology* (an extension of the then popular science theory of *physiognomy*) which postulated that a person's "character" could be determined by bumps and ridges on the skull, the idea being that mental faculties that were exercised would lead to growth of corresponding brain areas which could be detected by protrusions into overlying skull bone. Unfortunately, although many of Gall's ideas were prescient, they were not based on empirical data such as brain lesion studies, and in retrospect were naive and overly simplistic.

On the other hand, advocates of the *holistic* view of the brain, such as the physiologist Pierre Flourens, believed that brain computing was accomplished in a totally distributed manner, so that any part of the brain could perform any function, akin to the generic computer servers that comprise cloud computing. Flourens' theories (based on his work in the 1820s making focal lesions in animals) were carried into the twentieth century by advocates such as Karl Lashley, who noted in his experiments that rats who were given brain lesions and then had to learn to navigate a maze, appeared to have learning deficits that corresponded to the size of the lesion and not to the specific area of the lesion. Lashley concluded, in his theory of *mass action,* that it was the total mass of the brain that was important to accomplish mental functions, not specific brain areas [30].

Over time, however, converging evidence emerged favoring localist theories of brain function [30]. Broca and Wernicke's work on stroke patients in the mid-nineteenth century localized specific language deficits to specific areas of cortex (now referred to as *Broca's area* and *Wernicke's area*). Hughling Jackson's work on patients with focal epilepsy strongly suggested that motor and sensory functions were based on different areas of cortex. Painstaking work at the microscopic level by the anatomist Korbinian Brodmann elucidated at least 52 distinct brain areas (*Brodmann's areas*) distinguished by differences in cell morphology and spatial arrangement (*cytoarchitectonics*), supporting the idea that distinct cortical areas were specialized for distinct functions [31].

Meanwhile, electrophysiologists Gustav Fritsch and Eduard Hitzig demonstrated in 1870 that electrical stimulation in discrete areas of precentral gyrus in dogs caused characteristic limb movements—in effect, they discovered primary motor cortex and its topographical organization. Later, topographical maps of motor and somatosensory cortex were directly demonstrated in humans by neurosurgeons such as Wilder Penfield in the 1950s, who electrically stimulated discrete areas of cortex as part of functional mapping preceding epilepsy surgery [32]. Electrophysiological work by Hubel and Wiesel in the 1950s–1970s pushed mapping to the extreme, elucidating the exquisite retinotopic organization of cat visual cortex and its spatial segregation into ocular-dominance and orientation-selective columns [33].

The advent of functional brain imaging techniques revolutionized our understanding of how the human brain accomplishes tasks. Unlike lesion studies or electrophysiological stimulation studies, simultaneous activity patterns across the entire brain could be seen for the first time. Further, because functional brain imaging is noninvasive, extensive studies in humans became possible. The consensus model of brain function that has emerged from thousands of functional imaging studies is *distributed processing* [34], which integrates ideas from both localist and holistic paradigms. The distributed processing model of brain function acknowledges that brain areas are specialized for basic functions (*functional specialization* [35]) but extends localist models by positing that the brain accomplishes any given task by dynamically recruiting a set of localized/specialized cortical modules, which act in a coordinated, circuit-based fashion to produce the desired result. This is akin to how different apps on smartphones differentially activate computer chips, each specialized for various functions (GPS, graphics, sound, memory, etc.), which work as an ensemble to accomplish the tasks required by the app.

The distributed processing model explains both why lesions to specific areas can cause specific deficits, and why sometimes they do not—for example, some complex tasks such as speech generation rely heavily on specific cortical modules (e.g., Broca's area) which are critical to the task, while other tasks such as maze-learning are accomplished by a flexible network of modules with some redundancy, so that destruction of any one cortical module does not destroy the overall ability [36].

Key Neuroinnovation: Elucidation of the Human Connectome

Recent advances in neuroimaging techniques have allowed researchers to image connections between brain areas at a large scale, resulting in the elucidation of the *human connectome*, essentially a blueprint for the wiring diagram of the human brain, revealing the anatomical basis allowing for coordinated activities of distributed networks. This was a significant milestone in neuroscience, and the basis of a new approach in conceptualizing and treating brain disorders as "circuitopathies" [37].

Early attempts at elucidating brain connectivity were slow and laborious, involving the injection of radioactive, fluorescent, or viral tracers into specific brain areas in laboratory animals, waiting days to weeks for the tracer to diffuse down axonal pathways, then sacrificing the animal and visualizing pathways of tracer diffusion in brain slices, which could then be laboriously reconstructed to form a 3D image of a specific axonal pathway (for example, [38]). In a technical tour de force, investigators at the Allen Institute pushed this technique to the limit, generating a reasonably detailed whole-brain connectivity map of the mouse brain by injecting fluorescent viral tracers into hundreds of non-overlapping anatomical brain locations and reconstructing the resulting labeled fiber pathways in 3D [39].

Obviously, such tracer studies would not be possible in living human subjects. It is only recently that noninvasive neuroimaging techniques became refined enough to visualize brain connectivity in the intact human brain. One of these, *diffusion tensor imaging* (DTI), relies on the fact that while MRIs do not have the resolution to directly visualize axonal pathways, they do have the ability to track diffusion patterns of water molecules. Because water molecules in neurons are more likely to diffuse up and down the axon, rather than across it (a property referred to as *anisotropic diffusion*), tracking the diffusion of water molecules indirectly visualizes anatomical fiber tracts [2]. The other technique, *resting-state fMRI* (rs-fMRI), relies on statistical analysis of fMRI brain scans of subjects while at rest (in contrast to *task-based* fMRI, in which brain scans are recorded while subjects are engaged in a particular task of interest). Brain areas naturally fluctuate in activity over time, and by analyzing which brain areas fluctuate together (either correlated or anticorrelated), inferences can be drawn regarding functional connectivity [2].

Together, these two approaches, which map brain connectivity in complementary ways (DTI visualizing anatomical connectivity, rs-fMRI revealing functional connectivity), have formed the methodological basis of a large-scale, publicly funded (in part by the BRAIN initiative), multicenter collaborative effort known as the Human Connectome Project (HCP, http://www.humanconnectomeproject.org/), whose goal is to create highly accurate, high-resolution connectivity maps of the human brain based on thousands of high-quality brain scans [40]. At this point, the HCP has amassed data on over 1100 human subjects to form a brain connectivity map with unprecedented accuracy and resolution. The consortium has made all this data freely available to the research community, along with tools to help researchers navigate the data. This invaluable resource, a high-resolution map of brain connectivity, has played a vital role in advancing circuit-based understanding of psychiatric illnesses [36], opening up the potential for circuit-based approaches to treatment [37].

From Neuroinnovation to Neurotherapeutics for Depression

The increasing availability of neuroimaging tools in the 1980s and 1990s led to converging observations that informed initial models of the neural bases of major depression; these models then provided a road map, identifying potential targets of intervention by neuromodulatory tools such as DBS and TMS. Over time, two brain areas emerged as having particular significance in clinical depression: the prefrontal cortex (PFC) and subgenual cingulate cortex (sgCC).

Depression and Prefrontal Cortex Dysfunction

By the mid-1990s, a large body of literature had accumulated which convincingly associated depression with prefrontal cortex dysfunction (reviewed in [41–43]). Early structural brain MRI studies revealed that subjects with clinical depression have average smaller frontal lobe volumes [44]. Further, lesion-mapping studies in patients who developed depression after strokes [45–48], and multiple sclerosis [41, 42] strongly implicated the prefrontal lobes, with depression severity correlated with lesion burden in the left hemisphere [41, 42] and proximity to the left frontal pole [46, 47]. Although studies have been somewhat inconsistent in identifying a left vs right hemispheric preference for lesion locations leading to depression, taken as a whole, the research literature favors left hemisphere involvement over right (see [49] for recent meta-review).

Corroborating structural neuroimaging findings, functional neuroimaging studies consistently revealed that subjects with primary depression have lower prefrontal cortex activity compared with healthy controls (reviewed in [41, 42]), particularly on the left side [50–52]. Further, depression severity correlates with degree of prefrontal hypoactivity [50, 53, 54], and hypoactivity normalizes with recovery from depression, whether from antidepressant treatment [50, 51, 55], or placebo response [56]. Similarly, functional neuroimaging studies of subjects with secondary depression from psychiatric conditions such as OCD [50], or neurological conditions such as Parkinson's disease and Huntington's disease, revealed frontal lobe hypoactivity compared with non-depressed subjects with such conditions, irrespective of primary disease etiology [43].

Taken as a whole, while exceptions have been reported [53], both structural and functional brain studies have revealed that in general, clinically depressed subjects tend to have smaller and less active frontal lobes, particularly in the left prefrontal cortex, and that successful treatment of depression results in normalization of frontal lobe activity [41, 42, 50–52, 54–56].

Depression and the Subgenual Cingulate Cortex (sgCC)

In addition to prefrontal cortex dysfunction, similar lines of neuroimaging research comparing depressed and non-depressed patients revealed a central role for the subgenual cingulate cortex (sgCC), a key node of the brain linked to cortical, limbic, and paralimbic structures implicated in affective processes. Early volumetric studies reported a reduction in volume of the sgCC in those with clinical depression [57–59].

Functional studies strongly implicated sgCC activity with depression: the sgCC is activated by provoked sadness in healthy controls [56, 60–62] and by tryptophan depletion in remitted depressives [63] and in healthy controls [64]. The sgCC is hyperactive in subjects with clinical depression [65], particularly in those that later respond to treatment [66]. Further, depression severity correlates with sgCC activity levels [67], and recovery from clinical depression is associated with reduction of sgCC activity, whether by treatment with SSRIs [55, 68], placebo response [56], sleep deprivation [69], rTMS [52], or electroconvulsive therapy (ECT) [70].

Taken as a whole, based on its central anatomical connectivity to multiple brain areas relevant for emotional processing, hyperactivity in evoked sadness and in patients with clinical depression, and normalization associated with recovery from depression, the evidence appeared compelling that sgCC played an important role in the pathogenesis of depressive symptoms. Neuroimaging evidence is by nature only associative—were these neural patterns the result of depression, or the cause? The stage was set to test causality with a neuromodulatory intervention.

Deep Brain Stimulation for Depression

Helen Mayberg and colleagues were the first research group to attempt to treat severe depression with deep brain stimulation (DBS), extending the applications of this technology from neurological movement disorders such as essential tremor [71] and Parkinson's disease [72] to psychiatric diseases. DBS, which generally involves the delivery of high-frequency electrical pulses to an electrode surgically implanted into a targeted brain area, appears to effectively decrease neuronal activity at the target via multiple mechanisms that result in functional disruption of the target [73], creating what is essentially a "reversible lesion" while the stimulator is turned on [74].

In a seminal study reported in 2005 [4], Mayberg's group explored the effects of DBS stimulation at the sgCC (which they refer to as Brodmann's area 25, or BA25) of 6 patients with highly treatment-refractory depression (inclusion criteria required the failure of at least 4 prior antidepressant treatments; 5/6 patients had tried and failed prior ECT). Their results were striking and rapid; spontaneous intraoperative reports from patients during the placement and initial DBS tuning indicated an immediate effect of stimulation at this area, along the lines of "sudden calmness or lightness," increased interest, "disappearance of the void," and feelings of "connectedness." One week after implantation, 5/6 patients showed dramatic improvement in depressive symptoms, meeting criteria for response (>50% decrease in HDRS score). By the end of the trial at the 6-month time point, 4/6 patients had a sustained response to treatment (of these 4 responders, 2 had remitted).

Functional brain activation measured with FDG-PET revealed that prior to treatment, depressed subjects had increased sgCC activity and decreased dorsolateral prefrontal cortex (DLPFC) activity compared with healthy controls, similar to previously reported findings. After treatment, responders showed a reduction in sgCC activity (as expected, from direct effects of DBS stimulation) and revealingly, increased activity in remote sites such as the DLPFC—a pattern similar to depression recovery induced by SSRI antidepressants (Mayberg 2000). Although the study was not blinded nor placebo-controlled, placebo response appeared to be an unlikely explanation for clinical improvement, due to the treatment-refractory nature of the selected subjects (inclusion criteria required failure of at least four adequate antidepressant treatments), as well as the tight temporal relationship between symptomatic relief and DBS activation; for example, during intraoperative DBS placement, stimulation was delivered in a blinded and varying fashion, and behavioral improvements were noted to be time-locked to stimulation parameters. In addition, blinded discontinuation of DBS stimulation after 6 months in one subject who achieved sustained remission resulted in clear return of depressive symptoms within a month.

In a follow-up study, Mayberg's group [75] expanded the number of total subjects to 20 patients with treatment-resistant depression and reported similar results. Six months after DBS implantation into the sgCC, patients showed a 60% response rate (35% remission), and responders maintained benefits at 12-month follow-up; long-term follow-up studies of responders reported maintenance of benefits 6 years post-implant [76]. Extension of studies to bipolar depressed patients reported similar promising results [77], with sustained benefits for at least 8 years post-implant [78]. In comparison, results from the Sequenced Treatment Alternatives to Relieve Depression (STAR*D) trial indicate a 13% remission rate for antidepressant regimens in depressed patients who had failed to respond to three prior antidepressant treatments, with a relapse rate of 71% after 1 year of treatment [79].

The stage was set for a large, multicenter, blinded placebo-controlled trial for DBS, targeting sgCC to treat depression, dubbed the BROADEN (Broadman Area 25 Deep Brain Neuromodulation) trial [80]. Fifteen medical centers were involved in recruiting a planned 200 patients with treatment-resistant depression and tracking outcomes for a year after implantation. Unfortunately, when blinding was broken at the 6-month mark to conduct a futility analysis (90 patients had been enrolled by that point), no statistical differences were observed between sham (20% response rate) and active stimulation (22% response rate) groups, and enrollment of new patients was discontinued. Regardless, the existing implanted subjects continued to be followed in an open-label phase (with all patients receiving active stimulation) to the 2-year mark.

The BROADEN trial was widely deemed a failure and cast a chill over further DBS research. Still, Mayberg's group persevered, drilling into their data to explain the discrepancy between positive results from their early open-label studies and the apparent failure of the BROADEN randomized, placebo-controlled trial. Using connectomic tools such as DTI and rs-fMRI, Mayberg and colleagues demonstrated the utility of individualized targeting, thus advancing psychiatry into an era of personalized precision-medicine, akin to modern cancer treatment [81].

Previous positive results from open-label DBS studies did not appear likely to be due to placebo effects. As mentioned earlier, the patients enrolled were highly treatment-resistant, and improvements seen with active stimulation were acute and time-locked to stimulation during intraoperative test sessions. One could literally see individual patients' mood states changing on a minute-to-minute basis as stimulation was turned on and off [4]. In addition, blinded discontinuation of stimulation in responders [77] and naturalistic discontinuation (e.g., from battery failure) [78] consistently resulted in prompt relapse. Intriguingly, even in the BROADEN study, patients who continued into the open-label phase (with all patients receiving active stimulation at that point) continued to improve over the follow-up period, with 49% meeting criteria for response (26% remission) at the end of 2 years. To give these numbers context, it is useful to keep in mind that the patients enrolled in the BROADEN trial were highly treatment-refractory, having failed an average of eight adequate antidepressant treatments, including 82% with a prior trial of ECT.

What factors explained why some patients responded to DBS and why some did not? One obvious factor to examine was DBS placement. Did nonresponders receive DBS electrodes in the wrong place? Initial studies examining simple anatomical DBS lead placement did not differentiate responders from nonresponders [82]. Brain activation patterns from early pilot studies provided an initial clue: while both responders and nonresponders showed decreased sgCC activity (as expected from direct effects of DBS stimulation at that site), responders showed a broader pattern of activity changes at remote sites [4]. This suggested that even when the sgCC is accurately targeted using standardized anatomical methods (e.g., structural MRI guidance), individual variability in placement of the electrode *within* the sgCC could result in differential activation of connected circuits that could, in theory, explain response vs non-response.

In subsequent studies, Mayberg et al. [83] confirmed this hypothesis using a combination of high-resolution DTI and voltage field modeling to create probabilistic activation maps at the individual patient level which accurately differentiated responders from nonresponders (see Fig. 2.1). High-resolution DTI imaging was collected on 16 patients receiving DBS for depression and followed for 2 years; of

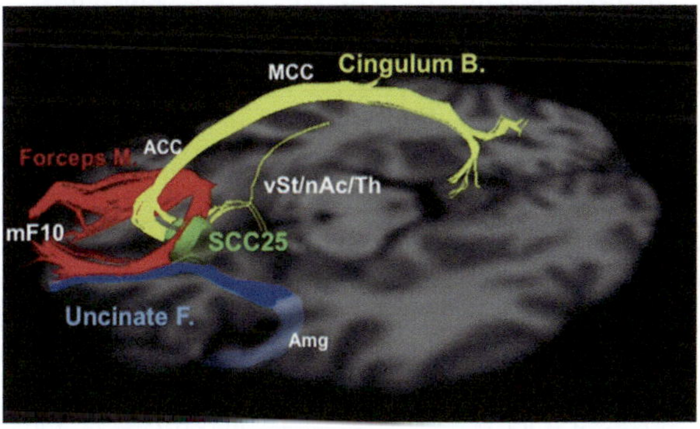

Fig. 2.1 Optimal subcallosal cingulate deep brain stimulation fiber bundle target template. Red: forceps minor. Blue: uncinate fasciculus. Yellow: cingulate bundle. *ACC* anterior cingulate cortex, *Amg* amygdala, *Cingulum B.* cingulum bundle, *Forceps M.* forceps minor, *MCC* middle cingulate cortex, *mF10* medial frontal (Brodmann area 10), *nAc* nucleus accumbens, *SCC25* subcallosal cingulate cortex (Brodmann area 25), *Th* thalamus, *Uncinate F.* uncinate fasciculus, *vSt* ventral striatum. From Riva-Posse P, Choi KS, Holtzheimer PE, et al. Defining critical white matter pathways mediating successful subcallosal cingulate deep brain stimulation for treatment-resistant depression. Biol Psychiatry. 2014;76(12):963–969. Used with permission from Elsevier

these 16 patients, 12 were responders at the 2-year mark, and 6 were not. DTI imaging was used to create circuit diagrams emanating from each individual's sgCC, with 4 distinct white matter tracts identified:

1. Uncinate fasciculus, connecting sgCC to ipsilateral mF10 (medial aspect of Brodmann's area 10),
2. Forceps minor, connecting sgCC to bilateral mF10,
3. Cingulum bundle, connecting sgCC to remote cingulate cortical areas such as the ACC (anterior cingulate cortex) and MCC (middle cingulate cortex),
4. Descending fronto-striatal fibers, connecting sgCC to deep brain structures such as the ventral striatum (containing the nucleus accumbens) and hypothalamus.

Similar to previous studies [82], simple inspection of anatomical localization of DBS leads did not differentiate responders from nonresponders; all appeared to have correct placement of the DBS lead into the sgCC. By using voltage field modeling to visualize the spread of electrical current from the DBS lead, combined with individual DTI connectivity information, researchers were able to create activation maps predicting which remote brain areas would be activated by the DBS lead in each patient; the researchers dubbed this technique "patient-specific activation volume tractography." Responders could be differentiated from nonresponders by virtue of activation of multiple circuits (and consequent neuromodulation of connected brain areas, such as mF10 and nucleus accumbens) that were not observed in nonresponders. In other words, because of its intricate circuitry, tiny changes in electrode placement within the sgCC could result in vastly different patterns of brain activation at remote sites, which appeared to explain why some patients responded and others did not.

Intriguingly, subsequent work monitoring awake depressed patients during intraoperative placement and stimulation of DBS leads [84] correlated specific types of symptom improvement with activation of specific circuits within the sgCC. Activating the cingulum bundle resulted in improvements in somatic symptoms ("I feel lighter" "the tension is gone"), whereas additional activation of mF10 (via uncinate fasciculus and forceps minor) and ventral striatum resulted in improved motivation and feelings of connectedness to others ("exteroceptive awareness"). This approach has recently been extended by Scangos and colleagues, who reported a case study of a highly depressed patient simultaneously implanted with deep brain electrodes in multiple brain areas thought to be involved in affect regulation, including the orbitofrontal cortex (OFC), amygdala, hippocampus, ventral striatum, and the sgCC bilaterally. By stimulating each area individually and monitoring patient feedback on changes in affective symptoms, researchers were able to map out distinct changes in affective symptoms by brain area. For example, OFC stimulation appeared to have an anxiolytic effect, while stimulation of the ventral striatum appeared to be activating. Further, whether such effects were interpreted as pleasurable or not depended on the mood state of the subject at the time; the anxiolytic effect of OFC stimulation felt pleasurable when the patient was in an anxious/agitated state of mind, but unpleasurable when the patient was in an anergic state [85].

Illustrating the power of their connectomic approach, in one of their most recent studies [86] Mayberg and colleagues demonstrated that precise individualized targeting using DTI imaging and voltage field modeling to guide optimal DBS lead placement to the exact location within that individual's sgCC (the location that intersected all four aforementioned fiber bundles), resulted in dramatically superior outcomes compared to previous conventional anatomic targeting approaches—an 82% response (55% remission) rate after 1 year of stimulation . This result is yet to be replicated in broad, placebo-controlled randomized trials—thus some caution is warranted in interpreting these initial findings—but nevertheless appears to be a promising path forward.

Mayberg's initial results launched a flurry of research activity investigating DBS stimulation of other targets for depression, such as the ventral striatum [87], nucleus accumbens [88], and medial forebrain bundle [89], with promising initial results, albeit in small non-blinded pilot studies thus far. Researchers are also investigating deep brain stimulation at other targets for psychiatric conditions with well-known circuitry such as obsessive-compulsive disorder (OCD) [90, 91] and addiction [92].

Currently, most DBS systems deliver constant, uniform stimulation, with occasional changes made manually by clinicians, e.g., to optimize treatment parameters. The development of *closed loop* systems, however, will allow future DBS implementations to make adaptive changes continuously. Such auto-sensing/auto-adapting systems are made possible by the fact that modern electrodes can both sense brain activity and deliver stimulation, as well as by advances in computational methods that can accurately analyze and interpret incoming signals. A closed loop system would be able to monitor brain activity to sense when stimulation is needed and deliver stimulation only during those times (for example, to abort an oncoming seizure [93]) or to continuously adapt stimulation parameters (for example, to optimize tremor reduction in patients with Parkinson's disease [93]). In Scangos et al.'s study [85], for example, OFC stimulation was deemed pleasurable when the patient was in an activated/agitated state of mind, but not when in a low-energy, anhedonic state. A closed loop system could in theory monitor the patient's affective state and deliver stimulation only when needed. [85] Of course, such continuous monitoring of brain activity also raises issues of privacy of mind, particularly as advances in computational machine-learning approaches allow for ever more accurate mind-reading from brain activity [94, 95].

Mayberg's work provided the first proof of principle that direct, focused neuromodulation could effectively treat a psychiatric condition. As a neuromodulatory intervention, DBS is able to target deep brain structures with exquisite temporal and spatial precision. Because of its inherently invasive nature, however, adverse events are common; the BROADEN trial reported 28/90 participants (31%) experienced at least one serious adverse effect [80]. Additionally, the need for extremely precise targeting using individualized, high-resolution connectomic approaches [86] is likely to be cost-prohibitive for all but the most serious, treatment-refractory patients. Fortunately, in recent years, noninvasive neuromodulatory technologies have made significant advances. We turn our attention now to one of the most promising, transcranial magnetic stimulation (TMS).

TMS: A Ground-Breaking, Noninvasive Neuromodulatory Treatment for Depression

Transcranial magnetic stimulation (TMS) was developed in its modern form by Anthony Barker, who first reported its use in 1985 [5] to directly stimulate motor cortex in humans, eliciting contralateral movements consistent with known topographic representation of body parts in primary motor cortex (the so-called motor homunculus [96]). Unlike its predecessor, transcranial electrical stimulation (TES) [97], which directly stimulates cortical tissue with high voltage electric shocks to the scalp (it's as painful as it sounds), TMS is non-painful and well tolerated.

TMS takes advantage of principles of electromagnetic induction—the same principles that make wireless recharging of electric toothbrushes possible—to allow for noninvasive stimulation of the brain. As the name implies, TMS machines generate brief intense pulses of electrical current which are delivered through a coiled loop of wire placed on the subject's head. Pulsed currents create fluctuating magnetic fields perpendicular to the coil (Ampere's law) that can freely pass through the skull and scalp, inducing the generation of electric fields (Faraday's law) within the brain, which, if powerful enough, can depolarize neuronal membranes and trigger action potentials. The size and magnitude of the induced electric field are dependent on the strength of the electrical pulse and coil geometry, but in all cases drops with increasing distance from the coil [98]. In practice, modern TMS machines and coil configurations can easily stimulate superficial cortical areas such as motor cortex (e.g., 2–3 cm from the surface of the scalp), but stimulation of deeper areas requires stronger currents and/or larger coils which increase the volume of brain stimulated (thus less focal; this is the so-called depth-focality tradeoff) and increases the risk for seizure induction [99].

Initial TMS studies focused on single-pulse TMS' ability to transiently excite or inhibit cortical areas (depending on the intensity of stimulation) in a safe, reversible, and noninvasive manner, allowing testing in healthy humans. Single-pulse TMS has been used to functionally map cortical areas (for example, motor cortex [5] and visual cortex [100]), to measure cortical excitability, and perhaps most significantly, to create reversible functional lesions that could finally test theories of causality in human subjects (reviewed in [101]).

Researchers studying repetitive sequences of TMS pulses (repetitive TMS or rTMS) discovered that, depending on stimulation frequency, rTMS could either enhance or inhibit cortical excitability for a period of time after the stimulation period [101]. High-frequency stimulation (5 Hz and above) applied to motor cortex appeared to facilitate cortical excitability [102], whereas low-frequency stimulation (e.g., 1 Hz) appeared to suppress it [103]. Such effects lasted on the order of minutes to hours after the end of stimulation and appear to be the result of cellular learning processes such as long-term potentiation (LTP) or long-term depression (LTD) (reviewed in [104]).

In summary, rTMS appeared to be a safe, noninvasive tool to focally modulate cortical activity in humans whose effects lasted beyond the period of stimulation, opening the door to therapeutic applications. As reviewed above, the left prefrontal cortex appeared to be an appealing target to treat depression, as this area of the brain was consistently found to be underactive in clinically depressed patients, and accessible to the magnetic fields generated by TMS coils. Could increasing the activity of this area with rTMS treat depression?

Mark George and colleagues were the first to demonstrate the utility of rTMS in treating depression in humans [60, 61]. Six subjects with treatment-resistant clinical depression received daily sessions high-frequency (20 Hz) TMS pulses (600 pulses per session) delivered to the left DLPFC for at least 5 days. Subjects showed significant averaged improvements in mood as measured by Hamilton Depression Rating Scores (HDRS), with 2 exhibiting robust responses (1 subject achieving remission for the first time in 3 years). Importantly, rTMS was also well tolerated; the main reported side effect was mild headache, which was treated effectively with over-the-counter NSAIDs. A flurry of other clinical trials followed, most with positive results (reviewed in [105]), albeit all non-blinded small studies.

The large, randomized, placebo-controlled trials that followed—one sponsored by industry (the "Neuronetics" trial) [106], the other funded by the NIH (the "OPT-TMS" trial) [107]—confirmed the effectiveness of high-frequency rTMS as a treatment for depression, with active treatment groups responding at 2–3 times the rate of sham groups, and with significant overall reductions in standardized measures of depression severity. Based on data from the O'Reardon 2007 study, the FDA approved rTMS in 2008 "for the treatment of Major Depressive Disorder in adult patients who have failed to achieve satisfactory improvement from one prior antidepressant medication at or above the minimal effective dose and duration in the current episode."

In a significant way, rTMS appeared to be safe and well tolerated, with low dropout rates (around 5% in both studies) due to side effects. The most common side effects were minor headaches and scalp tenderness (TMS coil firing contracts underlying scalp muscles) which generally resolved after the first week of treatment. Unlike electroconvulsive therapy (ECT), rTMS for depression did not cause cognitive impairment, and in fact multiple studies indicate cognitive improvement in depressed patients as well as healthy controls, presumably due to frontal lobe activation [108, 109].

Comprehensive meta-reviews indicate that the most serious adverse event caused by TMS appears to be seizure induction, but incidence is exceedingly rare, conservatively estimated to be 1:30,000 sessions [110]. As modern TMS protocols typically involve around 30 sessions for a treatment course, this translates to roughly 1 seizure per 1000 patients treated. In comparison, the estimated seizure risk from citalopram, based on meta-analyses of FDA phase II and phase III clinical trials, is roughly 1:300 patients, and bupropion SR is 1:1000 [111].

Further, careful analysis of the 16 cases of TMS-induced seizures reported in the medical literature up to 2009 [112] indicates that most events involved patients with known seizure risk factors, such as a seizure history, recent sleep deprivation, recent cessation of heavy alcohol use, or the use of medications that lower seizure threshold. In addition, based on review of clinical details, some reported cases were, in retrospect, likely vasovagal syncopal events rather than seizures (stimulation of the DLPFC reliably leads to parasympathetic activation, presumably due to connections with the vagal nerve [113]).

Since the publication of the industry-sponsored and NIH-sponsored RCTs, large naturalistic studies have been conducted which verified the effectiveness of rTMS for depression in "real-world" clinical settings. The largest naturalistic study, by Carpenter et al. [114], enrolled 307 patients with unipolar depression from 42 clinics, and reported response and remission rates of 58% and 37.1%. This is an impressive result, considering the treatment-resistant nature of the enrolled patients (averaging 2.5 previous adequate antidepressant trials). Regarding durability of benefits, a long-term follow-up to the Carpenter 2017 study reported that 62.5% of subjects who had responded to TMS treatment in that study maintained their gains after 12 months [115], in line with other long-term follow-up studies to OPT-TMS [116] and Neuronetics [117] RCTs.

In comparison, results from the Star*D study indicate that remission rates for subjects who have failed 2 antidepressant treatments and were being trialed for a third medication regimen (step 3) were 14%, with a 65% relapse rate after 1 year [79]. For subjects with 3 prior failed treatments (step 4), the chance of remission with a fourth medication was 13%, with a 71% relapse rate after 1 year [79]. With regard to effect size, large meta-analyses confirm that TMS is a highly effective treatment for depression, with effect sizes reported between 0.39 and 0.81, comparing favorably to mean effect size of 0.31 reported for antidepressant medication treatment in FDA trials (reviewed in [118]).

The Story of Targeting: Optimizing Coil Placement

TMS depression research has not stood still since receiving FDA approval in 2008. One of the most interesting stories to emerge from the recent literature concerns optimal targeting of the TMS coil. Early studies by George [119] and subsequent pivotal trials [106, 107] used a simple "5 cm rule" to attempt placement of the coil within the DLPFC: in each patient, the area of the left motor cortex controlling movement of the right abductor pollicis brevis was located and the coil was then moved 5 cm anteriorly along the curvature of the scalp. While this rule was easy to apply, it did not take into account considerable intersubject variability in skull size and prefrontal anatomy. As a result, rigorous neuroimaging analyses of cortical areas localized with the 5 cm rule found that this rule actually missed DLFPC more than half the time [120, 121], perhaps contributing to suboptimal response to TMS treatment.

This led to the rise of alternative targeting measures to localize DLPFC, such as localization of the electroencephalography F3 coordinate (based on the international 10–20 system), which better accounts for head size and thus more reliably targets DLPFC [122]. A variety of other investigators[123] have attempted to target DLPFC using neuronavigation relative to cortical anatomical landmarks or areas identified by functional studies (e.g., targeting the cortical area activated in working memory tasks [124]. Although neuronavigation and the use of EEG landmarks lead to better consistency in targeting compared with the 5 cm rule, few studies have compared depression outcomes with different targeting methods head-to-head, resulting in a lack of consensus regarding which target/targeting method is clinically superior.

One exception is Fitzgerald's 2009 study which directly compared depression outcomes for 51 patients treated with rTMS targeted either by the 5 cm rule ($n = 24$) vs a neuronavigation approach ($n = 27$) based on cortical anatomical landmarks (in this study, investigators targeted a particular location near the junction of BA 9 and 46). Fitzgerald found that patients targeted with neuronavigation had significantly better outcomes than patients targeted with the 5 cm rule, as measured by MADRS scores over 4 weeks. Further, comparison of cortical sites targeted with the two methods revealed that the neuronavigation site chosen in the study was on average 2–3 cm anterior to the average "5 cm rule" cortical location. In other words, it was possible that rTMS was being delivered in a non-optimal location in the majority of studies, which used the 5 cm rule to target, resulting in stimulation of cortical areas posterior to the DLPFC.

Other studies have examined clinical depression outcomes with regard to TMS coil position targeted with the 5 cm rule. As mentioned previously, because of inter-subject variability (e.g., varying head size), use of the 5 cm rule results in wide variation of cortical targeting relative to the DLPFC. Herbsman [125] took advantage of this natural variation to conduct a post-hoc analysis of treatment outcomes with variations in coil targeting ascertained using the 5 cm rule. He found, consistent with previous studies [121] that the 5 cm rule frequently missed DLPFC (in this study, only 68% of targeted regions were within DLPFC), often targeting cortical areas posterior and superior. Comparison of cortical targets of responders vs. non-responders revealed a consistent pattern: responders had coil targeting, on average, more anterior and lateral to nonresponders. This was consistent with Fitzgerald's finding of superior outcomes associated with cortical targeting anterior to the average 5 cm rule target.

Fitzgerald's and Herbsman's studies revealed that moving the cortical target anteriorly and laterally to the average 5 cm rule target led to better depression treatment outcomes but could not explain why. Fortunately, the advent of connectome mapping provided a method for Fox et al. [123] to elucidate a unifying theory. Similar to Mayberg's work analyzing the intricate circuitry within the sgCC, Fox and colleagues consulted a normative connectome dataset to reveal differential connectivity patterns within the DLPFC. By reanalyzing targeting and outcomes data in previous TMS studies by Herbsman [125], Fitzgerald [126], and Paillère Martinot et al. [127], Fox found that better treatment outcomes were associated with cortical locations that were more strongly anticorrelated to the sgCC. In other words, stimulation of cortical areas within the DLPFC that more strongly deactivated sgCC led to better antidepressant response. Using connectome data, Fox then generated a map of cortical areas functionally connected to the sgCC to identify an average optimal left DLPFC target that was maximally anticorrelated with the sgCC.

Fox's findings neatly complement results from Mayberg's DBS studies, which both converge on the central importance of the sgCC in regulating mood. As discussed above, Mayberg and colleagues found that DBS-induced suppression of the sgCC led to strong antidepressant effects, but only if properly targeted to subregions within the sgCC that were connected to remote cortical areas. Coming from the other direction, Fox's findings implied that the antidepressant effects of rTMS delivered to DLPFC were mediated by its ability to suppress sgCC activity. In other words, from the perspective of successful TMS treatment, the DLPFC could be regarded as an accessible "node" in a mood-modulating brain circuit involving the sgCC.

Fox's findings were prospectively validated in a recent rTMS study correlating depression outcomes with cortical targeting [128]. Again, more effective cortical targets were found to be more strongly anticorrelated with sgCC activity. Intriguingly, subgenual connectivity was associated with improvement in some symptoms of depression, such as sadness and anhedonia, but not with others, such as irritability, appetite, or fatigue, raising the question of whether distinct mood-regulating circuits exist that mediate different aspects of the clinical phenomenon of major depression. Weigand's findings are reminiscent of findings by Mayberg's group [84] and Scangos' group [85], correlating stimulation of specific subcircuits in the brain to distinct profiles of symptom improvement. For example, in Choi et al's study [84], activating the cingulum bundle resulted in improvements in somatic symptoms ("I feel lighter"), whereas activating mF10 (via uncinate fasciculus and forceps minor) and ventral striatum resulted in improved feelings of motivation.

Major depression has long been viewed as a heterogeneous clinical syndrome, based on early pioneering studies that established several clinical subtypes such as atypical, melancholic, seasonal and agitated (reviewed in [129]), raising the question of whether different subtypes of depression are caused by different kinds of underlying brain disturbances. The advent of connectome-based analysis has greatly accelerated our understanding of the contribution of different neural subcircuits to distinct depression symptoms by elucidating underlying circuitry and allowing analysis of coordinated whole-brain patterns of activity [130]. For example, the cingulum bundle, a white matter tract connecting brain areas that comprise the "default mode network" [131] is overactive in many subjects with depression and appears to mediate symptoms of negative rumination and pessimism. In contrast, abnormalities in the medial forebrain bundle, which connects cortical areas such as the DLPFC to limbic reward structures such as the nucleus accumbens, appear to mediate symptoms of anhedonia (reviewed in [130]).

TMS studies have confirmed and extended theories regarding neural bases of depression subtypes. For example, Downar et al. [132] treated 47 depressed subjects with rTMS (in this study, TMS was targeted to the dorsomedial prefrontal cortex) and analyzed outcomes with resting-state fMRI connectivity maps. Consistent with previous studies, rTMS was highly effective in treating depression, with a 51% response rate and 43% remission rate. In comparing responders to nonresponders, they found that nonresponders could be distinguished from responders clinically by the presence of anhedonia; the degree of anhedonia present before TMS treatment was a strong negative predictor of response. Further, analysis of resting-state connectivity revealed that nonresponders had significantly less functional connectivity of cortical areas to limbic reward areas such as ventral striatum.

Subsequent work by Drysdale et al. [133] extended neural biotyping to the extreme. Researchers analyzed resting-state functional connectivity maps in 1188 subjects with and without major depression and applied sophisticated machine-learning algorithms to discern 4 neurophysiological subtypes (which they dubbed "biotypes") of depression, each with a distinct neural signature, clinical-symptom profiles, and predicted targets for responsiveness to rTMS. For example, their models correctly predicted that patients in their study with biotype 1 depression (corresponding to the depression without anhedonia group in Downar's study) would have the best response to rTMS targeted to the DMPFC, compared to biotypes of depression with high levels of anhedonia, replicating Downar's finding.

The implications of these studies are clear: different types of depression are mediated by different patterns of neural dysfunction, which may require different TMS targets to optimally treat. Downar et al. [132] established that rTMS to the DMPFC was effective in treating depression with preserved hedonic function, but not depression with anhedonia. Fox et al. [123] established that targeting TMS to the subregion of the DLPFC most anticorrelated with sgCC activity maximized effectiveness in treating subjects with major depression, but subsequent detailed analysis by Weigand et al. [128] revealed that improvements were symptom-specific; for example, improvements were seen in symptoms such as sadness and anhedonia, but not in irritability, appetite, or fatigue. Could there be another TMS target best suited to treat the latter symptoms?

For the clinician, these results raise a vexing question: for any given patient, how should we match TMS coil placement with depression symptom profile? Recent work by Siddiqi [134] has started to address this question. Siddiqi and colleagues analyzed results of 109 subjects with major depression who were treated with rTMS targeted with the 5-cm rule, with depression severity measured over time by standardized measures such as the Beck Depression Inventory (BDI) and the Hamilton Depression Rating Scale (HAM-D).

As mentioned earlier, because of variability in brain size and cortical organization from person to person, the 5-cm rule generates a wide range of targeting relative to the DLPFC. Siddiqi took advantage of such targeting variability in order to discern whether certain locations are better than others in treating specific depressive symptoms. First, cortical targets were discerned for each patient receiving TMS, and magnetic coil field modeling was used to determine the extent of cortical tissue stimulated by the TMS coil at the target. This information was then plugged into a large normative connectome database in order to generate whole-brain maps of cortex stimulated by TMS at each site ("connectivity maps"). For each subject, activity of each voxel in the connectivity map was then compared with change in each depression symptom measured, and correlations graphed as a colored heat map for each symptom (the researchers dubbed these "symptom response maps"). Symptom response maps were then averaged across all subjects. Essentially what these maps revealed was the average degree to which stimulation at any given site in the brain was correlated (or anticorrelated) with change in a specific depression symptom in the study.

Remarkably, Siddiqi and colleagues found that in the 20 or so symptom response maps generated (one for each depressive symptom measured), many of the maps looked similar to each other and appeared to cluster overall into two distinct topographic patterns; formal clustering analysis (using Ward's hierarchical clustering method) revealed an optimal two-cluster solution that explained 73% of the variance between maps. For example, symptom response maps for sadness, decreased interest, and suicidality appeared similar, and were combined into a map the researchers dubbed the "dysphoric cluster." Similarly, symptom response maps for sleep, libido, and anxiety appeared similar to each other (and highly dissimilar to the dysphoric cluster map) and were combined into a map the researchers dubbed the "anxiosomatic cluster."

Notably, dysphoric and anxiosomatic cluster maps appeared to be largely non-overlapping, meaning that stimulation at any given site in the DLPFC tended to be correlated with change in dysphoric symptoms or anxiosomatic symptoms, but not both. Thus, by analyzing connectivity patterns of each voxel in the DLPFC and quantifying its similarity to either the dysphoria or anxiosomatic cluster map, the researchers generated a "targeting atlas," a color-coded brain map which could be used to predict whether (and to what degree) stimulation of any given spot in the DLPFC could be expected to modulate dysphoric symptoms vs anxiosomatic symptoms over the TMS treatment course. Predictions were tested and validated by comparing stimulation site and depressive symptom outcomes at an individual level in their study subjects, and at a group level by reanalyzing outcomes of 12 published rTMS trials that measured anxiety and depressive symptoms.

To sum up this section, studies of TMS targeting over time have revealed that the effectiveness of TMS treatment for depression is highly dependent on cortical targeting. The advent of whole-brain connectome analysis has provided a conceptual framework to unify disparate findings in the field and converges with findings from the DBS world. Cortical targets which are more connected with sgCC are more effective in treating depression, reiterating the importance of the sgCC in mood disorders, and reframing treatment of depression with TMS from simplistic models of cortical activity to circuit-based models involving neuroplasticity of cortical-subcortical mood circuits for which cortical sites serve as "entry nodes" accessible to the superficial magnetic fields of TMS. Further, connectome analysis has significantly advanced our understanding of depression as disorders of whole-brain connectivity, and advanced statistical and computing methods have revealed distinct neural subtypes of depression, each of which may be optimally treated by distinct stimulation targets.

Accelerated Protocols

Another area of intense interest amongst TMS researchers is development of highly potent, accelerated protocols. While TMS is a highly effective treatment for depression, receiving treatment can present logistical difficulties. A typical clinical course of TMS treatment involves daily treatments (5x/week) for 6 weeks, followed by a

3-week taper. While each treatment is brief (modern protocols typically employ sessions of 10 min or less) and noninvasive (patients stay awake throughout the procedure and are typically able to resume all activities afterwards including driving), making all the necessary appointments can be difficult. Further, similar to the time course of SSRI antidepressant response [79], patients who improve with TMS typically require 2–4 weeks of treatment for response [135], which can be difficult to tolerate for acutely depressed patients, particularly if suicidal.

At the same time, as the safety and tolerability of clinical rTMS became increasingly evident over time [112], researchers grew more comfortable exploring protocols delivering higher TMS doses, either by increasing the number of pulses per session, increasing the number of sessions, or both. An early study by Anderson et al. [136] explored the safety and tolerability of large doses of TMS in 63 healthy young men. They found that TMS delivered at 12,960 pulses per day for 3 days (total of 38,880 pulses), at the time the largest known exposure to TMS, was safe and tolerable. No seizures were recorded, and headaches occurred at the same statistical frequency in active TMS vs sham TMS sessions.

Other studies explored whether extending the typical treatment course by increasing the number of TMS sessions resulted in better outcomes. On the whole, such studies indicate that longer treatment courses result in higher remission/response rates. As a reference point, early TMS trials for depression, such as the multicenter RCT [107] that helped to establish TMS as an efficacious treatment for depression, delivered 3000 pulses/session, 5 sessions per week for 3–6 weeks, for a total of 45,000–90,000 total pulses per treatment course. This resulted in a 14% remission rate (13/92) for the active treatment group vs a 5% remission rate for the sham treatment group (remission rates in these early studies were lower than subsequent naturalistic studies such as Carpenter 2012 due to adherence to strict protocols). Interestingly, for remitters, only half (6/13) achieved remission by the 3-week point; the other half required up to an additional 3 weeks of treatment. An open-label extension of this study [137] delivered additional TMS treatments (3–6 weeks) to subjects that did not remit in the previous study, resulting in an additional 9 remissions. Similarly, extending treatment for subjects that did not respond to an acute 4-week course of deep TMS (rTMS delivered using a specially designed coil developed by Brainsway [138]) resulted in a 61% response rate after 4 additional weeks of treatment [139]. Finally, a recent study which analyzed trajectories of response to rTMS in major depression over 6 weeks of treatment [135] revealed that, while a minority of subjects were nonresponders at all time points ($n = 43/388$), all other subjects showed progressive improvements in depressive symptoms (measured by HAM-D) week by week.

Given the desire for more rapid response to TMS, and studies that indicated higher TMS doses result in higher response/remission rates, accelerated protocols have been developed which deliver a high number of TMS pulses, compressed over days instead of weeks. One early accelerated study conducted by Holzheimer et al. [140] delivered 15,000 rTMS pulses over 2 days (1000 pulses per session, 5 sessions on day 1, 10 sessions on day 2) to 14 depressed patients. No seizures were reported, but one subject discontinued treatment due to increased suicidal ideation. Results were dramatic: a 43% response rate (29% remission) on day 3, with improvements maintained at 3 week and 6-week follow-up. Similarly, another early accelerated study by Hadley et al. [141] which roughly doubled the number of pulses given for treatment of depression (6800 pulses per session, 5 sessions/week for a total of 34,000 pulses per week) resulted in rapid improvement within 1–2 weeks, with 67% of subjects reporting decrease in suicidal ideation after 1 week, with no serious adverse events reported in their 19 subjects.

Relevant to accelerated protocols, an important recent development in TMS has been the introduction of *theta-burst* protocols. Theta burst is a pulse pattern (high-frequency "bursts" of 3–5 pulses, repeated at low "theta" frequencies) derived from endogenous neuronal signaling patterns in the hippocampus that appears to facilitate neuroplasticity with high efficiency (reviewed in [142]). Early researchers established that rTMS delivered to human motor cortex in an intermittent theta-burst pattern (3 pulses given at 50 Hz, repeated every 200 ms, for 2 s followed by an 8 s rest period) could achieve LTP after only 190 seconds (600 pulses) [143]. This paved the way for the first randomized sham-controlled study of iTBS for depression [144], which demonstrated that 10 sessions of iTBS over 2 weeks (each session consisting of 1800 pulses delivered over 8.5 min) resulted in a 40% response rate (vs 13% for sham). Most recently, Blumberber et al. [145] demonstrated that a 3-min intermittent theta-burst protocol (600 total pulses per session) was non-inferior to the usual 37.5-min, non-patterned rTMS protocol (3000 total pulses per session) for treatment of depression. Subjects in both groups improved considerably to a statistically identical degree. In other words, delivering rTMS pulses in an intermittent theta-burst (iTBS) pattern sped up treatment sessions by a factor of 10 compared with conventional, non-patterned rTMS, thus opening the door to accelerated protocols capable of delivering 10 times the treatment dose of conventional rTMS in the same amount of time.

Putting It All Together: Stanford Accelerated Intelligent Neuromodulation Therapy (SAINT)

What is the upper limit of effectiveness of TMS in treating depression? As discussed above, previous researchers have established several factors that influence treatment outcomes:

1. Targeting location, with optimal results reported when targeting the subregion of left DLPFC maximally anticorrelated with sgCC,
2. Dosage, with higher dosages (e.g., more pulses) associated with better treatment response,
3. Theta burst, a patterned version of rTMS that is equally effective but 10 times more time-efficient than non-patterned rTMS, thus allowing for high potency protocols,
4. Compressing sessions to allow for an accelerated treatment course, which appears to result in a more rapid response.

Williams and colleagues devised a protocol combining these four factors (later dubbed Stanford Accelerated Intelligent Neuromodulation Therapy, or SAINT [146]) to create what may arguably be the most potent, effective, and safe treatment for depression currently in existence. Williams first tested the SAINT protocol on a small group ($n = 6$) of depressed patients considered to be at the highest level of treatment-refractoriness [147]. To be included in the study, subjects had to be highly depressed (score >20 on the HDRS17), tried and failed ECT, tried and failed standard rTMS, and scored at or above 14/15 on the Maudsley Staging Method (representing the highest level of treatment-refractoriness for non-psychotic depressions, indicating >10 failures of previous adequate medication trials). All subjects met criteria to be considered for deep brain stimulation and were profoundly functionally impaired. The average length of the current depressive episode at time of study was 14.8 years.

Each subject received resting-state fMRI scans from which coordinates were derived locating the precise region of left DLPFC maximally anticorrelated with sgCC activity. These coordinates, which were unique for each individual, were then used with neuronavigation methods to target the TMS coil to the optimal location with high precision and reliability. Subjects received the highest-intensity exposure to TMS of any treatment to date: 5 days of treatment, 10 sessions per day, each session lasting 9 min and delivering 1800 pulses in an intermittent theta-burst (iTBS) pattern. As the usual iTBS protocol for depression involves the delivery of 18,000 pulses over 30 sessions (600 pulses per session, 1 session per day, for 30 sessions) [145], subjects receiving the SAINT protocol essentially received the equivalent of a full "regular" iTBS treatment course (18,000 pulses) compressed into a single day, which was then repeated over 5 days, for a total of 90,000 pulses—the highest TMS dose ever attempted in humans to that point.

Remarkably, this high dosage appeared to be both safe and tolerable. No major adverse events, including seizures, were reported. Neuropsychological testing conducted before and after treatment indicated no impairment in any cognitive domain. Most remarkably, in this population of patients whose average length of current depressive episode was 14.8 years, a single 5-day course of TMS treatment resulted in a mean 76% drop in HDRS-17 scores, with 5/6 subjects meeting criteria for response after the fifth day (4/6 met criteria for remission). As can be expected for this highly treatment-refractory population, however, durability of gains was an issue: 2 weeks after treatment, only 2/6 subjects still met criteria for response, and after 4 weeks, all 6 subjects had relapsed. Still, Williams' study shows a glimpse of the potential power of TMS as a depression treatment.

Williams and colleagues followed up their 2018 study with a larger cohort of 21 subjects with treatment-resistant depression [146]. Although average level of treatment resistance was high (subjects had tried and failed an average of 5.9 adequate antidepressant trials and had an average Maudsley Staging Method score of 10.1), subjects were not as treatment-resistant as the 2018 study (those subjects all had Maudsley scores ≥14). In this slightly less treatment-resistant population, a single 5-day course of TMS using the SAINT protocol achieved stunning results: 19/21 patients (90.5%) achieved *remission* after the fifth day of treatment; after 1 month, 70% continued to meet criteria for response (60% met criteria for remission).

Again, no serious adverse events were reported, and treatments were well tolerated, with the main reported side effects being fatigue and some discomfort at the stimulation site and in facial muscles during stimulation. Again, neuropsychological testing revealed no impairment in any domain tested, and in fact improvements were noted in some domains such as cognitive inhibition, which is not unexpected given that rTMS applied to the DLPFC increases prefrontal cortical activity [109].

In a follow-up study [148], Williams and colleagues directly compared TMS delivered with the SAINT protocol to ECT for treatment of severe depression in a small sample ($n = 15$) of highly depressed, suicidal hospitalized patients. After just 5 days of TMS, 86% of patients achieved response (73% remission), which compared favorably with ECT which achieved 73% response (67% remission). In contrast, the effects of TMS treatment were much more rapid; response was achieved after an average 2.7 days of TMS treatment (3.5 days to remission). In contrast, response was achieved with ECT after an average of 18.5 days of treatment (31.3 days to remission). Reflecting these differential time-courses of improvement, TMS patients were hospitalized for much shorter periods of time, an average of 8.4 days, compared to 22.3 days for ECT patients.

These three studies were limited by the lack of randomized sham control groups. Thus it was possible, though unlikely (given the highly treatment-refractory nature of enrolled subjects), that positive results were due to the placebo effect. This concern has been laid to rest with the Williams group's latest study [149] a placebo-controlled double-blinded study which demonstrated a 78.6% remission rate in individuals with treatment-resistant depression after 5 days of SAINT treatment, compared with 13.3% remission rate in the sham treatment group. The effect size was so large and statistically significant that the trial was halted midway, at the planned interim analysis. Based on these remarkable results, on September 6, 2022 the FDA issued 510(k) clearance via its Breakthrough Device Designation pathway to Magnus Medical Inc. for the use of the SAINT protocol in the treatment of treatment-resistant Major Depressive Disorder in adults, opening the door to widespread clinical use. [https://www.magnusmed.com/press-releases/magnus-medical-receives-fda-clearance-for-the-saint-neuromodulation-system/]. It is worth mentioning that although TMS is a highly effective treatment for depression, it is not a cure. Durability of response can vary from subject to subject, and, similar to ECT, maintenance treatments may be necessary to maintain gains in some, a field of very active research currently [150]. As a whole, however, the TMS literature offers a tantalizing glimpse of a possible post-depression future—a future in which major depression is relegated from its current position as the fourth leading cause of disability worldwide [151], to a chronic but highly treatable condition characterized by long symptom-free intervals punctuated with brief relapses treated with neuromodulation.

This discussion of TMS has focused mainly on its remarkable qualities as a treatment for major depression. In theory, any brain-based illness involving dysfunction of circuits accessible to TMS magnetic fields should be amenable to neuromodulation and treatment. TMS-based approaches have already achieved FDA approval for the treatment of OCD [152] and smoking addiction, and other indications such as PTSD [153] appear close at hand based on promising preliminary studies.

A Next Big Thing: Focused Ultrasound

As discussed above, neuromodulatory approaches to psychiatric treatment hold enormous promise, particularly as researchers continue to refine maps of dysfunctional brain circuitry. The first of these approaches, DBS, offers excellent spatial and temporal precision, but applications will likely be limited by its invasive nature. The second approach discussed, TMS, has already proven itself as a highly effective treatment for a number of psychiatric conditions. Since FDA approval in 2008, TMS has emerged as arguably both the most effective and safest treatment for major depression, particularly in treatment-resistant populations. TMS coils, however, offer relatively poor spatial precision, and effects are limited to superficial (e.g., cortical) targets due to the physics of magnetic field decay. Although magnetic field properties and spatial resolution can be manipulated to a certain extent by coil design, in all cases fundamental physics dictate a *depth-focality tradeoff*—stimulation of deeper areas requires stronger currents and/or larger coils which increase the volume of brain stimulated (thus less focal) and increase seizure risk [99].

The ideal example of neuromodulation would be a technology which permits the ability to safely and noninvasively excite or inhibit any point(s) in 3 dimensions of brain-space with high spatial and temporal precision. Focused ultrasound, which combines the spatial and temporal precision of DBS with the noninvasive nature of TMS, appears poised to fulfill this role in the near future.

Diagnostic ultrasound imaging, which has been in wide clinical use for decades, generates images by analyzing reflections of low-intensity sound waves generated by a single transducer as they bounce off various bodily tissues and fluids. The FDA has established guidelines for ultrasound imaging [154] that limit acoustic energies to levels considered safe enough for routine fetal monitoring [155]. Focused ultrasound, on the other hand, typically employs multiple transducers arrayed around the head and focused to summate at a target of interest, similar to gamma-knife radiosurgery [156]. Focused ultrasound beams are capable of penetrating past the skull deep into the brain and can be marshalled at high intensity to ablate brain tissue (via heating) with millimeter accuracy [157]. The FDA has recently approved the use of such *high-intensity focused ultrasound (HIFU)* systems to treat medication-refractory essential tremor [158] and tremor-dominant Parkinson's disease [159]. For these conditions, high intensity ultrasound waves intersect to create tiny thermal lesions in the thalamus, thus avoiding the need for open brain surgery.

More relevant to neuromodulation, in the 1950s Fry and colleagues discovered that while high-intensity focused ultrasound could focally ablate tissue, lower-intensities could reversibly inhibit brain activity in a localized manner [160]. Decades of subsequent research in animal and human studies since have led to the discovery and refinement of ultrasound parameters and protocols (collectively dubbed *low-intensity focused ultrasound, or LIFU*) which appear to be capable of modulating brain activity safely, reversibly, and with millimeter localization (reviewed in [157]). In particular, the advent of *intermittent pulse protocols*, which deliver ultrasound waves in brief pulses rather continuously, have allowed researchers to safely modulate brain tissue with very low energies (e.g., less than 100 W/cm^2), similar or lower than used in ultrasound imaging (reviewed in [161]).

Focused ultrasound technology has currently reached the point where it can replicate a virtual DBS needle to focally stimulate motor and sensory brain areas, eliciting action potentials that give rise to discrete sensations and movements [163–165]. By manipulating pulse frequency and duration, researchers have discovered ways to focally increase or decrease brain excitability in a long-lasting manner [157], thus opening the door to therapeutic applications. For example, Beisteiner and colleagues recently reported the first patient study in which transcranial pulsed ultrasound (TPU) was demonstrated to improve cognition in subjects with Alzheimer's disease by focal stimulation of brain memory networks, an effect that lasted until the end of the study at 3 months [165]. Relevant to psychiatric illness, Sanguinetti and colleagues recently demonstrated [166] that transcranial ultrasound targeted to the right ventrolateral prefrontal cortex (R-VLPFC), a region of brain implicated in cognitive control and mood regulation, improved mood in healthy volunteers 30 min after stimulation. Examination of functional connectivity with rs-fMRI revealed changes in connectivity consistent with mood elevation, such as an increase in connectivity in cognitive control networks involving the DLPFC, and decrease in connectivity of the default mode network, which other studies have strongly implicated in depression [130].

While the mechanism of action of LIFU-induced neuromodulation is unclear, it is *not* thought to involve thermal effects, due to the very low energies involved, lack of evident heating when analyzed with MR-thermography [161], and lack of heat damage in histological studies which have examined this issue [157]. Rather, the mechanism is thought to be mainly related to mechanical stretching of neuronal membranes, which can activate and modify kinetics of voltage-gated ion channels, thereby affecting neuronal excitability [167]. Other contributory mechanisms may include the formation of bilayer sonophores, non-damaging cavitation, and modulation of neurotrophic factor activity (reviewed in [161]). Similar to TMS, long-lasting effects of ultrasonic pulse trains (inducing facilitation or inhibition) are presumably due to cellular learning processes such as LTP and LTD [168].

Just Add Nanoparticles

Although LIFU can directly modulate neural tissue, researchers have discovered that it's capabilities can be extended even further by adding nanoparticles. For example, Hynynen, McDannold et al. [169, 170] discovered a highly efficient method to focally open the blood–brain barrier (BBB) by combining LIFU with a microbubble contrast agent such as Optison (copyright GE Healthcare). Such agents, which are already FDA approved for use as contrast agents for diagnostic ultrasound, appear to concentrate ultrasound waves to the vascular wall, allowing focal and reversible disruption of the BBB at very low power levels that appear to be safe to surrounding tissues. Initial studies in humans have demonstrated the feasibility of this method to safely and temporarily open the BBB in patients with Alzheimer's [171] and amyotrophic lateral sclerosis (ALS) [172], which may one day allow targeted delivery of large molecule therapeutics that would otherwise be hindered by the BBB.

Airan and colleagues have developed another approach [173]: nanoparticles which can be loaded with bioactive drugs. When unperturbed, bioactive molecules are inert due to encapsulation within nanoparticles; however, when perturbed by a burst of focused ultrasound, nanoparticles can be induced to release their load, thus enabling focal delivery of bioactive drugs. Using this approach, Airan and colleagues demonstrated the ability to focally uncage propofol (a small molecule anesthetic that readily crosses the BBB), resulting in suppression of seizure activity in a mouse model of epilepsy [173] and neuromodulation at the local site of drug release—effects which propagated across whole-brain networks functionally connected to the site [174].

Finally, researchers are advancing methods that combine ultrasound approaches with other interventional modalities such as optogenetics, potentially allowing even more neuromodulatory precision. Optogenetics, a field pioneered by Karl Deisseroth [175], involves the insertion of light-sensitive ion channels into selective neuronal populations using genetic targeting techniques. Optogenetics has revolutionized the study of the brain by allowing the manipulation of individual brain circuits with exquisite precision, thereby allowing testing of causality in unprecedented detail. Modulating optogenetically modified brain circuits in vivo typically requires the insertion of fiber optics into the brain to deliver light to neurons of interest, obviously limiting its applications in humans.

Focused ultrasound offers a way to focally activate genetically targeted neurons noninvasively. For example, Ibsen and colleagues have devised mechanosensitive ion channels that can be inserted into neuronal populations of interest in *C. elegans* with genetic targeting [176]. Genetically targeted neurons can then be selectively activated with low-powered focused ultrasound; they dubbed this approach "sonogenetics." In another approach, Wu et al. [177] invented mechanoluminescent nanoparticles which are capable of storing and releasing light under controlled conditions. In theory, such nanoparticles can be introduced into the circulation, "charged" with light in the periphery, and induced to release light in the brain by application of focused ultrasound. The released light can then drive optogenetically modified circuits.

Such approaches, which combine focused ultrasound with genetic targeting, offer another layer of precision than focused ultrasound alone. Current focused ultrasound approaches can stimulate millimeter-sized regions of brain tissue, but in a relatively non-selective manner, akin to a DBS electrode [161]. By adding genetic targeting, the effects of focused ultrasound can be further limited to selected neuronal populations within the stimulated area, giving rise to a method with exquisite control at temporal, spatial, and neuronal population levels.

One of the exciting features of focused ultrasound approaches is that clinical translation from the lab to human patients appears highly feasible. Clinically oriented ultrasound systems can be built in a modular fashion by modifying and incorporating technologies that are already available and FDA approved. For example, MR-guided focused ultrasound systems for treatment of essential tremor already exist and have received FDA approval for this purpose [178]. The FDA has already established guidelines outlining safe operating parameters for diagnostic ultrasound imaging [154], which LIFU can readily adopt. Microbubble contrast agents such as Optison (copyright GE Healthcare) are FDA approved and widely used in diagnostic ultrasound imaging. Drug-carrying nanoparticles invented by Airan and colleagues were constructed with components and methods already approved by the FDA for clinical use in other contexts [173].

Further, human engineers have had centuries of experience in the manipulation of waves, for example, in the field of optics [179]. Such experience is currently being leveraged to produce breakthroughs in acoustic metamaterials (artificial materials constructed at the atomic level to produce materials with radically unconventional properties) to create acoustic hyper-lenses capable of focusing sound waves with unprecedented precision [179]. Advances in the manufacture of microscale ultrasonic transducers, combined with advances in computer modeling to control the behaviors of multiple arrays of such transducers, will enable precise adaptive neuromodulation of multiple brain foci simultaneously (reviewed in [161]).

Closing Thoughts: What Are the Ethical Limits of Neuromodulation?

From a historical perspective, advances in science have fundamentally been driven by the invention of new tools (see Table 2.1). The invention of the telescope in the early 1600s enabled astronomers such as Galileo to map the heavens in detail and provide experimental validation of Copernicus' heliocentric model of the solar system [183]. The invention of the microscope enabled Cajal and Golgi at the turn of the nineteenth century to elucidate the neuron as the fundamental unit of the brain [8]. In the twentieth century, humans have advanced from crude, X-ray based visualizations of the brain (pneumoencephalography) to extremely high-resolution MRI imaging, whose most powerful machines are currently capable of safely imaging the human brain at 10.5 T with 0.4 mm resolution [184].

In the twenty-first century, further advances in MRI imaging methods have enabled visualization of brain activity (fMRI) and brain connectivity (DTI,

Table 2.1 Historical examples of tools for neuroscience research and neurotherapeutics

Year discovered/invented	Structural imaging	Activity imaging	Neurointervention
1895, Roentgen	X-ray imaging First imaging modality able to visualize hard vs soft tissues noninvasively Poor resolution for soft tissues such as the brain		
1918, Walter Dandy	Air ventriculography X-ray visualization of ventricles by injecting air into ventricles. Invasive, painful, low resolution		
1927, Egaz Moniz	Cerebral angiography X-ray visualization of cerebral vasculature by injection of radiopaque contrast materials		
1960s, Godfrey Houndsfield	CT scan First noninvasive modality able to image human brain tissue in vivo Relatively low resolution, high radiation dose (technique based on reconstruction from multiple X-rays)		
1975, Ter-Pogossian et al. [15]	–	PET scan Requires injection of radioactive particles	
1970s, Lauterbur and Mansfield	MRI Better resolution than CT, no radiation		
1990, Ogawa et al. [24]	–	fMRI Higher resolution than PET, radioactive particles not required; hemoglobin acts as contrast agent	
1990, Moseley et al. [180]	DTI Visualizes anatomical connectivity by tracking diffusion of water molecules		

(continued)

Table 2.1 (continued)

Year discovered/ invented	Structural imaging	Activity imaging	Neurointervention
1995, Biswal et al. [181]		Rs-fMRI Visualizes functional connectivity by correlating activity between brain regions	
1995, George et al. ([60], [61])			TMS for depression Moderate spatial precision, high temporal precision, limited to superficial (i.e., cortical) targets; noninvasive
2005, Deisseroth et al. [3]			Optogenetics Insertion of light-sensitive ion channels into genetically targeted neuronal population Requires insertion of fiber optic cable into the brain for neuromodulation
2005, Mayberg et al. [4]			DBS for depression High spatial and temporal precision, but invasive
2014, Legon et al. [182]			Focused ultrasound for neuromodulation in humans High spatial and temporal precision, able to hit deep targets. Noninvasive precision can be enhanced by combining with optogenetics and nanoparticles

rs-fMRI), allowing the compilation of high-resolution maps of brain circuitry. Large-scale consortiums such as the Human Connectome Project, funded in part by the BRAIN initiative, have collected thousands of high-quality human connectome maps and made them freely accessible to the public. Such connectome maps represent a fundamental new tool for the research community, and have already enabled the discovery of invaluable insights, such as optimal targeting of focal neurointerventions. With focused ultrasound, medicine is poised on the brink of a neuromodulation tool capable of safely targeting and tuning the activity of multiple brain loci simultaneously, sculpting neural pathways with efficiency and precision… which raises some interesting ethical questions.

Just as LASIK (laser in-situ keratomileusis) revolutionized the treatment of refractive eye conditions by mapping, modeling, and correcting the optical properties of the eye by resculpting the cornea with a laser [185], one day in the near future it may be possible to map out an individual's brain activity and connectivity, compare such maps to normative databases, then correct aberrations by resculpting neural pathways with targeted neuromodulation—in essence, "LASIK for the brain." As discussed earlier, elementary forms of these approaches are already being used by interventional psychiatrists with success to treat major depression. The potential of neuromodulation, however, extends far beyond treatment of disease; in principle, targeted modification of neural pathways could be used to change any thought or behavior of interest.

Such capabilities raise the question of when medical intervention is warranted. There is little disagreement about the appropriateness of treating generally accepted medical conditions that are highly impairing and cause intense suffering, such as major depression, but what about milder conditions?

As one example, the neurodiversity movement has emphasized the idea that, to a certain extent, unusual variations in human brains (and the thoughts/behaviors arising from such brains) can be reframed in a positive light, as features rather than defects of the brain that ultimately enhance survival and enrich the collective human condition [186]. Recent research in conditions such as autism reinforces the general concept of human conditions and abilities as lying on a spectrum [187], often with no clear bright line dividing "disease" from unusual-but-potentially-beneficial variations, particularly when benefits are considered in a broader context. Neuromodulation has the potential to resculpt brains to fit a predetermined ideal, but overzealous pursuit towards a simplistic ideal risk creating a cognitive monoculture, with all the attendant risks of monocultures. What would have happened to Beethoven, and his music, if he was treated with Prozac?

As neuromodulatory approaches become safer and more widely available, the temptation to use such technologies beyond disease treatment will inevitably arise, and with them, a slew of ethical considerations related to issues of coercion, fairness, integrity, and moral acceptability ("the yuck factor") [188]. Ethical considerations raised by neuromodulation are the latest instantiation of such concerns, which were raised previously by the advent of safe(r) antidepressants such as Prozac in treating subclinical conditions (what the psychiatrist Peter Kramer dubbed "cosmetic psychopharmacology"[189]), the increasing availability of purported cognitively enhancing pharmaceuticals such as Ritalin [190], and increasing sophistication of gene editing techniques such as CRISPR [191], to name a few.

Ethical issues raised by neuromodulation go beyond previous biological interventions due to its potentially greater safety, specificity, and effectiveness, as well as the special status accorded to the brain as the organ of the mind. Conundrums will likely arise from sometimes conflicting goals between individuals and the greater society. As explored in Chap. 11, on the one hand, individuals may want to utilize neuromodulation to enhance their abilities. Will this give such individuals an unduly unfair advantage over others, or is this simply another manifestation of "the usual" unfair advantages accorded to individuals with high socioeconomic status, such as access to better nutrition, education, and training?

On the other hand, as explored in Chap. 10, societies may want to forcibly impose neuromodulatory treatments on individuals in certain situations, for example, to increase empathy in criminal psychopaths [192], or to decrease pedophilic drives in sexual offenders. Such speculations are no longer the realm of science fiction; the strategy of selecting and inhibiting specific behaviors and drives by targeted neuromodulation is already being utilized clinically—and with high effectiveness—to treat specific compulsions with TMS in patients with OCD [152]. Is neuromodulatory modification of behavior an extension of other types of involuntary treatments already in widespread use, such as forcible medication of psychotic individuals, or chemical castration of sexual offenders [193] or, as legal scholars such as Nita Farahany have argued [194], is neuromodulation something fundamentally different because it directly infringes on an individual's freedom of thought, which is arguably the basis of all other liberties?

In the twenty-first century, humans have become increasingly adept in learning *how* to manipulate our environment, including our bodies and minds, but the thorny issue of *what* is appropriate to manipulate, and *when*, continue to be ongoing and open-ended questions that science alone will be unable to answer. Neuromodulation is a tool with great potential, but the same features that make it so compelling as a medical treatment (precision, safety, effectiveness) are the same features that make it so attractive for enhancement or, more darkly, coercion, and mind control. How each society decides to use such tools ultimately rests on the moral intuitions of individuals in that society, and their collective expression in the rich overlapping tapestry of public policies, regulations, laws, and culture that bind society together.

Key Points

1. **Diffusion tensor imaging (DTI)** is an MRI-based imaging approach that visualizes *anatomical* connectivity in the brain by tracking diffusion of water molecules.
2. **Resting-state functional magnetic resonance imaging (rs-fMRI)** is an MRI-based imaging approach that visualizes *functional* connectivity in the brain by identifying brain regions whose activity levels are temporally correlated.
3. **Distributed processing** is a currently widely accepted model of brain function which posits that the brain accomplishes complex tasks by dynamically recruiting specialized brain modules which act in a coordinated, circuit-based manner via functional connections.
4. The **Human Connectome Project** is a publicly funded neuroscience consortium which aims to create definitive high-resolution maps of human brain circuitry pooled from large populations with DTI and rs-fMRI imaging.

Questions to Consider

1. What are the clinical implications of recent findings that demonstrate the existence of different neural subtypes of major depression?
2. What are the relative advantages and disadvantages of neuromodulation approaches such as optogenetics, DBS, TMS, and focused ultrasound?
3. How should neuromodulation be regulated for purposes other than disease treatment?
4. Is neuromodulation fundamentally different from other forms of biological intervention, such as medications and gene therapy? How is it different? How is it similar?
5. Under what circumstances should society allow forcible neuromodulation of individuals? How is forcible neuromodulation different from involuntary treatment with medications?

References

1. Kaku M. The Golden age of neuroscience has arrived. Wall Street J. 2014;20
2. Behrens TE, Sporns O. Human connectomics. Curr Opin Neurobiol. 2012;22(1):144–53.
3. Boyden ES, Zhang F, Bamberg E, Nagel G, Deisseroth K. Millisecond-timescale, genetically targeted optical control of neural activity. Nat Neurosci. 2005;8(9):1263–8.
4. Mayberg HS, Lozano AM, Voon V, McNeely HE, Seminowicz D, Hamani C, Schwalb JM, Kennedy SH. Deep brain stimulation for treatment-resistant depression. Neuron. 2005;45(5):651–60.
5. Barker AT, Jalinous R, Freeston IL. Non-invasive magnetic stimulation of human motor cortex. Lancet. 1985;325(8437):1106–7.
6. Williams NR, Taylor JJ, Kerns S, Short EB, Kantor EM, George MS. Interventional psychiatry: why now? J Clin Psychiatry. 2014;75(8):895.
7. Lapidus KA, Kopell BH, Ben-Haim S, Rezai AR, Goodman WK. History of psychosurgery: a psychiatrist's perspective. World Neurosurg. 2013;80(3–4):S27–e1.
8. Glickstein M. Golgi and Cajal: the neuron doctrine and the 100th anniversary of the 1906 Nobel prize. Curr Biol. 2006;16(5):R147–51.
9. Stahl SM. Mechanism of action of serotonin selective reuptake inhibitors: serotonin receptors and pathways mediate therapeutic effects and side effects. J Affect Disord. 1998;51(3):215–35.
10. Vaswani M, Linda FK, Ramesh S. Role of selective serotonin reuptake inhibitors in psychiatric disorders: a comprehensive review. Prog Neuro-Psychopharmacol Biol Psychiatry. 2003;27(1):85–102.
11. Miller DD, Caroff SN, Davis SM, Rosenheck RA, McEvoy JP, Saltz BL, Riggio S, Chakos MH, Swartz MS, Keefe RS, Stroup TS. Extrapyramidal side-effects of antipsychotics in a randomised trial. Br J Psychiatry. 2008;193(4):279–88.
12. Sinyor M, Schaffer A, Levitt A. The sequenced treatment alternatives to relieve depression (STAR* D) trial: a review. Can J Psychiatry. 2010;55(3):126–35.
13. Rutten GJ. The Broca-Wernicke doctrine: a historical and clinical perspective on localization of language functions. Berlin: Springer; 2017.
14. Shorvon SD. A history of neuroimaging in epilepsy 1909–2009. Epilepsia. 2009;50:39–49.
15. Ter-Pogossian MM, Phelps ME, Hoffman EJ, Mullani NA. A positron-emission transaxial tomograph for nuclear imaging (PETT). Radiology. 1975;114(1):89–98.
16. Small SA, Heeger DJ. Functional imaging of cognition. In: Kandel ER, Schwartz JH, Jessell TM, editors. Principles of neural science. New York: McGraw-Hill; 2012. p. 426–42.

17. Raichle ME. In: Cabeza R, Kingstone A, editors. Functional neuroimaging: a historical and physiological perspective. In handbook of functional neuroimaging of cognition. Cambridge: MIT Press; 2006. p. 3–20.
18. Logothetis NK. What we can do and what we cannot do with fMRI. Nature. 2008;453(7197):869–78.
19. Savage N. The world's most powerful MRI takes shape. IEEE Spectr. 2013;50(11):11–2.
20. Fox PT, Mintun MA, Raichle ME, Herscovitch P. A noninvasive approach to quantitative functional brain mapping with $H_2{}^{15}O$ and positron emission tomography. J Cereb Blood Flow Metab. 1984;4(3):329–33.
21. Raichle ME. Behind the scenes of functional brain imaging: a historical and physiological perspective. Proc Natl Acad Sci. 1998;95(3):765–72.
22. Rosen BR, Belliveau JW, Aronen HJ, Kennedy D, Buchbinder BR, Fischman A, Gruber M, Glas J, Weisskoff RM, Cohen MS, Hochberg FH. Susceptibility contrast imaging of cerebral blood volume: human experience. Magn Reson Med. 1991;22(2):293–9.
23. Belliveau JW, Kennedy DN, McKinstry RC, Buchbinder BR, Weisskoff R, Cohen MS, Vevea JM, Brady TJ, Rosen BR. Functional mapping of the human visual cortex by magnetic resonance imaging. Science. 1991;254(5032):716–9.
24. Ogawa S, Lee TM, Kay AR, Tank DW. Brain magnetic resonance imaging with contrast dependent on blood oxygenation. Proc Natl Acad Sci. 1990;87(24):9868–72.
25. Pauling L, Coryell CD. The magnetic properties and structure of hemoglobin, oxyhemoglobin and carbonmonoxyhemoglobin. Proc Natl Acad Sci. 1936;22(4):210–6.
26. Bandettini PA, Wong EC, Hinks RS, Tikofsky RS, Hyde JS. Time course EPI of human brain function during task activation. Magn Reson Med. 1992;25(2):390–7.
27. Frahm J, Bruhn H, Merboldt KD, Hänicke W. Dynamic MR imaging of human brain oxygenation during rest and photic stimulation. J Magn Reson Imaging. 1992;2(5):501–5.
28. Kwong KK, Belliveau JW, Chesler DA, Goldberg IE, Weisskoff RM, Poncelet BP, Kennedy DN, Hoppel BE, Cohen MS, Turner R. Dynamic magnetic resonance imaging of human brain activity during primary sensory stimulation. Proc Natl Acad Sci. 1992;89(12):5675–9.
29. Ogawa S, Tank DW, Menon R, Ellermann JM, Kim SG, Merkle H, Ugurbil K. Intrinsic signal changes accompanying sensory stimulation: functional brain mapping with magnetic resonance imaging. Proc Natl Acad Sci. 1992;89(13):5951–5.
30. Kandel ER, Hudspeth AJ. The brain and behavior. In: Kandel ER, Schwartz JH, Jessell TM, editors. Principles of neural science. New York: McGraw-Hill; 2012. p. 5–20.
31. Brodmann K. Vergleichende Lokalisationslehre der Grosshirnrinde in ihren Prinzipien dargestellt auf Grund des Zellenbaues. Barth, Lafayette; 1909.
32. Penfield W, Jasper H. Epilepsy and the functional anatomy of the human brain. Neurol. 1954;4(6):1891–1976.
33. Hubel DH, Wiesel TN. Receptive fields, binocular interaction and functional architecture in the cat's visual cortex. J Physiol. 1962;160(1):106–54.
34. Mesulam MM. Large-scale neurocognitive networks and distributed processing for attention, language, and memory. Ann Neurol. 1990;28(5):597–613.
35. Zeki S, Watson JD, Lueck CJ, Friston KJ, Kennard C, Frackowiak RS. A direct demonstration of functional specialization in human visual cortex. J Neurosci. 1991;11(3):641–9.
36. Fornito A, Zalesky A, Breakspear M. The connectomics of brain disorders. Nat Rev Neurosci. 2015;16(3):159–72.
37. Lozano AM, Lipsman N. Probing and regulating dysfunctional circuits using deep brain stimulation. Neuron. 2013;77(3):406–24.
38. Bishop KM, Goudreau G, O'Leary DD. Regulation of area identity in the mammalian neocortex by Emx2 and Pax6. Science. 2000;288(5464):344–9.
39. Oh SW, Harris JA, Ng L, Winslow B, Cain N, Mihalas S, Wang Q, Lau C, Kuan L, Henry AM, Mortrud MT. A mesoscale connectome of the mouse brain. Nature. 2014;508(7495): 207–14.
40. Glasser MF, Smith SM, Marcus DS, Andersson JL, Auerbach EJ, Behrens TE, Coalson TS, Harms MP, Jenkinson M, Moeller S, Robinson EC. The human connectome project's neuroimaging approach. Nat Neurosci. 2016;19(9):1175.

41. George MS, Ketter TA, Post RM. Prefrontal cortex dysfunction in clinical depression. Depression. 1994a;2(2):59–72.
42. George MS, Kellner CH, Bernstein H, Goust JM. A magnetic resonance imaging investigation into mood disorders in multiple sclerosis: a pilot study. J Nerv Ment Dis. 1994b;182:410–2.
43. Mayberg HS. Frontal lobe dysfunction in secondary depression. J Neuropsychiatry Clin Neurosci. 1994;6(4):428.
44. Coffey CE, Wilkinson WE, Weiner RD, Djang WT, Webb MC, Figiel GS, Spritzer CE. Quantitative cerebral anatomy in depression: a controlled magnetic resonance imaging study. Arch Gen Psychiatry. 1993;50(1):7–16.
45. Morris PLP, Robinson RG, Raphael B, Hopwood MJ. Lesion location and poststroke depression. J Neuropsychiatry Clin Neurosci. 1996;8(4):399–403. https://doi.org/10.1176/jnp.8.4.399.
46. Robinson RG, Szetela B. Mood change following left hemispheric brain injury. Ann Neurol. 1981;9(5):447–53.
47. Robinson RG, Kubos KL, Starr LB, Rao K, Price TR. Mood disorders in stroke patients: importance of location of lesion. Brain. 1984;107(1):81–93.
48. Vataja R, Leppävuori A, Pohjasvaara T, Mäntylä R, Aronen HJ, Salonen O, Erkinjuntti T. Poststroke depression and lesion location revisited. J Neuropsychiatry Clin Neurosci. 2004;16(2):156–62.
49. Zhang Y, Zhao H, Fang Y, Wang S, Zhou H. The association between lesion location, sex and poststroke depression: meta-analysis. Brain Behav. 2017;7:e00788.
50. Baxter LR, Schwartz JM, Phelps ME, Mazziotta JC, Guze BH, Selin CE, Gerner RH, Sumida RM. Reduction of prefrontal cortex glucose metabolism common to three types of depression. Arch Gen Psychiatry. 1989;46(3):243–50.
51. Martinot JL, Hardy P, Feline A, Huret JD, Mazoyer B, Attar-Levy D, Pappata S, Syrota A. Left prefrontal glucose hypometabolism in the depressed state: a confirmation. Am J Psychiatry. 1990;147(10):1313–7.
52. Mottaghy FM, Keller CE, Gangitano M, Ly J, Thall M, Parker JA, Pascual-Leone A. Correlation of cerebral blood flow and treatment effects of repetitive transcranial magnetic stimulation in depressed patients. Psychiatry Res Neuroimaging. 2002;115(1–2):1–4.
53. Drevets WC, Videen TO, Price JL, Preskorn SH, Carmichael ST, Raichle ME. A functional anatomical study of unipolar depression. J Neurosci. 1992;12(9):3628–41.
54. Drevets WC. Prefrontal cortical-amygdalar metabolism in major depression. Ann N Y Acad Sci. 1999;877(1):614–37.
55. Mayberg HS, Brannan SK, Tekell JL, Silva JA, Mahurin RK, McGinnis S, Jerabek PA. Regional metabolic effects of fluoxetine in major depression: serial changes and relationship to clinical response. Biol Psychiatry. 2000;48(8):830–43.
56. Mayberg HS, Liotti M, Brannan SK, McGinnis S, Mahurin RK, Jerabek PA, Silva JA, Tekell JL, Martin CC, Lancaster JL, Fox PT. Reciprocal limbic-cortical function and negative mood: converging PET findings in depression and normal sadness. Am J Psychiatr. 1999;156(5):675–82.
57. Botteron KN, Raichle ME, Drevets WC, Heath AC, Todd RD. Volumetric reduction in left subgenual prefrontal cortex in early onset depression. Biol Psychiatry. 2002;51(4):342–4.
58. Drevets WC, Price JL, Simpson JR, Todd RD, Reich T, Vannier M, Raichle ME. Subgenual prefrontal cortex abnormalities in mood disorders. Nature. 1997;386(6627):824–7.
59. Öngür D, Drevets WC, Price JL. Glial reduction in the subgenual prefrontal cortex in mood disorders. Proc Natl Acad Sci. 1998;95(22):13290–5.
60. George MS, Ketter TA, Parekh PI, Horwitz B, Herscovitch P, Post RM. Brain activity during transient sadness and happiness in healthy women. Am J Psychiatr. 1995a;152(3):341–51.
61. George MS, Wassermann EM, Williams WA, Callahan A, Ketter TA, Basser P, Hallett M, Post RM. Daily repetitive transcranial magnetic stimulation (rTMS) improves mood in depression. Neuroreport. 1995b;6(14):1853–6.
62. Pardo JV, Pardo PJ, Raichle ME. Neural correlates of self-induced dysphoria. Am J Psychiatry. 1993;150(5):713–9.
63. Neumeister A, Nugent AC, Waldeck T, Geraci M, Schwarz M, Bonne O, Bain EE, Luckenbaugh DA, Herscovitch P, Charney DS, Drevets WC. Neural and behavioral responses to tryptophan

depletion in Unmedicated patients with remitted major depressive disorder and controls. Arch Gen Psychiatry. 2004;61(8):765–73.
64. Talbot PS, Cooper SJ. Anterior cingulate and subgenual prefrontal blood flow changes following tryptophan depletion in healthy males. Neuropsychopharmacology. 2006;31(8):1757–67.
65. Greicius MD, Flores BH, Menon V, Glover GH, Solvason HB, Kenna H, Reiss AL, Schatzberg AF. Resting-state functional connectivity in major depression: abnormally increased contributions from subgenual cingulate cortex and thalamus. Biol Psychiatry. 2007;62(5):429–37.
66. Mayberg HS, Brannan SK, Mahurin RK, Jerabek PA, Brickman JS, Tekell JL, Silva JA, McGinnis S, Glass TG, Martin CC, Fox PT. Cingulate function in depression: a potential predictor of treatment response. Neuroreport. 1997;8(4):1057–61.
67. Osuch EA, Ketter TA, Kimbrell TA, George MS, Benson BE, Herscovitch MW, Post RM. Regional cerebral metabolism associated with anxiety symptoms in affective disorder patients. Biol Psychiatry. 2000;48(10):1020–3.
68. Drevets WC, Bogers W, Raichle ME. Functional anatomical correlates of antidepressant drug treatment assessed using PET measures of regional glucose metabolism. Eur Neuropsychopharmacol. 2002;12(6):527–44.
69. Wu JC, Gillin JC, Buchsbaum MS, Hershey T, Johnson JC, Bunney WE. Effect of sleep deprivation on brain metabolism of depressed patients. Am J Psychiatry. 1992;149(4):538–43.
70. Nobler MS, Oquendo MA, Kegeles LS, Malone KM, Campbell C, Sackeim HA, Mann JJ. Decreased regional brain metabolism after ECT. Am J Psychiatr. 2001;158(2):305–8.
71. Benabid AL, Pollak P, Gao D, Hoffmann D, Limousin P, Gay E, Payen I, Benazzouz A. Chronic electrical stimulation of the ventralis intermedius nucleus of the thalamus as a treatment of movement disorders. J Neurosurg. 1996;84(2):203–14.
72. Benabid AL. Deep brain stimulation for Parkinson's disease. Curr Opin Neurobiol. 2003;13(6):696–706.
73. Jakobs M, Fomenko A, Lozano AM, Kiening KL. Cellular, molecular, and clinical mechanisms of action of deep brain stimulation—a systematic review on established indications and outlook on future developments. EMBO Mol Med. 2019;11(4):e9575.
74. Herrington TM, Cheng JJ, Eskandar EN. Mechanisms of deep brain stimulation. J Neurophysiol. 2016;115(1):19–38.
75. Lozano AM, Mayberg HS, Giacobbe P, Hamani C, Craddock RC, Kennedy SH. Subcallosal cingulate gyrus deep brain stimulation for treatment-resistant depression. Biol Psychiatry. 2008;64(6):461–7.
76. Kennedy SH, Giacobbe P, Rizvi SJ, Placenza FM, Nishikawa Y, Mayberg HS, Lozano AM. Deep brain stimulation for treatment-resistant depression: follow-up after 3 to 6 years. Am J Psychiatr. 2011;168(5):502–10.
77. Holtzheimer PE, Kelley ME, Gross RE, Filkowski MM, Garlow SJ, Barrocas A, Wint D, Craighead MC, Kozarsky J, Chismar R, Moreines JL. Subcallosal cingulate deep brain stimulation for treatment-resistant unipolar and bipolar depression. Arch Gen Psychiatry. 2012;69(2):150–8.
78. Crowell AL, Riva-Posse P, Holtzheimer PE, Garlow SJ, Kelley ME, Gross RE, Denison L, Quinn S, Mayberg HS. Long-term outcomes of subcallosal cingulate deep brain stimulation for treatment-resistant depression. Am J Psychiatr. 2019;176(11):949–56.
79. Warden D, Rush AJ, Trivedi MH, Fava M, Wisniewski SR. The STAR*D project results: a comprehensive review of findings. Curr Psychiatry Rep. 2007;9(6):449–59.
80. Holtzheimer PE, Husain MM, Lisanby SH, Taylor SF, Whitworth LA, McClintock S, Slavin KV, Berman J, McKhann GM, Patil PG, Rittberg BR. Subcallosal cingulate deep brain stimulation for treatment-resistant depression: a multisite, randomised, sham-controlled trial. Lancet Psychiatry. 2017;4(11):839–49.
81. Chin L, Andersen JN, Futreal PA. Cancer genomics: from discovery science to personalized medicine. Nat Med. 2011;17(3):297.
82. Hamani C, Mayberg H, Snyder B, Giacobbe P, Kennedy S, Lozano AM. Deep brain stimulation of the subcallosal cingulate gyrus for depression: anatomical location of active contacts in clinical responders and a suggested guideline for targeting. J Neurosurg. 2009;111(6):1209–15.

83. Riva-Posse P, Choi KS, Holtzheimer PE, McIntyre CC, Gross RE, Chaturvedi A, Crowell AL, Garlow SJ, Rajendra JK, Mayberg HS. Defining critical white matter pathways mediating successful subcallosal cingulate deep brain stimulation for treatment-resistant depression. Biol Psychiatry. 2014;76(12):963–9.
84. Choi KS, Riva-Posse P, Gross RE, Mayberg HS. Mapping the "depression switch" during intraoperative testing of subcallosal cingulate deep brain stimulation. JAMA Neurol. 2015;72(11):1252–60.
85. Scangos KW, Makhoul GS, Sugrue LP, Chang EF, Krystal AD. State-dependent responses to intracranial brain stimulation in a patient with depression. Nat Med. 2021;27:229–31.
86. Riva-Posse P, Choi KS, Holtzheimer PE, Crowell AL, Garlow SJ, Rajendra JK, McIntyre CC, Gross RE, Mayberg HS. A connectomic approach for subcallosal cingulate deep brain stimulation surgery: prospective targeting in treatment-resistant depression. Mol Psychiatry. 2018;23(4):843–9.
87. Malone DA Jr, Dougherty DD, Rezai AR, Carpenter LL, Friehs GM, Eskandar EN, Rauch SL, Rasmussen SA, Machado AG, Kubu CS, Tyrka AR. Deep brain stimulation of the ventral capsule/ventral striatum for treatment-resistant depression. Biol Psychiatry. 2009;65(4):267–75.
88. Bewernick BH, Hurlemann R, Matusch A, Kayser S, Grubert C, Hadrysiewicz B, Axmacher N, Lemke M, Cooper-Mahkorn D, Cohen MX, Brockmann H. Nucleus accumbens deep brain stimulation decreases ratings of depression and anxiety in treatment-resistant depression. Biol Psychiatry. 2010;67(2):110–6.
89. Schlaepfer TE, Bewernick BH, Kayser S, Mädler B, Coenen VA. Rapid effects of deep brain stimulation for treatment-resistant major depression. Biol Psychiatry. 2013;73(12):1204–12.
90. Abelson JL, Curtis GC, Sagher O, Albucher RC, Harrigan M, Taylor SF, Martis B, Giordani B. Deep brain stimulation for refractory obsessive-compulsive disorder. Biol Psychiatry. 2005;57(5):510–6.
91. Greenberg BD, Gabriels LA, Malone DA, Rezai AR, Friehs GM, Okun MS, Shapira NA, Foote KD, Cosyns PR, Kubu CS, Malloy PF. Deep brain stimulation of the ventral internal capsule/ventral striatum for obsessive-compulsive disorder: worldwide experience. Mol Psychiatry. 2010;15(1):64–79.
92. Luigjes JV, Van Den Brink W, Feenstra MV, Van Den Munckhof P, Schuurman PR, Schippers R, Mazaheri A, De Vries TJ, Denys D. Deep brain stimulation in addiction: a review of potential brain targets. Mol Psychiatry. 2012;17(6):572–83.
93. Swann NC, de Hemptinne C, Thompson MC, Miocinovic S, Miller AM, Ostrem JL, Chizeck HJ, Starr PA. Adaptive deep brain stimulation for Parkinson's disease using motor cortex sensing. J Neural Eng. 2018;15(4):046006.
94. Haynes JD. A primer on pattern-based approaches to fMRI: principles, pitfalls, and perspectives. Neuron. 2015;87(2):257–70.
95. Shen FX. Neuroscience, mental privacy, and the law. Harv J Law Pub Poly. 2013;36:653.
96. Schott GD. Penfield's homunculus: a note on cerebral cartography. J Neurol Neurosurg Psychiatry. 1993;56(4):329.
97. Merton PA, Morton HB. Stimulation of the cerebral cortex in the intact human subject. Nature. 1980;285(5762):227.
98. Deng ZD, Lisanby SH, Peterchev AV. Electric field depth–focality tradeoff in transcranial magnetic stimulation: simulation comparison of 50 coil designs. Brain Stimul. 2013;6(1):1–3.
99. Deng ZD, Lisanby SH, Peterchev AV. Coil design considerations for deep transcranial magnetic stimulation. Clin Neurophysiol. 2014;125(6):1202–12.
100. Amassian VE, Cracco RQ, Maccabee PJ, Cracco JB, Rudell A, Eberle L. Suppression of visual perception by magnetic coil stimulation of human occipital cortex. Electroencephalogr Clin Neurophysiol. 1989;74(6):458–62.
101. Hallett M. Transcranial magnetic stimulation and the human brain. Nature. 2000;406(6792):147–50.
102. Pascual-Leone A, Valls-Solé J, Wassermann EM, Hallett M. Responses to rapid-rate transcranial magnetic stimulation of the human motor cortex. Brain. 1994;117(4):847–58.

103. Chen RM, Classen J, Gerloff C, Celnik P, Wassermann EM, Hallett M, Cohen LG. Depression of motor cortex excitability by low-frequency transcranial magnetic stimulation. Neurology. 1997;48(5):1398–403.
104. Chen R, Udupa K. Measurement and modulation of plasticity of the motor system in humans using transcranial magnetic stimulation. Motor control. 2009;13(4):442–53.
105. George MS. Whither TMS: a one-trick pony or the beginning of a Neuroscientific revolution? Am J Psychiatr. 2019;176(11):904–10.
106. O'Reardon JP, Solvason HB, Janicak PG, Sampson S, Isenberg KE, Nahas Z, McDonald WM, Avery D, Fitzgerald PB, Loo C, Demitrack MA. Efficacy and safety of transcranial magnetic stimulation in the acute treatment of major depression: a multisite randomized controlled trial. Biol Psychiatry. 2007;62(11):1208–16.
107. George MS, Lisanby SH, Avery D, McDonald WM, Durkalski V, Pavlicova M, Anderson B, Nahas Z, Bulow P, Zarkowski P, Holtzheimer PE. Daily left prefrontal transcranial magnetic stimulation therapy for major depressive disorder: a sham-controlled randomized trial. Arch Gen Psychiatry. 2010;67(5):507–16.
108. Guse B, Falkai P, Wobrock T. Cognitive effects of high-frequency repetitive transcranial magnetic stimulation: a systematic review. J Neural Transm. 2010;117(1):105–22.
109. Luber B, Lisanby SH. Enhancement of human cognitive performance using transcranial magnetic stimulation (TMS). NeuroImage. 2014;85:961–70.
110. Najib U, Horvath JC. Transcranial magnetic stimulation (TMS) safety considerations and recommendations. In: Transcranial magnetic stimulation. New York: Humana Press; 2014. pp. 15–30.
111. Alper K, Schwartz KA, Kolts RL, Khan A. Seizure incidence in psychopharmacological clinical trials: an analysis of Food and Drug Administration (FDA) summary basis of approval reports. Biol Psychiatry. 2007;62(4):345–54.
112. Rossi S, Hallett M, Rossini PM, Pascual-Leone A. Safety of TMS consensus group. Safety, ethical considerations, and application guidelines for the use of transcranial magnetic stimulation in clinical practice and research. Clin Neurophysiol. 2009;120(12):2008–39.
113. Iseger TA, Padberg F, Kenemans JL, Gevirtz R, Arns M. Neuro-cardiac-guided TMS (NCG-TMS): probing DLPFC-sgACC-vagus nerve connectivity using heart rate–first results. Brain Stimul. 2017;10(5):1006–8.
114. Carpenter LL, Janicak PG, Aaronson ST, Boyadjis T, Brock DG, Cook IA, Dunner DL, Lanocha K, Solvason HB, Demitrack MA. Transcranial magnetic stimulation (TMS) for major depression: a multisite, naturalistic, observational study of acute treatment outcomes in clinical practice. Depress Anxiety. 2012;29(7):587–96.
115. Dunner DL, Aaronson ST, Sackeim HA, Janicak PG, Carpenter LL, Boyadjis T, Brock DG, Bonneh-Barkay D, Cook IA, Lanocha K. A multisite, naturalistic, observational study of transcranial magnetic stimulation for patients with pharmacoresistant major depressive disorder: durability of benefit over a 1-year follow-up period. J Clin Psychiatry. 2014;75(12):1394–401.
116. Mantovani A, Pavlicova M, Avery D, Nahas Z, McDonald WM, Wajdik CD, Holtzheimer PE III, George MS, Sackeim HA, Lisanby SH. Long-term efficacy of repeated daily prefrontal transcranial magnetic stimulation (tms) in treatment-resistant depression. Depress Anxiety. 2012;29(10):883–90.
117. Janicak PG, Nahas Z, Lisanby SH, Solvason HB, Sampson SM, McDonald WM, Marangell LB, Rosenquist P, McCall WV, Kimball J, O'Reardon JP. Durability of clinical benefit with transcranial magnetic stimulation (TMS) in the treatment of pharmacoresistant major depression: assessment of relapse during a 6-month, multisite, open-label study. Brain Stimul. 2010;3(4):187–99.
118. Moyer ML, Cristancho MA, O'Reardon JP. Clinical efficacy of Transcranial magnetic stimulation in depression. In: A clinical guide to transcranial magnetic stimulation; 2014:17.
119. George MS, Wassermann EM, Williams WA, Callahan A, Ketter TA, Basser P, Post RM. Daily repetitive transcranial magnetic stimulation (rTMS) improves mood in depression. Neuroreport: Int J Rapid Commun Res Neurosci. 1995.

120. Ahdab R, Ayache SS, Brugières P, Goujon C, Lefaucheur JP. Comparison of "standard" and "navigated" procedures of TMS coil positioning over motor, premotor and prefrontal targets in patients with chronic pain and depression. Neurophysiol Clin. 2010;40(1):27–36.
121. Herwig U, Padberg F, Unger J, Spitzer M, Schönfeldt-Lecuona C. Transcranial magnetic stimulation in therapy studies: examination of the reliability of "standard" coil positioning by neuronavigation. Biol Psychiatry. 2001;50(1):58–61.
122. Herwig U, Satrapi P, Schönfeldt-Lecuona C. Using the international 10–20 EEG system for positioning of transcranial magnetic stimulation. Brain Topogr. 2003;16(2):95–9.
123. Fox MD, Buckner RL, White MP, Greicius MD, Pascual-Leone A. Efficacy of transcranial magnetic stimulation targets for depression is related to intrinsic functional connectivity with the subgenual cingulate. Biol Psychiatry. 2012;72(7):595–603.
124. Rusjan PM, Barr MS, Farzan F, Arenovich T, Maller JJ, Fitzgerald PB, Daskalakis ZJ. Optimal transcranial magnetic stimulation coil placement for targeting the dorsolateral prefrontal cortex using novel magnetic resonance image-guided neuronavigation. Hum Brain Mapp. 2010 Nov;31(11):1643–52.
125. Herbsman T, Avery D, Ramsey D, Holtzheimer P, Wadjik C, Hardaway F, Haynor D, George MS, Nahas Z. More lateral and anterior prefrontal coil location is associated with better repetitive transcranial magnetic stimulation antidepressant response. Biol Psychiatry. 2009;66(5):509–15.
126. Fitzgerald PB, Hoy K, McQueen S, Maller JJ, Herring S, Segrave R, Bailey M, Been G, Kulkarni J, Daskalakis ZJ. A randomized trial of rTMS targeted with MRI based neuro-navigation in treatment-resistant depression. Neuropsychopharmacology. 2009;34(5):1255–62.
127. Paillère Martinot ML, Galinowski A, Ringuenet D, Gallarda T, Lefaucheur JP, Bellivier F, Picq C, Bruguière P, Mangin JF, Rivière D, Willer JC. Influence of prefrontal target region on the efficacy of repetitive transcranial magnetic stimulation in patients with medication-resistant depression: a [18F]-fluorodeoxyglucose PET and MRI study. Int J Neuropsychopharmacol. 2010;13(1):45–59.
128. Weigand A, Horn A, Caballero R, Cooke D, Stern AP, Taylor SF, Press D, Pascual-Leone A, Fox MD. Prospective validation that subgenual connectivity predicts antidepressant efficacy of transcranial magnetic stimulation sites. Biol Psychiatry. 2018;84(1):28–37.
129. Drysdale AT, et al. "Resting-state connectivity biomarkers define neurophysiological subtypes of depression." Nat Med. 2017,23(1):s28–38.
130. Anderson RJ, Hoy KE, Daskalakis ZJ, Fitzgerald PB. Repetitive transcranial magnetic stimulation for treatment resistant depression: re-establishing connections. Clin Neurophysiol. 2016;127(11):3394–405.
131. Fox MD, Snyder AZ, Vincent JL, Corbetta M, Van Essen DC, Raichle ME. The human brain is intrinsically organized into dynamic, anticorrelated functional networks. Proc Natl Acad Sci. 2005;102(27):9673–78.
132. Downar J, Geraci J, Salomons TV, Dunlop K, Wheeler S, McAndrews MP, Bakker N, Blumberger DM, Daskalakis ZJ, Kennedy SH, Flint AJ. Anhedonia and reward-circuit connectivity distinguish nonresponders from responders to dorsomedial prefrontal repetitive transcranial magnetic stimulation in major depression. Biol Psychiatry. 2014;76(3):176–85.
133. Drysdale AT, Grosenick L, Downar J, Dunlop K, Mansouri F, Meng Y, Fetcho RN, Zebley B, Oathes DJ, Etkin A, Schatzberg AF. Resting-state connectivity biomarkers define neurophysiological subtypes of depression. Nat Med. 2017;23(1):28.
134. Siddiqi SH, Taylor SF, Cooke D, Pascual-Leone A, George MS, Fox MD. Distinct symptom-specific treatment targets for circuit-based neuromodulation. Am J Psychiatr. 2020;177(5):435–46.
135. Kaster TS, Downar J, Vila-Rodriguez F, Thorpe KE, Feffer K, Noda Y, Giacobbe P, Knyahnytska Y, Kennedy SH, Lam RW, Daskalakis ZJ. Trajectories of response to dorsolateral prefrontal rTMS in major depression: a THREE-D study. Am J Psychiatr. 2019;176(5):367–75.
136. Anderson B, Mishory A, Nahas Z, Borckardt JJ, Yamanaka K, Rastogi K, George MS. Tolerability and safety of high daily doses of repetitive transcranial magnetic stimulation in healthy young men. J ECT. 2006;22(1):49–53.

137. McDonald WM, Durkalski V, Ball ER, Holtzheimer PE, Pavlicova M, Lisanby SH, Avery D, Anderson BS, Nahas Z, Zarkowski P, Sackeim HA. Improving the antidepressant efficacy of transcranial magnetic stimulation: maximizing the number of stimulations and treatment location in treatment-resistant depression. Depress Anxiety. 2011;28(11):973–80.
138. Levkovitz Y, Isserles M, Padberg F, Lisanby SH, Bystritsky A, Xia G, Tendler A, Daskalakis ZJ, Winston JL, Dannon P, Hafez HM. Efficacy and safety of deep transcranial magnetic stimulation for major depression: a prospective multicenter randomized controlled trial. World Psychiatry. 2015;14(1):64–73.
139. Yip AG, George MS, Tendler A, Roth Y, Zangen A, Carpenter LL. 61% of unmedicated treatment resistant depression patients who did not respond to acute TMS treatment responded after four weeks of twice weekly deep TMS in the Brainsway pivotal trial. Brain Stimul. 2017;10(4):847–9.
140. Holtzheimer PE III, McDonald WM, Mufti M, Kelley ME, Quinn S, Corso G, Epstein CM. Accelerated repetitive transcranial magnetic stimulation for treatment-resistant depression. Depress Anxiety. 2010;27(10):960–3.
141. Hadley D, Anderson BS, Borckardt JJ, Arana A, Li X, Nahas Z, George MS. Safety, tolerability, and effectiveness of high doses of adjunctive daily left prefrontal repetitive transcranial magnetic stimulation for treatment-resistant depression in a clinical setting. J ECT. 2011;27(1):18–25.
142. Suppa A, Huang YZ, Funke K, Ridding MC, Cheeran B, Di Lazzaro V, Ziemann U, Rothwell JC. Ten years of theta burst stimulation in humans: established knowledge, unknowns and prospects. Brain Stimul. 2016;9(3):323–35.
143. Huang YZ, Edwards MJ, Rounis E, Bhatia KP, Rothwell JC. Theta burst stimulation of the human motor cortex. Neuron. 2005;45(2):201–6.
144. Li CT, Chen MH, Juan CH, Huang HH, Chen LF, Hsieh JC, Tu PC, Bai YM, Tsai SJ, Lee YC, Su TP. Efficacy of prefrontal theta-burst stimulation in refractory depression: a randomized sham-controlled study. Brain. 2014;137(7):2088–98.
145. Blumberger DM, Vila-Rodriguez F, Thorpe KE, Feffer K, Noda Y, Giacobbe P, Knyahnytska Y, Kennedy SH, Lam RW, Daskalakis ZJ, Downar J. Effectiveness of theta burst versus high-frequency repetitive transcranial magnetic stimulation in patients with depression (THREE-D): a randomised non-inferiority trial. Lancet. 2018;391(10131):1683–92.
146. Cole EJ, Stimpson KH, Bentzley BS, Gulser M, Cherian K, Tischler C, Nejad R, Pankow H, Choi E, Aaron H, Espil FM. Stanford accelerated intelligent Neuromodulation therapy for treatment-resistant depression. Am J Psychiatr. 2020;177(8):716–26.
147. Williams NR, Sudheimer KD, Bentzley BS, Pannu J, Stimpson KH, Duvio D, Cherian K, Hawkins J, Scherrer KH, Vyssoki B, DeSouza D. High-dose spaced theta-burst TMS as a rapid-acting antidepressant in highly refractory depression. Brain. 2018;141(3):e18.
148. Jensen E, Nejad R, Cole E, Williams N. Benchmarking aiTBS with the gold standard. Biol Psychiatry. 2020;87(9):S129.
149. Cole EJ, Phillips AL, Bentzley BS, Stimpson KH, Nejad R. Barmak F, Williams NR. Stanford neuromodulation therapy (SNT): a double-blind randomized controlled trial. Am J Psychiatry. 2022:179(2):132–41.
150. Sackeim HA. Acute continuation and maintenance treatment of major depressive episodes with transcranial magnetic stimulation. Brain Stimulat. 2016;9(3):313–9.
151. Kessler RC. The costs of depression. Psychiatr Clin. 2012;35(1):1–4.
152. Carmi L, Tendler A, Bystritsky A, Hollander E, Blumberger DM, Daskalakis J, Ward H, Lapidus K, Goodman W, Casuto L, Feifel D. Efficacy and safety of deep transcranial magnetic stimulation for obsessive-compulsive disorder: a prospective multicenter randomized double-blind placebo-controlled trial. Am J Psychiatr. 2019;176(11):931–8.
153. Philip NS, Barredo J, Aiken E, Larson V, Jones RN, Shea MT, Greenberg BD, van't Wout-Frank M. Theta-burst transcranial magnetic stimulation for posttraumatic stress disorder. Am J Psychiatr. 2019;176(11):939–48.

154. Barnett SB, Ter Haar GR, Ziskin MC, Rott HD, Duck FA, Maeda K. International recommendations and guidelines for the safe use of diagnostic ultrasound in medicine. Ultrasound Med Biol. 2000;26(3):355–66.
155. U.S. Food and Drug Administration. Ultrasound imaging. [online]; 2020. https://www.fda.gov/radiation-emitting-products/medical-imaging/ultrasound-imaging. Accessed 3 Jul 2020.
156. Wu A, Lindner G, Maitz AH, Kalend AM, Lunsford LD, Flickinger JC, Bloomer WD. Physics of gamma knife approach on convergent beams in stereotactic radiosurgery. Int J Rad Oncol Biol Phys. 1990;18(4):941–9.
157. Bowary P, Greenberg BD. Noninvasive focused ultrasound for neuromodulation: a review. Psychiatr Clin. 2018;41(3):505–14.
158. Fishman PS. Thalamotomy for essential tremor: FDA approval brings brain treatment with FUS to the clinic. J Ther Ultrasound. 2017;5:19.
159. Bond AE, Shah BB, Huss DS, Dallapiazza RF, Warren A, Harrison MB, Sperling SA, Wang XQ, Gwinn R, Witt J, Ro S. Safety and efficacy of focused ultrasound thalamotomy for patients with medication-refractory, tremor-dominant Parkinson disease: a randomized clinical trial. JAMA Neurol. 2017;74(12):1412–8.
160. Fry FJ, Ades HW, Fry WJ. Production of reversible changes in the central nervous system by ultrasound. Science. 1958;127(3289):83–4.
161. Tyler WJ, Lani SW, Hwang GM. Ultrasonic modulation of neural circuit activity. Curr Opin Neurobiol. 2018;50:222–31.
162. Kim H, Chiu A, Lee SD, Fischer K, Yoo SS. Focused ultrasound-mediated non-invasive brain stimulation: examination of sonication parameters. Brain Stimul. 2014;7(5):748–56.
163. Lee W, Kim H, Jung Y, Song IU, Chung YA, Yoo SS. Image-guided transcranial focused ultrasound stimulates human primary somatosensory cortex. Sci Rep. 2015;5:8743.
164. Lee W, Kim HC, Jung Y, Chung YA, Song IU, Lee JH, Yoo SS. Transcranial focused ultrasound stimulation of human primary visual cortex. Sci Rep. 2016;6(1):1–2.
165. Beisteiner R, Matt E, Fan C, Baldysiak H, Schönfeld M, Philippi Novak T, Lohse-Busch H. Transcranial pulse stimulation with ultrasound in Alzheimer's disease—a new navigated focal brain therapy. Adv Sci. 2020;7(3):1902583.
166. Sanguinetti JL, Hameroff S, Smith EE, Sato T, Daft CM, Tyler WJ, Allen JJ. Transcranial focused ultrasound to the right prefrontal cortex improves mood and alters functional connectivity in humans. Front Hum Neurosci. 2020;14:52.
167. Kubanek J, Shukla P, Das A, Baccus SA, Goodman MB. Ultrasound elicits behavioral responses through mechanical effects on neurons and ion channels in a simple nervous system. J Neurosci. 2018;38(12):3081–91.
168. Yoo SS, Yoon K, Croce P, Cammalleri A, Margolin RW, Lee W. Focused ultrasound brain stimulation to anesthetized rats induces long-term changes in somatosensory evoked potentials. Int J Imaging Syst Technol. 2018;28(2):106–12.
169. Hynynen K, McDannold N, Vykhodtseva N, Jolesz FA. Noninvasive MR imaging–guided focal opening of the blood-brain barrier in rabbits. Radiology. 2001;220(3):640–6.
170. McDannold N, Vykhodtseva N, Raymond S, Jolesz FA, Hynynen K. MRI-guided targeted blood-brain barrier disruption with focused ultrasound: histological findings in rabbits. Ultrasound Med Biol. 2005;31(11):1527–37.
171. Lipsman N, Meng Y, Bethune AJ, Huang Y, Lam B, Masellis M, Herrmann N, Heyn C, Aubert I, Boutet A, Smith GS. Blood–brain barrier opening in Alzheimer's disease using MR-guided focused ultrasound. Nat Commun. 2018;9(1):1–8.
172. Abrahao A, Meng Y, Llinas M, Huang Y, Hamani C, Mainprize T, Aubert I, Heyn C, Black SE, Hynynen K, Lipsman N. First-in-human trial of blood–brain barrier opening in amyotrophic lateral sclerosis using MR-guided focused ultrasound. Nat Commun. 2019;10(1):1–9.
173. Airan RD, Meyer RA, Ellens NP, Rhodes KR, Farahani K, Pomper MG, Kadam SD, Green JJ. Noninvasive targeted transcranial neuromodulation via focused ultrasound gated drug release from nanoemulsions. Nano Lett. 2017;17(2):652–9.
174. Wang JB, Aryal M, Zhong Q, Vyas DB, Airan RD. Noninvasive ultrasonic drug uncaging maps whole-brain functional networks. Neuron. 2018;100(3):728–38.

175. Deisseroth K. Optogenetics: 10 years of microbial opsins in neuroscience. Nat Neurosci. 2015;18(9):1213–25.
176. Ibsen S, Tong A, Schutt C, Esener S, Chalasani SH. Sonogenetics is a non-invasive approach to activating neurons in Caenorhabditis elegans. Nat Commun. 2015;6(1):1–2.
177. Wu X, Zhu X, Chong P, Liu J, Andre LN, Ong KS, Brinson K, Mahdi AI, Li J, Fenno LE, Wang H. Sono-optogenetics facilitated by a circulation-delivered rechargeable light source for minimally invasive optogenetics. Proc Natl Acad Sci. 2019;116(52):26332–42.
178. Ghanouni P, Pauly KB, Elias WJ, Henderson J, Sheehan J, Monteith S, Wintermark M. Transcranial MRI-guided focused ultrasound: a review of the technologic and neurologic applications. Am J Roentgenol. 2015;205(1):150–9.
179. Cummer SA, Christensen J, Alù A. Controlling sound with acoustic metamaterials. Nat Rev Mater. 2016;1(3):1–3.
180. Moseley ME, Cohen Y, Kucharczyk J, Mintorovitch J, Asgari HS, Wendland MF, Tsuruda J, Norman D. Diffusion-weighted MR imaging of anisotropic water diffusion in cat central nervous system. Radiology. 1990;176(2):439–45.
181. Biswal B, Zerrin Yetkin F, Haughton VM, Hyde JS. Functional connectivity in the motor cortex of resting human brain using echo-planar MRI. Magn Reson Med. 1995;34(4):537–41.
182. Legon W, Sato TF, Opitz A, Mueller J, Barbour A, Williams A, Tyler WJ. Transcranial focused ultrasound modulates the activity of primary somatosensory cortex in humans. Nat Neurosci. 2014;17(2):322–9.
183. Harris JC. Galileo Galilei: scientist and artist. Arch Gen Psychiatry. 2010;67(8):770–1. https://doi.org/10.1001/archgenpsychiatry.2010.95.
184. Sadeghi-Tarakameh A, DelaBarre L, Lagore RL, Torrado-Carvajal A, Wu X, Grant A, Adriany G, Metzger GJ, Van de Moortele PF, Ugurbil K, Atalar E. In vivo human head MRI at 10.5 T: a radiofrequency safety study and preliminary imaging results. Magn Reson Med. 2020;84(1):484–96.
185. Lawless M, Hodge C. Lasik. Int Ophthalmol Clin. 2013;53(1):111–28.
186. Silberman S. Neurotribes: the legacy of autism and the future of neurodiversity. New York: Penguin; 2015.
187. Lord C, Cook EH, Leventhal BL, Amaral DG. Autism spectrum disorders. Neuron. 2000;28(2):355–63.
188. Greely HT. Remarks on human biological enhancement. U Kan Law Rev. 2007;56:1139.
189. Kramer P. Listening to prozac: a psychiatrist explores antidepressants drugs and the remaking of the self. Viking: New York; 1993.
190. Farah MJ, Illes J, Cook-Deegan R, Gardner H, Kandel E, King P, Parens E, Sahakian B, Wolpe PR. Neurocognitive enhancement: what can we do and what should we do? Nat Rev Neurosci. 2004;5(5):421–5.
191. Brokowski C, Adli M. CRISPR ethics: moral considerations for applications of a powerful tool. J Mol Biol. 2019;431(1):88–101.
192. Canavero S. Criminal minds: neuromodulation of the psychopathic brain. Front Hum Neurosci. 2014;8:124.
193. Greely HT, Farahany NA. Neuroscience and the criminal justice system. Ann Rev Criminol. 2019;2:451–71.
194. Farahany NA. Incriminating thoughts. Stan Law Rev. 2012;64:351.

Further Reading

Airan R. Focused ultrasound with nanoparticles: neuromodulation with nanoparticles. Science. 2017;357(6350):465.

Deisseroth K. Optogenetics review: optogenetics. Nat Methods. 2011;8(1):26–9.

Fox MD. Mapping symptoms to brain networks with the human connectome. N Eng J Med. 2018;379(23):2237–45. Connectomic approaches to neurotherapeutics.

George MS, Whither TMS. TMS review: a one-trick pony or the beginning of a neuroscientific revolution? Am J Psychiatry. 2019;176(11):904–10.

Greely HT. Ethics of neuroenhancement: remarks on human biological enhancement. U Kan Law Rev. 2007;56:1139.

Holtzheimer PE, Mayberg HS. DBS review: deep brain stimulation for psychiatric disorders. Ann Rev Neurosci. 2011;34:289–307.

Tyler WJ, Lani SW, Hwang GM. Focused ultrasound review: ultrasonic modulation of neural circuit activity. Curr Opinion Neurobiol. 2018;50:222–31.

Williams NR, Taylor JJ, Kerns S, Short EB, Kantor EM, George MS. Interventional psychiatry: interventional psychiatry: why now? J Clin Psychiatry. 2014;75(8):895.

Chapter 3
Clinical Neuroinnovation: Ethical Frameworks and Emerging Issues

Max Kasun, Laura B. Dunn, Barton W. Palmer, Jane Paik Kim, and Laura Weiss Roberts

Introduction

Brain-based illnesses (i.e., psychiatric, neurological, neurodevelopmental, neurodegenerative, and substance use disorders) contribute to immense personal, familial, societal, and economic costs worldwide, including premature mortality and high levels of disability [1, 2]. The great personal and societal devastation attributable to these illnesses—combined with the reality that even the best available treatments are inadequate in the treatment of many individuals—together create the imperative for further advancement in neuroscience to enhance our models of disease etiologies and their underlying mechanisms, test disease-modifying treatments, and develop interventions for complex symptom patterns (e.g., behavioral, cognitive, and neurological comorbidities). Extraordinary neuroinnovations—defined as new approaches and tools that contribute to the treatment of brain-based illness—have already improved the lives of patients [3, 4]. Future generations of clinical neuroinnovations hold immense promise for identifying the concrete origins and presentations of brain-based diseases and for improving humanity's capacity to reduce their great burden.

M. Kasun · J. P. Kim · L. W. Roberts
Department of Psychiatry and Behavioral Sciences, Stanford University School of Medicine, Stanford, CA, USA
e-mail: maxkasun@stanford.edu; janepkim@stanford.edu; robertsl@stanford.edu

L. B. Dunn (✉)
Department of Psychiatry, University of Arkansas for Medical Sciences, Little Rock, AR, USA
e-mail: Ldunn@uams.edu

B. W. Palmer
UC San Diego, La Jolla, CA, USA

VASDHS, La Jolla, CA, USA
e-mail: bpalmer@ucsd.edu

L. W. Roberts (ed.), *Ethics and Clinical Neuroinnovation*,
https://doi.org/10.1007/978-3-031-14339-7_3

The shared and primary moral and intellectual mission of clinicians and biomedical researchers is to reduce the emotional, social, functional, and economic hardships associated with mental and physical disorders. One of the foundations for this mission is the public's trust in medical professionals and institutions [5]. However, the history of medicine in general, and of psychiatry and neurology specifically, provide many examples of excitement, hype, and apparent promise—followed by disappointment, unintended consequences, and outright harms [6, 7]. Even the most prescient of ethics codes designed to prevent ethical failures can gradually lose prescriptive utility as their historical, cultural, and intellectual contexts elapse. Public trust is therefore hard-earned, conferred on medicine only through habits of active and systematic ethical analysis.

Modern bioethical inquiry includes multiple approaches that are jointly essential for establishing a coherent sense of what is good and right in the act of caring for human beings, including in situations that create ethical dilemmas. Neuroethical inquiry makes use of these same approaches because it shares this overarching purpose. Three foremost approaches are empirical (or descriptive) ethics, normative (or prescriptive) ethics, and principlism [8]. Empirical ethics is an approach that describes and studies, through observation and experimentation, the broad array of ethical issues and questions that accompany clinical research. Many ethical issues in clinical neuroinnovation can be anticipated and resolved through empirical attunement to the individuals who assume their societal, professional, and individual benefits and risks. For moral questions for which sufficient answers may not be available empirically—namely, questions of *how* to act, given our observations—normative ethics can offer ways of reaching a justifiable action. Normative ethics extend from traditions in moral philosophy; some of them—though outside the scope of this text—are Kantianism, utilitarianism, and deontology. Complementary to empirical and normative approaches are bioethical principles that have guided research through the past half-century, such as nonmaleficence, beneficence, justice, respect for persons, and autonomy. Bioethical principles represent key outcomes of normative thought to which the medical community has ascribed special and enduring value. These topics are discussed further in the following sections, "Historical background to ethical issues: principles and norms" and "Assessing ethical aspects of clinical neuroinnovation."

Neuroinnovations of the past several decades represent cultural, intellectual, technological, and pragmatic transformations that cause them to differ from neuroinnovations of the twentieth century. This earlier period included the birth of basic neuroimaging tools and device-based therapies that are still central to research and treatment of brain-based illness, including magnetic resonance imaging (MRI), electroencephalography (EEG), deep brain stimulation (DBS), transcranial magnetic stimulation (TMS), pharmacological therapies, and sense-restoring implants. The current generation of neuroscience, as conceptualized by the NIH BRAIN Initiative, is a more integrated effort to discover cell diversity, map circuits at multiple scales, dynamically visualize the functioning brain, demonstrate neural circuit causality, identify fundamental principles of mental processes, support neuroscience research networks, and develop comprehensive models of cognition, emotion, perception, and action [9].

Many recent transformations in neuroinnovation have been driven by advancements in computational utilities and approaches that enable us to exceed the capabilities of natural cognitive systems. Important achievements in this category include the capability to collect, model, and run optimization and analysis functions on big, multi-modal data, as well as developments in artificial intelligence (especially computer vision, machine learning (ML), and natural language processing (NLP)) to enhance diagnoses, prognoses, and treatments [9]. Emerging applications of computational methods have been proposed and studied for high-dimensional systems problems (e.g., distinguishing the neural correlates of specific behaviors, or finding optimal solutions for day-to-day resource allocation objectives in healthcare systems) [10, 11]. Achievements already made in deriving improved taxonomies of neural circuit types may soon lead to more finely personalized circuit-based interventions [12].

Artificial intelligence may soon help us more effectively navigate the explore-exploit tradeoffs of neuroscientific inquiry itself—as evidenced by promising hypothesis-testing simulators of brain disorder pathologies that are orders of magnitude larger and more resource-efficient than human-run trials [13, 14]. Artificial intelligence is being explored as a means of enhancing knowledge production, including through the systematic capturing of knowledge from research studies and automation of the labor-intensive synthesis of research findings [15]. Collectively, neuroinnovations are rapidly transforming the culture of research, raising many new ethical concerns. Some concerns relate to novel challenges for traditional ethical safeguards (e.g., informed consent and personal data protections), while others relate to theoretical questions, such as the epistemic capabilities and medically acceptable roles of machine agents. Examples of such ethical concerns are given at the end of this chapter in the section titled "Application of modern ethical models in neuroinnovation."

The social science dimension of ethical inquiry on neuroinnovation is quickly evolving as well. Many of the individuals who collaborate in the creation and use of neuroinnovative tools now work outside of public and/or academic institutions, may be committed to or familiar with varied ethics codes (or none at all), and may hold different views about their enterprises' primary problems and goals. There is therefore major ethical interest in increasing attunement to the perspectives of the diverse stakeholder groups who impact and are impacted by modern research, policy, and translation, including neuroscientists, engineers, ethicists, legal scholars, policymakers, clinicians, and patients. A more empirically informed and coherent foundation for ethics in neuroinnovation is essential to ensure that mutualism prevails among disciplines, to sustain the public trust, and ultimately to maximally protect the rights and welfare of patients and the wider public.

The immense and boldly ambitious NIH BRAIN Initiative—whose goal is to "to produce a revolutionary new dynamic picture of the brain that, for the first time, shows how individual cells and complex neural circuits interact in both time and space"—explicitly recognized the relevance and importance of contemplating and addressing ethical issues in innovative neuroscience as this groundbreaking research proceeds [16]. With an extraordinarily rich and diverse range of research projects, the BRAIN Initiative has laid the foundation for revolutionary advances in clinical applications of innovative neurotechnologies. In addition, outside of the BRAIN Initiative itself, numerous other research entities are also making inroads into understanding and manipulating the brain.

As these phenomenal research advances proceed from animal and organoid models to human trials, attention to the numerous ethical issues accompanying them will remain vital to reinvigorating and sustaining the public's trust in the research community. In this chapter, we describe the complementary prescriptive, principles-based, and empirical (descriptive) approaches to analyzing the ethical aspects of clinical neuroinnovation.

Historical Background to Neuroethics: Principles and Norms

Neuroethics extends from the long intellectual and legal tradition of bioethics which has elevated a range of core principles. While ethical principles can be formulated as ideals, they function in everyday health care to guide how we reason in circumstances where the particulars may be complex, unfamiliar, or even new, the principles may exist in tension, and the best course of action may be especially hard to determine. The 1979 "Belmont Report" of the National Commission for the Protection of Human Subjects of Biomedical and Behavioral Research remains a landmark document guiding ethical human research [17]. The three core ethical principles identified and described therein are *respect for persons, beneficence,* and *justice*. Of note, *respect for persons* includes two components that are themselves sometimes in tension—respect for individuals as autonomous agents and protection of those with diminished autonomy. *Beneficence* also has two components, i.e., one proscriptive (nonmaleficence) and one prescriptive (promotion of good). Although sometimes overlooked in operationalization of research ethics, Beauchamp & Childress [18] note that "principles of beneficence potentially demand more than the principle of nonmaleficence because agents must take positive steps to help others, not merely refrain from harmful acts" (p. 165). Similar prescriptive beneficence considerations were also noted within the Belmont Report in regard to the societal obligation to support sound and ethical research that has potential to enhance our understanding of disease, and to lead to more effective and safer prevention and intervention methods.

Similarly, the principle of *justice* is not solely proscriptive (the avoidance of exploiting segments of the population due to vulnerability or convenience); it also requires equitable distribution of the opportunity to benefit from the insights and tools generated by research. In short, a major conceptual contribution of the Belmont Report was to affirm that ethics are not merely a list of prohibitions: they also strongly demand development and support of otherwise ethically and scientifically sound research that holds promise to promote and advance human wellbeing. Much of the current and foreseeable neuroinnovative research meet this latter description. In addition, the Belmont Report emphasized that there are often not straightforward answers to ethical dilemmas; rather, resolving them requires thoughtful balancing of ethical principles that are in tension.

The so-called Common Rule (i.e., common to many different U.S. federal agencies), instituted in the United States in 1981, built upon the foundation of ethical principles laid out in the Belmont Report to establish an explicit system of regulations for research involving human subjects. The main effects of the Common Rule were to define a human subject (i.e., as any living person about whom a researcher obtains data), and to define categories of "vulnerable" individuals requiring additional protection due to the potential for greater risks related to physical health and/or informed consent, including prisoners, pregnant women, fetuses, newborns, and children [19]. A major revision to the Common Rule was added in 2018. This revision broadened the initial definition of "obtaining data" to "using, studying, or analyzing individuals' information or biospecimens or generating identifiable private information or identifiable biospecimens," reflecting ethical concerns raised by advances in computing and many advances in biomedical innovation. The revision also newly excluded several data-collecting activities from the definition of "research" requiring IRB approval, including public health surveillance, criminal justice or investigative functions, certain national security purposes, and scholarly and journalistic activities that utilize biographical data [20].

In the time since the advent of the Common Rule, a semantic shift has taken place—in which the language of "human subjects" has moved increasingly towards a more participatory and volunteer-centric tone [21]. This shift reflects a growing awareness of the importance of the perspectives of individuals who participate in research regarding a broad range of ethical issues in research [22]. Accompanying this evolving awareness, a growing body of empirical work has illuminated the individual factors at play in research decision-making, including factors that can threaten fully authentic and voluntary participation in research. Some of this work has overturned prior assumptions in medicine and in broader society about the relative strengths and weaknesses of populations considered vulnerable due to mental or physical health conditions.

For example, although individuals with mental illnesses such as schizophrenia had often been presumed to lack the capacity to make research decisions, a body of work that emerged in the 1990s and subsequent decades documented the wide range of interindividual differences in decision-making capacity among individuals with mental illnesses [23–30]. Reflecting increased understanding of this heterogeneity and awareness of ethical risks associated with presuming impairments in decisional capacity without holistic and sustained assessment of individual patients, the term "individuals with impaired decision-making capacity" has replaced "handicapped or mentally disabled persons" as a vulnerable category in the Revised Common Rule [20].

Assessing Aspects of Clinical Neuroinnovation

As discussed further in Chap. 15, "Qualitative findings: A focus on professional stakeholder perspectives on the environments and challenges of innovative neuroscience research," numerous aspects of neuroinnovation are shaped not only by scientific contexts and technological capabilities, but also by regulatory and industrial pressures. Research related to the brain and all other research in the clinical context is within the jurisdiction of ethical regulations set forth by the Department of Health and Human Services (the entity responsible for the Common Rule), the NIH, the Food and Drug Administration (FDA), and other federal health institutions.

In contrast, privately funded research involving human volunteers is not subject to the provisions of the Common Rule [23], but rather is subject to the FDA's Code of Federal Regulations (CFR), which contains provisions somewhat analogous to those of the Common Rule. These provisions apply to research related to drugs, biological products (including organoids, biospecimens, and derived data), and medical devices, including neurotechnologies [31]. Increased attention has recently been called to areas where these two regulatory codes diverge, particularly where there may be ambiguities that could reduce safeguards for protection of human participants [32]. Emerging technologies raise new contexts where ethical concerns play out; for example, issues raised by digital mental health applications and their use and retention of personal health data remain largely unaddressed in these codes [33, 34]. Numerous additional ethical concerns will continue to emerge in the context of human trials for neurotechnologies such as brain–machine interfaces—for example, as prefigured by prototypes developed by researchers at Neuralink [35]. To the extent that such issues remain unaddressed (or inadequately addressed), neuroinnovations extend outside the public trust (Fig. 3.1).

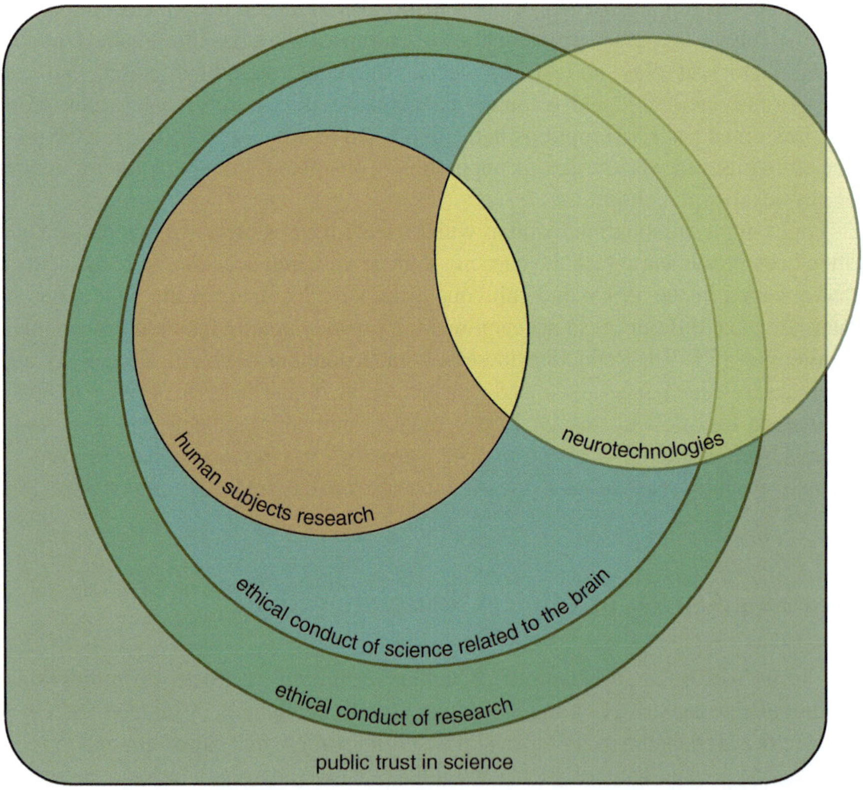

Fig. 3.1 Overlapping domains of ethical interest in neuroinnovation. Copyright 2019, Roberts Ethics Lab. Used with permission

Value (Clinical Need)

First and foremost, clinical interventions research with human volunteers requires a clear, valuable, clinical need. Greater than minimal risk research on humans cannot be justified solely on the existence of an interesting scientific question; it requires sufficient prior evidence that the research will uniquely address a health problem not already resolvable through standard means and will present lower risk than the problem itself. The decision tree in Fig. 3.2 outlines key components of this process.

Informed Consent

To date, much of the literature on ethical issues in neuropsychiatric research has focused on informed consent, especially in the context of schizophrenia or dementias [36, 37]. This was a logical starting point for empirical research on ethics. The notion of informed consent, to begin with, needed to be articulated [38]. The research community has largely accepted a primarily cognitive model of informed

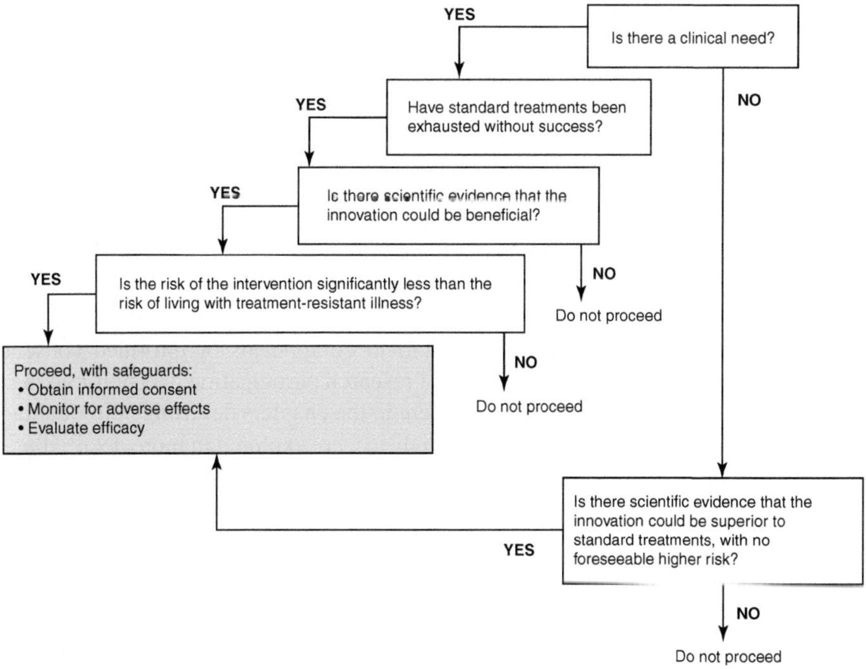

Fig. 3.2 Decision tree for ethical treatment using clinical neuroinnovations. Adapted from Hoop JG, Layde J, Roberts LW: "Ethical Considerations in Psychopharmacological Treatment and Research," in Textbook of Psychopharmacology, fourth Edition. Edited by Schatzberg AF, Nemeroff CB. Washington, DC, American Psychiatric Publishing, 2009, p. 1490

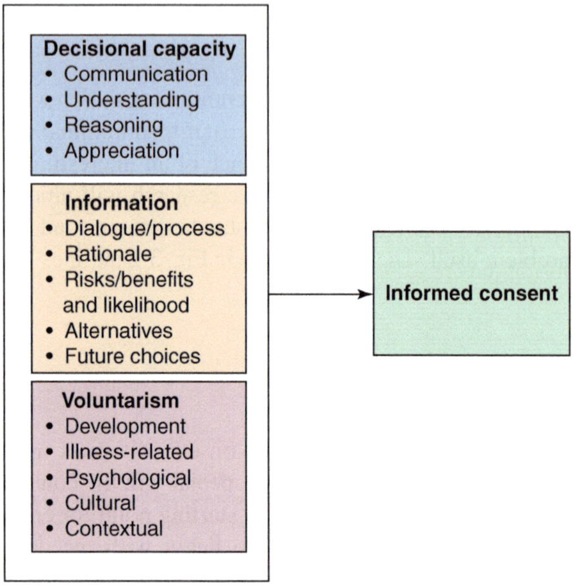

Fig. 3.3 Elements of informed consent. Adapted from Roberts LW, Dyer AR (eds): "Informed Consent and Decisional Capacity," in *Concise Guide to Ethics in Mental Health Care*. Washington, DC, American Psychiatric Publishing, 2004, p. 52

consent (Fig. 3.3), which has at its center the idea of decision-making capacity as a prerequisite for autonomous informed consent [39, 40]. In turn, decision-making capacity has been operationalized as comprising four component abilities—understanding, appreciation, reasoning, and expression of a choice [40, 41]. For over four decades, this model of consent and capacity has dominated discussions of ethical issues in clinical research. The model has meanwhile evolved, and now accommodates two additional components that are considered generative of true informed consent: information (related to the quality and comprehensiveness of knowledge transmitted to prospective volunteers about study content and purpose, risk, and benefit) and voluntarism (predicated on attunement to prospective volunteers' coherent senses of identity and experience, self-efficacy, and value systems).

Threats to the information and voluntarism components of informed consent include a number of less-studied aspects of research participation decision-making. These include several broad domains, which, in the chapters describing our qualitative research findings (i.e., Chaps. 14 through 16), are explored in more detail, illustrating the notable themes with remarks from the interview participants themselves. Here we describe each of these domains in consideration of the ethical questions and issues raised by neuroinnovation.

Information

Ethical engagement of potentially vulnerable volunteers in human studies, especially neuroinnovative research with as-yet poorly understood risks and benefits, is predicated on rigorous yet genuine informed consent processes that enable positive influences on participation decisions and safeguard against negative influences toward participation [42]. Informed consent consists in a continuous dialogue and is not simply a form transmitted to a potential volunteer for their signature [43]. The problematic focus on legalistic consent forms has been noted for decades. As noted four decades ago by Roth et al. [44], "obtaining informed consent through consent forms too frequently becomes a 'rite' rather than a right." And as observed in a 2003 Institute of Medicine (IOM) report "consent forms have been hijacked as 'disclosure documents' for the risk management purposes of research organizations" [45]. In this regard, the past 25 years of research on improving participant comprehension during the consent has been of critical importance [46].

Voluntarism

Voluntarism is the principle that individuals should be able to contemplate participation in research through their free exercise of reason and judgment, with maximal preservation and respect of the coherent set of values, beliefs, and preferences that makes up each individual's understanding of self. The full rational conception of voluntarism thus also presupposes also an accurate sense of the features of a given study, including risk and benefit. Voluntarism has received relatively little attention in the literature [47–50]. In the interviews conducted as part of the study described in Chaps. 14 through 16, however, voluntarism emerged as a crucial concern of our participants. For instance, although concerns have been raised for many years about the potential that individuals may enroll in research out of desperation or "false" or misplaced hope [51], minimal work has empirically examined whether such negative valence factors actually outweigh more positive valence factors, such as a desire to help others with the same condition, or a robust informed consent process. Prior work, including ours, identified factors that influence research participation decision-making [52, 53]. Some affect a potential participant's decision-making favorably and appropriately; we call these "positive valence" factors, e.g., altruism; salience of the condition under study; accurate understanding of study procedures, risks, and benefits. Other influences are more ethically problematic; they may sway an individual toward participation by factors that have "negative valence," e.g., desperation; lack of resources; threats to voluntarism [48, 49, 54] (Fig. 3.4). Finally, it is not yet fully understood how (or whether) the valence factors currently appreciated by research ethics are commensurable to an overall ideal of voluntarism, given their highly individual, dynamic, and context-sensitive nature and the ultimately reductive outcome salient to prospective volunteers (whether or not to participate).

It is expected that a decision to participate in research will be influenced by a number of factors, some "positive" and some "negative." The presence of negative valence factors is not in and of itself ethically problematic, but overly weighted negative valence factors are problematic. In ethically sound decision-making, negative valence factors will be at least balanced by positive valence factors. Ideally, positive valence factors will shape the decision to participate in research. Researchers can "tip the scale" through robust study-specific safeguards that ensure that positive factors outweigh negative factors.

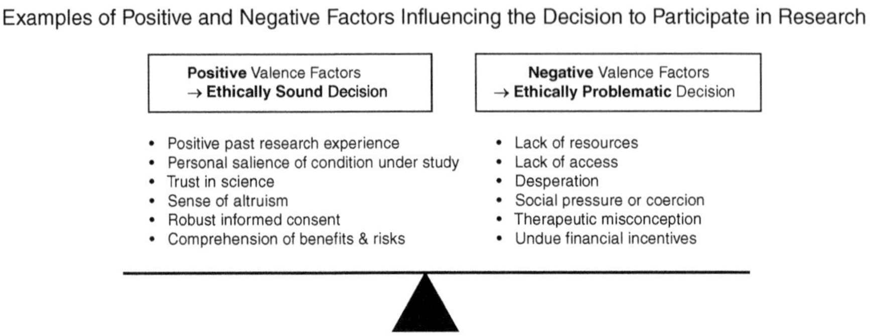

Fig. 3.4 Roberts Valence Model with examples of positive and negative factors influencing the decision to participate in research. Adapted from Roberts LW, Kasun M, Termuehlen G. Ethics in the mental health professions. IN: Roberts LW, Termuehlen G, eds. Professionalism and Ethics Q&A Self Study Guide for Mental Health Professionals, second Edition. Washington, DC: American Psychiatric Association Publishing: 2021, pg 112. Used with permission

Decisional Capacity

As described in Chaps. 12–16 of this book, our group conducted a series of interviews with clinical researchers engaged in innovative neuroscience research, with the specific goal of obtaining their perspectives on the primary/priority ethical challenges faced in current and foreseeable innovative neuroscience research. One of the already-common challenges identified by the interviewed researchers was the issue of obtaining true informed consent. Specifically, due to the potential cognitive effects of many brain-based disorders, the very nature of these conditions, such as dementias, may increase the risk of impaired decision-making capacity. And yet, it is now known that decisional capacity varies among people with mental illness—and even between people who have the same disorder. For this reason, assessing informed consent bears critically on the expertise and careful attunement of clinician-researchers to the research volunteers who entrust them to provide good care.

As an example relevant to neuroinnovative research, consider one unique ethical tension in play in researching promising disability reduction interventions for brain injury and serious mental illness that entail loss of decisional capacity; for example, comatose patients in an emergency room setting. The promise of such research to reduce this specific burden necessarily requires research on people who are not capable of reaching decisions. Although specific regulatory requirements, such as proxy and deferred consent, have been specified for emergency research under such conditions, the ethical tension remains. Prohibition of such research could simultaneously promote the protective component of respect for persons, while, at least at a societal level, challenge the prescriptive aspects of beneficence and justice. It is because of the fact that the weight given to such considerations is inherently subjective, and that even well-meaning researchers and IRB reviewers are not free from implicit biases, that regulations for emergency research generally require community consultation [55].

Return of Sensitive Findings

In addition to issues of capacity to consent to research, a key issue that arose in virtually every interview was on the theme of sensitive findings. These could involve either incidental findings (i.e., unexpected discovery of clinically relevant information within research assessments), or foreseeable information that might not be "actionable" (or at least not currently so)—such as a biomarker indicating that an otherwise cognitively healthy individual may be in the pre-clinical stages of a neurodegenerative condition. While the terminology for such findings has not always been consistent in the literature, here we use the terms "secondary" or "sensitive" findings to include those that may be primary to the research question (e.g., biomarker assessment in studies of neurodegenerative disorders), and which are anticipated to be found in some proportion within the study population, but which have not been validated or approved for clinical use and/or that have no currently actionable implications.

The primary (potential) ethical tension with return of incidental and secondary findings rests between respect for persons and beneficence (including nonmaleficence). Respect for persons compels researchers to pre-warn potential participants about the possibility of incidental and secondary findings, including the potential nature of those findings; plans for management of any such findings; and determination of participant preferences [56]. However, sometimes due to advances in technology or analytic methods, the nature of secondary findings may be such that they were not known even to the investigators during the consent process. Several of the investigators whom we interviewed noted that ongoing and rapid methodological advances in clinical neuroscience may result in a higher number of previously unimagined forms of incidental or secondary findings in the context of neuroinnovative research.

Researchers may be ethically compelled by the prescriptive component of beneficence to disclose any serious findings when potentially actionable (e.g., a potentially treatable tumor found during MRI). On the other hand, nonmaleficence requires researchers to reduce harm, including potential psychological harms (anxiety, depression, and suicidal ideation). Yet, even the determination of what is "actionable" may be more uncertain than anticipated. Advances in treatment following study participation may change a finding to readily actionable. Moreover, several of the investigators in our interviews noted that even in the context of a perfectly sensitive and specific biomarker for pre-clinical Alzheimer's disease, the definition of "actionable" might include not only potential treatments, but also lifestyle choices and planning for the future. The latter case illustrates the potential presence of subjectivity and role of personal preferences in determinations of what is deemed "actionable."

Secondary analyses of data in biorepositories or electronic data repositories, as is generally now required for all federally funded research [57, 58], further complicate the issue of unforeseen sensitive information. The ability to identify certain clinically relevant findings (such as identification of a biomarker of Alzheimer's disease risk with high sensitivity and specificity) may emerge in analyses completely removed from the purpose of the original study to which the participant consented. This issue has received substantial attention in the past 15 years in the context of genomic and imaging research [59, 60].

Novel Ethical Issues in Neuroinnovative Approaches

Machine Learning

Of the diverse subfields of artificial intelligence, machine learning and its more recent subbranch, deep learning, have demonstrated the greatest promise in the past two decades in clinical neuroinnovation and medicine more broadly [61, 62; see 63, 64]. Remarkably, machine learning and deep learning are inspired by the neural architecture and heuristic-building processes of the brain. These algorithms can be generally classified as utilizing unsupervised, supervised, or reinforcement learning. In the first category, unsupervised learning, the algorithm's goal is to learn patterns in unstructured data with no labels. The problem of understanding patterns between brain, behavior, and genetics, by identifying novel groupings of symptoms or brain circuits, is one that has been well addressed by unsupervised learning techniques. In supervised learning, an algorithm relies on symbolic, statistical, or probabilistic methods of learning from training data with the goal of predicting labels on new and unseen data. At the heart of deep learning algorithms, for example, is a subset of algorithms called artificial neural networks, which are composed of a layer of input nodes, multiple hidden layers of nodes, and an output layer. Each node, which acts as its own regression function, takes in its input through an activation function, and results in an output and either "fires" the neuron (i.e., passes on data to the next layer) or does not. Reinforcement learning is concerned with an agent who makes sequential decisions with the goal of learning a task and receiving rewards and is often used to optimize decision-making in certain settings. For example, reinforcement learning has been used in healthcare to develop optimal, tailored treatments for epilepsy and as a tool to help with clinical decision-making related to sepsis management [65].

Deep learning methods that can be accomplished without importing a priori, human-developed classifications belong to feature learning or representation learning. Deep learning algorithms for medical image segmentation outperform traditional methods, allowing physicians to divide medical images into semantically meaningful regions for the purpose of diagnosis, disease monitoring, prognosis, and treatment planning [66]. Computer vision technologies applied to hospital settings have demonstrated promise in detecting patient mobilization activities (such as getting out of bed) in the ICU and may be used to minimize negative outcomes related to frequent movement [67, 68]. In neuroinnovation, deep learning has enabled computer vision algorithms to quantify and detect patterns in key biomarkers such as brain tissue atrophy in brain scans, and then to use these patterns to build multidimensional characterizations of brain disorders including schizophrenia, multiple sclerosis, and Alzheimer's disorder [69]. Some deep learning approaches have also demonstrated the potential to identify and characterize structural or circuit-related neurobiological features of brain-based disorders, which may be used to distinguish between individuals with and without brain-based disorders or to predict treatment response (e.g., [70, 71]).

Several major ethical concerns have emerged in applications of machine learning. First, there are bias issues that can compromise the ethical status of innovation on principles including beneficence, nonmaleficence, and justice. It is increasingly well understood that machine learning and deep learning algorithms can automate biases and perpetuate health disparities that already exist in the world. Such biases can originate in sampling biases, biased labeling of training data that can function as proxies for preexisting health inequities, and many other biases, as enumerated by Rajkomar et al. [72]. For example, Hutchinson et al. [73] found that machine learning models trained on natural language text tend to classify texts which mention disability as more negative, regardless of their actual semantic content—and that these undesirable biases were partly explained by underlying topical biases incorporating homelessness, gun violence, and addiction [73]. During translation, clinical machine learning involves many potential biases at virtually every stage where value-laden judgments take place (e.g., sampling, labeling, and algorithm design) (see Fig. 3.5).

A second major concern relates to the use and retention of data. New machine learning applications involve questions about what privacy rights belong to individuals, and what obligations such rights entail. The Revised Common Rule and National Defense Authorization Act of 2020 both contain provisions that seek to clarify how best to ensure the right to control how and when one's personal data is collected and used [74]. The latter document provides federal funding to address privacy risks in emerging applications of artificial intelligence while also creating programs that may pose new privacy risks, such as shared data computing resources for researchers. There are specific privacy risks introduced by ambient intelligence, which uses machine learning to track and understand behavioral data to improve clinical workflows and health outcomes [75]. Ambient intelligence relies on sensors embedded in the clinical environment that often have the potential to track and use data that are incidental to the target phenomena for which individuals may have originally given consent.

A third major concern relates to assuring informed consent. Applications of machine learning raise questions including what kinds of activity in the clinical environment should require patients' consent, whether patients must be aware of and consent to a machine's influence on a health decision, when and how to obtain consent within clinical workflows, the duration of consent, and whether consent should be occasionally reaffirmed. These questions are likely to become more complex if clinical environments grow more saturated with data collection activities. Therefore, applications of machine learning likely will require enhanced and additional ethical safeguards [76]. While these concerns do not represent the exhaustive

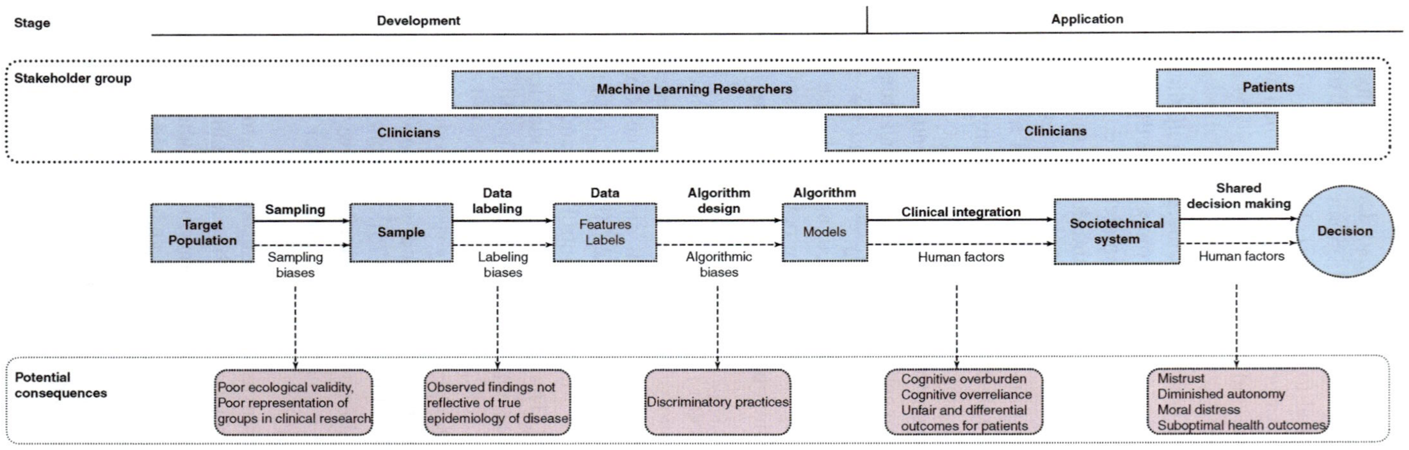

Fig. 3.5 Translational continuum of clinical machine learning systems. Copyright 2020, Jane Paik Kim and Max Kasun. Used with permission

list of concerns, they are reflective of the emergent issues that are circulating in the research community.

Failure to thoroughly investigate the potential limitations and risks of machine learning applications can compromise their ethical use (see Kim [77]). Overtrust in automation is another specific concern [77, 78]. Overtrust as a dispositional attribute is different from automation bias (cognitive overreliance on technology) [77]. Recent research into automation bias has shown that it appears to stem from the adaptive features of cognition that enable multitasking [79]. Additional research into not only the cognitive constraints on human reasoning processes, but also on factors better characterized as rational or dispositional and that occur during clinical integration and shared decision-making, will help to assure the moral agency of clinicians and to develop adequate safeguards for clinical care.

An additional class of ethical concerns relates to the properties of machine learning approaches themselves, and the question of how they can contribute to an ethically acceptable medical decision (see Table 3.1). If human clinicians are to be understood as the arbiters of acceptability, machine behavior may need to comport with our best understanding of our own rational processes. For this reason, clinically desired properties of machine behavior tend to resemble our conceptions of the properties of human intelligence, though they may differ mechanistically.

Two concepts in machine learning that are considered most important in evaluating their ethical acceptability are transparency and explainability. *Transparency* considers the extent to which the full series of states of a machine model (or decision-making processes of a clinician) are observable; a model without transparency is often called a "black box" model or approach. Transparency concerns may

Table 3.1 Common ethical concerns for the evaluation of clinical machine learning algorithms

Concept	Meaning
Transparency	How much are the full series of states of a machine model (or decision-making processes of a clinician) observable to an entrusted supervisor?
Explainability/ interpretability	How much does a model's output or clinician's explanation for a decision comport with an accepted model of rationality that is understandable to its entrusted supervisor?
Reliability	How frequently does a (clinician or machine) give the same output when using the same input data?
Validity	How much does the (clinician or machine)'s output accord to a valid biological or medical understanding of reality (across time and in each individual case)?
Accuracy	How frequently does a (clinician or machine) give the correct output on a given category of task, based on our choice of definition of what is "correct"?
Attributability	To what extent do the features of an explanation that contribute to its validity have explicit referents in the data used to make the explanation?
Interoperability	How well does a (clinician or machine) cooperate with another agent or agents in the service of patient care?
Generalizability	How accurate is a (clinician or machine) on a given category of task when moved to a new environment, such as a new patient population?

not be ethically relevant in a utilitarian sense (such as in deep learning models that are opaque, but nonetheless consistently outperform human experts). Yet, they are relevant in the sense that they interrogate how we create knowledge and why we think of it as clinically essential (i.e., what can an inscrutable, black box machine be said to "know"?), as well as the conditions on which a machine could hypothetically replace the critical role of human expertise.

Second, the concept of *explainability* (also referred to as *interpretability)* relates to how much the reason for a decision or output comports with an accepted model of rationality that is understandable to the person or team entrusted with using it. Therefore, a model can give a valid explanation without its explanation necessarily leading to the most accurate output, just as a human clinician can give a correct explanation for a symptom and still arrive at an inaccurate diagnosis. Explainability is a key factor in determining an approach's ethical acceptability in clinics where doctors will need to understand the reasons for its outputs and communicate these reasons to patients. Explainability also helps ensure that these approaches' "findings" can translate into scientific value for humans—that is, knowledge external to the utility of a machine agent in itself. Each of these concerns is critical to our sense of intellectual and moral responsibility as clinicians and users of machine learning approaches. Solutions for transparency and explainability issues in clinical machine learning algorithms have been proposed [80, 81]. In the coming years, society's embrace of artificial intelligence in health care will depend on greater pluralism between researchers, ethicists, clinicians, designers of intelligent systems, and other professionals, including those who have competencies in multiple domains.

The prevailing view regarding clinical machine agents is that they are not yet sophisticated enough to be moral agents in their own right; thus, clinicians retain the moral agency accorded to them on ethical principles and commitments that come with their professions. As noted by Wallach and Vallor, the current predominant way of thinking ethically about the design of machine agents uses "computationally friendly" approaches that aim to resemble overt utility-seeking and optimizing behaviors, which should not supplant the role of human behaviors that seem distinct from them, such as virtue [82]. Though the limits of artificial intelligence are not known, one can make an intuitive account of moral difference: machines can be trained to recognize signs and symptoms of disability and suffering, while human beings uniquely bear it, contemplate it, and can strive to envision a world in which it is not the case. As artificial intelligence evolves, explaining such differences may become more relevant in evaluating the extent to which the sense-making capacities of the human clinician are uniquely valuable and central to the identity of medicine.

Brain–Machine Interfaces

First prototyped in the 1920s, brain–machine interfaces (also known as neural interfaces or neuroprosthetics) were aspirational breakthrough interventions for brain-based disorders for decades, with ethically problematic study designs and disappointing results [83]. This class of innovative devices broadly aims to allow people to interact with their external environments via the systematic processing of neural activity. Contemporary brain–machine interfaces, which can interface with precise regions of the brain and even individual neurons via microfiber arrays, have come a long way since the use of scalp-embedded metal wires that processed low-resolution EEG signals. Several investigators, such as the Neuralink group, have announced that they are approaching human trials of this new generation of brain–machine interface products [35].

These new products do not come without new ethical concerns. More advanced devices tend to require greater physical proximity to the brain, creating higher resolution signals at the cost of greater invasiveness and, often, greater risks [84]. As with any clinical research, these higher risk devices necessitate an even higher ratio of benefit to be ethically acceptable. Careful attunement to the motivations of prospective volunteers is needed to ensure that they approach such studies with an informed appraisal of their potential risks and benefits, and with minimal negatively valenced factors (such as desperation, in the case of very serious illness). A number of examples of ethics research-focused studies conducted with individuals undergoing brain surgical interventions as part of early-phase clinical trials (e.g., deep brain stimulation for treatment-resistant depression) provided a range of intriguing findings about the motivations and concerns of these participants [85, 86]. However, the nature of newer brain–machine interfaces and the wide range of potential applications to brain-based illnesses and conditions will necessitate additional studies examining diverse ethical challenges in these studies [87].

Research on human volunteers in the context of brain–machine interfaces, as with any other technology, is only ethically sound if there exists compelling evidence that the research will uniquely address an immediate clinical need not already resolvable through standard treatments (refer to Fig. 3.2 for the relevant decision tree). Brain–machine interfaces that fit these ethical criteria include those that can be characterized as sense- or control-restoring implants or treatments for treatment-resistant illness (such as computer-modulated deep brain stimulation for Parkinson's disease [88]). Acquiring knowledge about non-clinical use cases may not be ethically problematic if they are secondary to an ethically sound research question; however, sustained ethical evaluation is required to ensure that conflicts of interest do not arbitrarily or intentionally influence the scientific questions being asked in ways that deprioritize or undermine volunteers' health-related interests. Moreover, this focus on immediate clinical need guards against the problematic use of utilitarian logic to justify forms of experimentation that pose unnecessary risk and minimal or no direct benefit to patients on the proposition that scientific value could generate benefit, in the abstract sense, to unidentified others in the future.

Another ethical concern in the use of brain–machine interfaces is the need to ensure that research contributes to generalizable knowledge and benefit to humanity. Sullivan and Illes note that human-centered motivating rationales are scarce in recent brain–machine interface literature (with rationales focused on technological advancement being predominant), and that this is an ethical issue in its own right [89]. The advancement of these devices is sensible as a social "good" only to the extent that it reduces the burden of illness and/or directly promotes human wellbeing. Ideal research rationales tend to demonstrate appropriate motivation through the selection of a specific target population and identification of everyday health needs, rather than general appeals to illness (e.g., "individuals with advanced ALS who require improved communication and control modalities to reduce health-related financial burdens and improve independence" as opposed to "individuals with neurological disorders").

Conclusion

Inspired and enabled by the work of generations of researchers striving to reduce the burdens of brain-based and neurological disease, neuroinnovations are developing promising avenues in biomedicine. Built on an inspired and interdisciplinary research community, clinical neuroinnovation serves jointly to accelerate discovery in the basic sciences and to transform new knowledge into interventions and therapies. In order not to turn away from its own promise, neuroinnovation will continue to depend on a commensurate effort to anticipate and resolve potential ethical flaws, and minimize potential harms for research volunteers, vulnerable and marginalized populations, and society more broadly. Normative ethical thought and empirical ethics tools continue to develop and branch out from traditional modes of inquiry, principles, and norms, while recognizing the importance of their foundations, and the culture of research continues to register such changes through increased recognition of the moral and intellectual value of stakeholders' diverse voices. Together, these transformations are building a more vibrant, humane, and promising future for the care of human health.

Key Points

1. Modern bioethics benefit from multiple approaches that are jointly essential for establishing a coherent sense of how best to provide health care for human beings and resolve ethical dilemmas (when two or more values, norms, or principles are in tension).
2. Because neuroinnovative research and its potential to reduce human suffering depend on the participation of volunteers, it benefits from the wider tradition of bioethics and key documents including the Belmont Report and Common Rule.
3. How exactly to honor bioethical principles in the clinical deployment of neuroinnovative tools is a context-specific question, requiring active and systematic attention to the relevant experiences of health care professionals, patients, and research volunteers.

Questions to Consider

1. What values, beliefs, and other perceptions are unexplored or underexplored as potential influences on research volunteers?
2. Structural factors such as financial compensation and the availability of alternative treatments have received more attention in the literature over the past several

decades. What other structural factors will play a role in the emerging research context?

3. Given disparities in access to existing health care resources, how can modern bioethics work to ensure that the benefits of neuroinnovation are justly distributed throughout historically vulnerable and/or marginalized populations and society at large?

References

1. Whiteford H, Ferrari A, Degenhardt L. Global burden of disease studies: implications for mental and substance use disorders. Health Aff. 2016;35(6):1114–20.
2. Bloom DE, Cafiero E, Jané-Llopis E, Abrahams-Gessel S, Bloom LR, Fathima S, Feigl AB, Gaziano T, Hamandi A, Mowafi M, O'Farrell D. The global economic burden of noncommunicable diseases. Program on the Global Demography of Aging; 2012
3. Krack P, Martinez-Fernandez R, Del Alamo M, Obeso JA. Current applications and limitations of surgical treatments for movement disorders. Mov Disord. 2017;32(1):36–52. https://doi.org/10.1002/mds.26890.
4. Bilge MT, Gosai AK, Widge AS. Deep brain stimulation in psychiatry: mechanisms, models, and next-generation therapies. Psychiatr Clin. 2018;41(3):373–83. https://doi.org/10.1016/j.psc.2018.04.003. Epub 2018 Jul 9. PMID: 30098651; PMCID: PMC6092041.
5. McCullough LB, Coverdale JH, Chervenak FA. Trustworthiness and professionalism in academic medicine. Acad Med. 2020;95(6):828–32.
6. López-Muñoz F, Alamo C, Dudley M, Rubio G, Garcia-Garcia P, Molina JD, Okasha A. Psychiatry and political–institutional abuse from the historical perspective: the ethical lessons of the Nuremberg trial on their 60th anniversary. Prog Neuro-Psychopharmacol Biol Psychiatry. 2007;31(4):791–806.
7. Schüklenk U. Protecting the vulnerable: testing times for clinical research ethics. Soc Sci Med. 2000;51(6):969–77.
8. Beauchamp TL. The theory, method, and practice of principlism. In: The oxford handbook of psychiatric ethics. Oxford: Oxford University; 2015.
9. Koroshetz W, Gordon J, Adams A, Beckel-Mitchener A, Churchill J, Farber G, et al. The state of the NIH BRAIN initiative. J Neurosci. 2018;38(29):6427–38. pmid:29921715.
10. Topol EJ. High-performance medicine: the convergence of human and artificial intelligence. Nat Med. 2019;25(1):44–56.
11. Young C, Harati S, Ball T, Williams L. S35. Using machine learning to characterize circuit-based subtypes in mood and anxiety disorders. Biol Psychiatry. 2019;85(10):S310.
12. Graham B, Bond R, Quinn M, Mulvenna M. Using data mining to predict hospital admissions from the emergency department. IEEE Access. 2018;6:10458–69.
13. Williams LM. Defining biotypes for depression and anxiety based on large-scale circuit dysfunction: a theoretical review of the evidence and future directions for clinical translation. Depress Anxiety. 2017;34(1):9–24.
14. Xia Y, Li L. Matrix graph hypothesis testing and application in brain connectivity alternation detection. Stat Sin. 2019;29:303–28.
15. Marshall IJ, Wallace BC. Toward systematic review automation: a practical guide to using machine learning tools in research synthesis. Syst Rev. 2019;8(1):163.
16. Koroshetz WJ, Ward J, Grady C. NeuroEthics and the BRAIN initiative: where are we? Where are we going? AJOB Neurosci. 2020;11(3):140–7. https://doi.org/10.1080/21507740.2020.1778119.
17. United States. National Commission for the Protection of Human Subjects of Biomedical, Behavioral Research. The Belmont report: ethical principles and guidelines for the protection

of human subjects of research. The Commission; 1978. http://www.hhs.gov/ohrp/humansubjects/guidance/belmont.html.
18. Beauchamp T, Childress J. Principles of biomedical ethics. 5th ed. New York: Oxford University Press; 2001.
19. Department of Health and Human Services. Federal policy for the protection of human subjects ("common rule"); 1981. http://www.hhs.gov/ohrp/humansubjects/commonrule/index.html.
20. Department of Health and Human Services. Federal Policy for the protection of human subjects. Final rule. Fed Regist; 2017. https://www.govinfo.gov/content/pkg/FR-2017-01-19/pdf/2017-01058.pdf.
21. Corrigan O, Tutton R. What's in a name? Subjects, volunteers, participants and activists in clinical research. Clin Ethics. 2006;1(2):101–4.
22. Roberts LW, Kim JP. Giving voice to study volunteers: comparing views of mentally ill, physically ill, and healthy protocol participants on ethical aspects of clinical research. J Psychiatr Res. 2014;56:90–7.
23. Jeste DV, Depp CA, Palmer BW. Magnitude of impairment in decisional capacity in people with schizophrenia compared to normal subjects: an overview. Schizophr Bull. 2006;32(1):121–8.
24. Palmer BW, Savla GN, Roesch SC, Jeste DV. Changes in capacity to consent over time in patients involved in psychiatric research. Br J Psychiatry. 2013;202:454. https://doi.org/10.1192/bjp.bp.112.121160.
25. Appelbaum PS, Grisso T. The MacArthur treatment competence study. I: mental illness and competence to consent to treatment. Law Hum Behav. 1995;19:105–26.
26. Appelbaum PS, Grisso T. The MacArthur treatment competence study. II: measures of abilities related to competence to consent to treatment. Law Hum Behav. 1995;19:127–48.
27. Appelbaum PS, Grisso T. The MacArthur treatment competence study. III: abilities of patients to consent to psychiatric and medical treatments. Law Hum Behav. 1995;19:149–74.
28. Dunn LB, Candilis PJ, Roberts LW. Emerging empirical evidence on the ethics of schizophrenia research. Schizophr Bull. 2006;32(1):47–68.
29. Appelbaum PS, Grisso T, Frank E, O'Donnell S, Kupfer DJ. Competence of depressed patients for consent to research. Am J Psychiatr. 1999;156(9):1380–4.
30. Other Research. Health and Human Services Office of Human Research Protections website. Updated January 28 2020. https://www.hhs.gov/ohrp/education-and-outreach/about-research-participation/protecting-research-volunteers/other-research/index.html
31. FDA U. Code of Federal Regulations (CFR), Title 21.
32. Meyer MN. There Oughta be a law: when does (n't) the US common rule apply? J Law Med Ethics. 2020;48(1_suppl):60–73.
33. Klugman CM, Dunn LB, Schwartz J, Cohen IG. The ethics of smart pills and self-acting devices: autonomy, truth-telling, and trust at the dawn of digital medicine. Am J Bioeth. 2018;18(9):38–47.
34. Torous J, Roberts LW. Needed innovation in digital health and smartphone applications for mental health: transparency and trust. JAMA Psychiat. 2017;74(5):437–8. https://doi.org/10.1001/jamapsychiatry.2017.0262.
35. Musk E. An integrated brain-machine interface platform with thousands of channels. J Med Internet Res. 2019;21(10):e16194.
36. Hostiuc S, Rusu MC, Negoi I, Drima E. Testing decision-making competency of schizophrenia participants in clinical trials. A meta-analysis and meta-regression. BMC Psychiatry. 2018;18(1):2.
37. Spencer BWJ, Shields G, Gergel T, Hotopf M, Owen GS. Diversity or disarray? A systematic review of decision-making capacity for treatment and research in schizophrenia and other non-affective psychoses. Psychol Med. 2017;47(11):1906–22.
38. Faden RR. Informed consent and clinical research. Kennedy Inst Ethics J. 1996;6(4):356–9.
39. Appelbaum PS, Roth LH. Competency to consent to research: a psychiatric overview. Arch Gen Psychiatry. 1982;39(8):951–8.
40. Meisel A, Roth LH, Lidz CW. Toward a model of the legal doctrine of informed consent. Am J Psychiatr. 1977;134(3):285–9.

41. Berg JW, Appelbaum PS, Lidz CW, Parker LS. Informed consent: legal theory and clinical practice. 2nd ed. New York: Oxford University Press; 2009.
42. Emanuel EJ, Wendler D, Grady C. What makes clinical research ethical? JAMA. 2000;283(20):2701–11.
43. Office for Protection from Research Risks. Informed Consent Tips: Department of Health and Human Services, Office of Human Research Protections; 1993. https://www.hhs.gov/ohrp/regulations-and-policy/guidance/informed-consent-tips/index.html.
44. Roth LH, Lidz CW, Meisel A, Soloff PH, Kaufman K, Spiker DG, et al. Competency to decide about treatment or research: an overview of some empirical data. Int J Law Psychiatry. 1982;5(1):29–50.
45. Federman D, Hannam K, Lyman RL. Responsible research: a systems approach to protecting research participants. Washington, D.C.: Institute of Medicine, National Academies Press; 2003.
46. Tam NT, Huy NT, Thoa LT, Long NP, Trang NT, Hirayama K, Karbwang J. Participants' understanding of informed consent in clinical trials over three decades: systematic review and meta-analysis. Bull World Health Organ. 2015;93:186–98H.
47. Roberts LW. Informed consent and the capacity for voluntarism. Am J Psychiatr. 2002;159(5):705–12.
48. Roberts LW. Mental illness and informed consent: seeking an empirically derived understanding of voluntarism. Report. Lippincott Williams & Wilkins; 2003. Report No.: 545; 16.
49. Misra S, Ganzini L. Capacity to consent to research among patients with bipolar disorder. J Affect Disord. 2004;80(2–3):115–23.
50. Appelbaum PS, Roth LH, Lidz CW, Benson P, Winslade W. False hopes and best data: consent to research and the therapeutic misconception. Hast Cent Rep. 1987;17(2):20–4.
51. Kaminsky A, Roberts LW, Brody JL. Influences upon willingness to participate in schizophrenia research: an analysis of narrative data from 63 people with schizophrenia. Ethics Behav. 2003;13(3):279–302.
52. Roberts LW, Hammond KG, Hoop J. An inverse relationship between perceived harm and participation willingness in schizophrenia research protocols. Am J Psychiatr. 2006;163(11):2002–4.
53. Grady C. Money for research participation: does it jeopardize informed consent? Am J Bioeth. 2001;1(2):40–4.
54. Carnahan SJ. Promoting medical research without sacrificing patient autonomy: legal and ethical issues raised by the waiver of informed consent for emergency research. Oklahoma Law Rev. 1999;52:566–91.
55. Gutman A, Wagner JW, Allen AL, Hauser SL, Arras JD, Kucherlapati RS, Atkinson BF, Michael NL, Farahany NA, Sulmasy DP. Anticipate and communicate: ethical management of incidental and secondary findings in the clinical, research and direct-to-consumer contexts. Presidential Commission for the Study of Bioethical Issues. Am J Epidemiol. 2013;2013:1–46.
56. Majumder MA, Guerrini CJ, Bollinger JM, Cook-Deegan R, McGuire AL. Sharing data under the 21st century cures act. Genet Med. 2017;19(12):1289–94.
57. Memorandum for the heads of executive departments and agencies: increasing access to the results of federally funded scientific research [press release]. Washington, DC: executive Office of the President - Office of Science and technology policy, February 22, 2013 February 23, 2013.
58. Ewuoso C. A systematic review of the Management of Incidental Findings in genomic Research. J West Afr Bioethics Train Prog. 2016;3(1):1–21.
59. Christenhusz GM, Devriendt K, Van Esch H, Dierickx K. Ethical signposts for clinical geneticists in secondary variant and incidental finding disclosure discussions. Med Health Care Philos. 2015;18(3):361–70.
60. Kaul V, Enslin S, Gross SA. The history of artificial intelligence in medicine. Gastrointest Endosc. 2020;92(4):807–12.
61. Graham S, Depp C, Lee EE, Nebeker C, Tu X, Kim HC, Jeste DV. Artificial intelligence for mental health and mental illnesses: an overview. Curr Psychiatry Rep. 2019;21(11):116. https://doi.org/10.1007/s11920-019-1094-0. PMID: 31701320; PMCID: PMC7274446.

62. Friedman J, Hastie T, Tibshirani R. The elements of statistical learning, vol. 1(10). New York: Springer; 2001.
63. James G, Witten D, Hastie T, Tibshirani R. An introduction to statistical learning, vol. 112. New York: Springer; 2013. p. 18.
64. Gottesman O, Johansson F, Komorowski M, Faisal A, Sontag D, Doshi-Velez F, Celi LA. Guidelines for reinforcement learning in healthcare. Nat Med. 2019;25:16–8.
65. Gottesman O, Johansson F, Komorowski M, Faisal A, Sontag D, Doshi-Velez F, Celi LA. Guidelines for reinforcement learning in healthcare. Nature medicine. 2019;25(1):16–8.
66. Yeung S, Rinaldo F, Jopling J, Liu B, Mehra R, Downing NL, Guo M, Bianconi GM, Alahi A, Lee J, Campbell B, Deru K, Beninati W, Fei-Fei L, Milstein A. A computer vision system for deep learning-based detection of patient mobilization activities in the ICU. NPJ Digit Med. Springer Nature. 2019;2(1):11.
67. Yeung S, Downing NL, Fei-Fei L, Milstein A. Bedside computer vision — moving artificial intelligence from driver assistance to patient safety. N Engl J Med [Internet]. 2018;378(14):1271–3. [cited 2020 Jan 24]; http://www.nejm.org/doi/10.1056/NEJMp1716891.
68. Akkus Z, Galimzianova A, Hoogi A, Rubin DL, Erickson BJ. Deep learning for brain MRI segmentation: state of the art and future directions. J Digit Imaging. 2017;30(4):449–59.
69. Baker JT, Holmes AJ, Masters GA, Thomas Yeo BT, Krienen F, Buckner RL, Öngür D. Disruption of cortical association networks in schizophrenia and psychotic bipolar disorder. JAMA Psychiat. 2014;71(2):109–18.
70. Kaiser RH, Andrews-Hanna JR, Wager TD, Pizzagalli DA. Large-scale network dysfunction in major depressive disorder: a meta-analysis of resting-state functional connectivity. JAMA Psychiat. 2015;72(6):603–11.
71. Ichikawa N, Lisi G, Yahata N, Okada G, et al. Primary functional brain connections associated with melancholic major depressive disorder and modulation by antidepressants. Sci Rep. 2020;10(1):3542.
72. Rajkomar A, Hardt M, Howell MD, Corrado G, Chin MH. Ensuring fairness in machine learning to advance health equity. Ann Intern Med. 2018;169(12):866–72.
73. Hutchinson B, Prabhakaran V, Denton E, Webster K, Zhong Y, Denuyl S. Unintended machine learning biases as social barriers for persons with disabilities. Accessibil Comput. 2020;125:1.
74. National Defense Authorization act for Fiscal Year 2020 U.S.C. § 1790 (2020). https://www.congress.gov/bill/116th-congress/senate-bill/1790/text.
75. Martinez-Martin N, Luo Z, Kaushal A, Adeli E, Haque A, Kelly SS, Wieten S, Cho MK, Magnus D, Fei-Fei L, Schulman K. Ethical issues in using ambient intelligence in health-care settings. Lancet Digital Health. 2021;3(2):E115–23.
76. Cohen IG. Informed consent and medical artificial intelligence: what to tell the patient? Geo LJ. 2019;108:1425.
77. Kim JP. Letter to the editor: machine learning and artificial intelligence in psychiatry: balancing promise and reality. J Psychiatr Res. 2021;136:244–5.
78. Kaur H, Nori H, Jenkins S, Caruana R, Wallach H, Wortman Vaughan J. Interpreting Interpretability: Understanding Data Scientists' Use of Interpretability Tools for Machine Learning. In Proceedings of the 2020 CHI conference on human factors in computing systems; 2020. pp. 1–14.
79. Lyell D, Coiera E. Automation bias and verification complexity: a systematic review. J Am Med Inform Assoc. 2017;24(2):423–31.
80. Ahmad MA, Eckert C, Teredesai A. Interpretable machine learning in healthcare. In: Proceedings of the 2018 ACM international conference on bioinformatics, computational biology, and health informatics; 2018. pp. 559–560.
81. Roscher R, Bohn B, Duarte MF, Garcke J. Explainable machine learning for scientific insights and discoveries. IEEE Access. 2020;8:42200–16.
82. Wallach W, Vallor S. Moral machines: from value alignment to embodied virtue. In Ethics of artificial intelligence (pp. 383-412). Oxford University Press: Oxford.
83. Mohiddin M, Premalatha M, Kedarnath B, Kumar KS, Prasad KV. Implementation of brain–computer interface technology using arduino. Int J Electric Eng Technol. 2017;8(2):25–35.

84. Lebedev MA, Nicolelis MA. Brain-machine interfaces: from basic science to neuroprostheses and neurorehabilitation. Physiol Rev. 2017;97(2):767–837.
85. Leykin Y, Christopher PP, Holtzheimer PE, Appelbaum PS, Mayberg HS, Lisanby SH, Dunn LB. Participants' perceptions of deep brain stimulation Research for treatment-resistant depression: risks, benefits, and therapeutic misconception. AJOB Prim Res. 2011;2(4):33–41. https://doi.org/10.1080/21507716.2011.627579. PMID: 26225215; PMCID: PMC4516276.
86. Christopher PP, Leykin Y, Appelbaum PS, Holtzheimer PE 3rd, Mayberg HS, Dunn LB. Enrolling in deep brain stimulation research for depression: influences on potential subjects' decision making. Depress Anxiety. 2012;29(2):139–46. https://doi.org/10.1002/da.20916. Epub 2011 Nov 17.
87. Wu H, Adler S, Azagury DE, Bohon C, Safer DL, Barbosa DAN, Bhati MT, Williams NR, Dunn LB, Tass PA, Knutson BD, Yutsis M, Fraser A, Cunningham T, Richardson K, Skarpaas TL, Tcheng TK, Morrell MJ, Roberts LW, Malenka RC, Lock JD, Halpern CH. Brain-responsive neurostimulation for loss of control eating: early feasibility study. Neurosurgery. 2020;87:1277–88. https://doi.org/10.1093/neuros/nyaa300. Epub ahead of print.
88. Bronte-Stewart HM, Petrucci MN, O'Day JJ, Afzal MF, Parker JE, Kehnemouyi YM, Wilkins KB, Orthlieb GC, Hoffman SL. Perspective: evolution of control variables and policies for closed-loop deep brain stimulation for Parkinson's disease using bidirectional deep-brain-computer interfaces. Front Hum Neurosci. 2020;14:353.
89. Sullivan LS, Illes J. Ethics in published brain–computer interface research. J Neural Eng. 2018;15(1):013001.

Further Reading

Department of Health and Human Services. Federal Policy for the Protection of Human Subjects. Fed Regist: Final rule; 2017. https://www.govinfo.gov/content/pkg/FR-2017-01-19/pdf/2017-01058.pdf

Greely HT, Grady C, Ramos KM, Chiong W, Eberwine J, Farahany NA, Johnson LS, Hyman BT, Hyman SE, Rommelfanger KS, Serrano EE. Neuroethics guiding principles for the NIH BRAIN initiative. J Neurosci. 2018;38(50):10586–8.

Gutman A, Wagner JW, Allen AL, Hauser SL, Arras JD, Kucherlapati RS, Atkinson BF, Michael NL, Farahany NA, Sulmasy DP. Anticipate and communicate: Ethical management of incidental and secondary findings in the clinical, research and direct-to-consumer contexts. Presidential Comission for the Study of Bioethical. 2013;2013:1–46.

Kim JP. Letter to the editor: machine learning and artificial intelligence in psychiatry: balancing promise and reality. J Psychiatric Res. 2021;136:244–5.

Kim JP, Kasun M. Pluralistic empirical research for the ethical advancement of algorithmic medicine. BMJ J Med Ethics. Under Review.

Majumder MA, Guerrini CJ, Bollinger JM, Cook-Deegan R, McGuire AL. Sharing data under the twenty-first Century Cures Act. Genetic Med. 2017;19(12):1289–94.

Topol EJ. High-performance medicine: the convergence of human and artificial intelligence. Nat Med. 2019;25(1):44–56.

United States. National Commission for the Protection of Human Subjects of Biomedical, Behavioral Research. The Belmont report: ethical principles and guidelines for the protection of human subjects of research. The Commission; 1978. http://www.hhs.gov/ohrp/humansubjects/guidance/belmont.html.

Chapter 4
Changing Contexts of Neuroinnovation: Societal Considerations

Mildred K. Cho

Introduction

The ethics of clinical research and biomedical innovation are in part determined by general principles but also subject to contextual influences. The changing context in which neuroinnovation occurs impacts the associated ethical considerations and implications. General ethical principles for neuroinnovation are derived from established ethical frameworks for clinical care, the ethics of research, and professional ethics. However, to the extent that these ethical frameworks are based on power relationships, policies, and expectations around the existence and perception of risks and benefits of technological innovations, context matters.

Established Ethical Frameworks and Changing Contexts of Neuroinnovation

The ethics of innovation in biomedicine have largely been framed around addressing the tension between the need for innovation to create medical benefit (principle of beneficence), minimizing the risks of doing so (principle of nonmaleficence), while allowing people to make well-informed, voluntary decisions about participating (principle of autonomy). Furthermore, ethical innovations are fairly and broadly accessible (principle of justice).

The ethics of medical innovation also has recognized the potential for conflicting interests when clinicians conduct research on their patients. In these situations, traditional professional ethics that are based on fiduciary relationships between doctor

M. K. Cho (✉)
Stanford Center for Biomedical Ethics, Stanford University, Stanford, CA, USA
e-mail: micho@stanford.edu

L. W. Roberts (ed.), *Ethics and Clinical Neuroinnovation*,
https://doi.org/10.1007/978-3-031-14339-7_4

and patient are challenged by the duties of researchers to create generalizable knowledge [1, 2]. That is, when conducting research conflicts with the best interests of a patient who is enrolled in a study, what is the right thing to do? For example, is it ethical to do placebo controls that involve sham surgery with burr holes drilled into the skull and general anesthesia, if such controls are necessary to rigorously establish effectiveness of the experimental intervention? To address such role conflicts, regulations and processes have been established to mitigate the conflicts of interest in several ways [3]. The primary mitigation strategy is ethical review by a third party, such as an institutional review board, that has neither a care-giving nor a research relationship with potential subjects. The second mitigation strategy is disclosure of the risks and benefits of research participation, in the form of written informed consent and a process of discussing the consent form with knowledgeable experts. Other commonly employed strategies are Data Safety Monitoring Boards, which provide third-party scientific and ethical review by third parties once clinical studies have begun collecting and analyzing data. Similar mitigation strategies are employed for clinical researchers who have financial conflicts of interest, for example, if they receive funding from or are employed by industry sponsors of their clinical studies [2].

These ethical frameworks were built on the assumption that patients and participants in medical innovation have interests that are in need of protection, and that there is a power imbalance between them and the clinicians and scientists conducting research. Increasingly, however, these assumptions are being challenged by technological and social changes that are shifting the power dynamics in subtle and radical ways, alike.

Over the last few decades, several ethically relevant changes in the social and technological context of neuroinnovation have become apparent. These changes include:

- Emergence and convergence of technologies to capture, analyze, and act upon digital data.
- Blurring of lines between clinical care, research, and commerce.
- Increased empowerment of patients and patient advocacy organizations.
- Growing involvement of non-scientist communities in the innovation process.

Each of these contextual changes and their ethical implications will be discussed.

Emergence and Convergence of Technologies to Capture, Analyze, and Act upon Digital Data

It is clear that neuroinnovation is and will be greatly affected by the emergence and convergence of technologies that can collect vast amounts of biological and clinical data in digital form. Digitization of data affords greater data sharing but poses greater threats to privacy. Democratization of access and technology that allows medically relevant data to be acquired by individuals and companies rather than clinicians shifts control of information away from physicians and researchers. The volume of big data that can be captured and linked not only facilitates but requires new forms of analysis such as machine learning because it exceeds the capability of human cognitive processes. The shift of analysis to artificial intelligence systems, which lack transparency and accountability [4], also shifts control of information and analysis away from physicians and research. With these shifts of control come uncertainty about responsibility and accountability about decision-making.

The growing availability of highly granular data facilitates identification of individuals even if it is deidentified to be compliant with privacy regulations about data sharing, such as the Health Insurance Portability and Accountability Act (HIPAA) in the US and the General Data Protection Regulation (GDPR) in the European Union [5]. Some data, such as magnetic resonance images (MRI) of the brain can be used to identify individuals using facial software [6]. Other data such as DNA sequences might not be identifying in themselves but can be linked to readily available or public data that contains identifiers, such as from social media or motor vehicle records. Individuals can be uniquely re-identified, even from incomplete datasets without overt identifiers [7].

The availability of high-resolution yet non-specific data without full understanding of its meaning can lead to risks. For example, whole body MRI can be very sensitive at detecting pathology such as malignancy because it is very granular, but also lead to a high rate of false positives. Similarly, whole genome sequencing generates high-resolution data but interpretation can be ambiguous and of uncertain clinical significance before large population datasets are acquired and analyzed. The capability of gathering large amounts of data in a non-targeted fashion creates incidental findings (IFs). While IFs are not a new phenomenon, the likelihood of generating them has been enhanced by the lowered barriers to collecting data in a non-specific manner, such as through whole body scanning or whole genome sequencing [8, 9].

Mobile and wearable devices have literally put powerful data collection and analysis tools into the hands of billions. Thus, innovation now readily occurs outside of the health care or medical research environment. For example, data from smart phones is used by companies for digital phenotyping to diagnose psychiatric conditions, and web-based chat bots serve as mental health counselors [10, 11]. Analysis of personal but non-health data can now be correlated with neurological and psychiatric diseases. For example, the speed of typing on a smart phone, prosody of voice calls, the specific choices of words used in social media, and speech analysis are used to detect conditions such as Parkinson's disease, dementia, schizophrenia, or depression. While some of the companies involved in such innovation are in the biotechnology, pharmaceutical or health industries, others are in consumer electronics, social media, and computing industries.

Broad availability of large amounts of digital biological and health data, in concert with software and hardware, can facilitate medical decisions or control of biological processes in unprecedented ways. However, these technologies, especially artificial intelligence, can also diminish autonomy of patients, clinicians, research participants, and researchers to the extent that decisions or actions based on data become automated or are implemented without informed consent. For example, decisions about prescribing treatments or making diagnoses are increasingly aided or conducted by computers, often without the knowledge of the patient [12]. Another example is closed-loop deep brain stimulation, in which stimulation parameters are automatically determined by sensors to detect biomarkers rather than manually by a neurologist [13].

Blurring of Lines Between Clinical Care, Research, and Commerce

The low barriers to collecting large amounts of health data and using it for research are blurring lines between clinical care, medical research, and commerce. Electronic medical records are now the norm, and they now can include a diversity of data from procedures such as CT scans or whole genome sequencing. The explosion of clinical data is a driving force towards the conversion of hospitals into learning health care systems that, in essence, turn each patient into a research subject. The hope is that the research facilitated by learning health systems can be fed back to change health care practices and improve care of individual patients. The diminishing distinctions between medical practice and medical research is important for neuroinnovation because their ethical and regulatory frameworks differ. When research is conducted on patients and patient data in the course of clinical care, the usual protections of human subjects such as informed consent may not apply [14].

Health data are also collected and used for research by companies. For example, the DNA testing company 23andMe generates genome data from its customers but also uses it, with consent, for research on Parkinson's disease and other conditions [15]. The data have been sold to pharmaceutical companies and used by 23andMe to develop drugs directly [16]. Furthermore, the majority of clinical research and biomedical innovation is now funded and conducted by industry. This is an important contextual feature of neuroinnovation for several reasons. First, regulations for protection of human subjects do not apply to much of the research conducted in the private sector. Second, the findings of such research, unlike most government-funded work, are usually proprietary. Third, private sector control of drugs, devices, or other technologies means that there may be limited incentive for making them broadly and fairly accessible.

Medical care and research are also increasingly conducted outside of traditional health care organizations and by non-professionals. Examples abound, such as crowdsourcing the identification of unsolved illnesses of real people in the Netflix documentary series *Diagnosis*. So-called do-it-yourself biologists are attempting to treat diseases such as *Clostridioides difficile* bacterial infections by fecal transplants [17]. Biohackers are implanting devices into their own brains or other body parts to augment functions such as inserting near-field communications chips in a fingertip to enable contactless access, replacing the use of keys or passwords, and kits can be purchased on websites such as dangerousthings.com in order to do so. Others have used implants to convert visual images into sounds to help a person with colorblindness "hear" colors in the visible light spectrum and perceive infrared and ultraviolet light [18]. These experiments are usually, but not always, self-administered, without the involvement of professional researchers or clinicians. While proponents argue that the democratization of science and medicine increases access to these activities, and therefore enhances fairness, there remain substantial questions about whether de-professionalization introduces more harms than benefits [19].

Biohacking, DIY bio, crowdsourced research, and citizen science cross over the ambiguous boundary between treatment and enhancement, as well as the boundary between professional and non-professional activity. Aside from the lack of regulation and clear obligations around activities conducted by non-professionals, biohacking raises ethical questions about the limits of self-experimentation and social questions about what constitutes enhancement and who is entitled access to them. While many neurotechnologies are not widely available for DIY use, some, such as transcranial direct current stimulation, a non-invasive neuromodulatory device to treat chronic pain, depression, or improve cognitive functions such as long-term memory or mathematical ability, have been used on a DIY basis [20]. However, they have been predominantly used by people with higher education higher income, and technical backgrounds [21]. This raises questions about whether greater access to such a technology actually does lead to more fair access.

Increased Empowerment of Patients and Patient Advocacy Organizations

An important contextual shift for biomedical innovation in general is the emergence of patient advocacy and advocacy organizations over the last few decades, heavily influenced by patients with breast cancer and those with AIDS [22, 23], but now encompassing virtually all conditions ranging from the very common to the very rare. Organized groups and individual patients and families have played major roles in biomedical innovation, including collection of biospecimens and data, assistance with recruiting to clinical trials and research design, and funding. Patient advocates have increasing, and sometimes controversial influence on the regulatory process of drug and device approval [24–26].

The ethical implications of patient advocacy for neuroinnovation are numerous and diverse. The ethical obligations and regulations around research were designed with the imbalance of power of professionals such as physicians and scientists as compared to patient advocates and research participants, who were viewed as vulnerable and in need of protection. Professional obligations are embedded in fiduciary duties, which entail putting the needs of others ahead of one's own, thus serving as guard rails on professional power.

Like crowdsourcing, the phenomenon of patient advocacy, however, reflects the shift in the balance of power away from professionals. With this shift, increased power of patient advocacy groups, the majority of which are funded by pharmaceutical and biotechnology companies, is not scaffolded by the ethical frameworks that are built around fiduciary duties. For example, professional physicians and researchers are subject to conflict of interest regulations which are designed to surface and mitigate potential interests that could interfere with the professional acting in the best interest of their patients or research subjects. Patient advocates could also have conflicts of interest in the roles they play in funding, conducting, and facilitating research, but are not necessarily mandated to disclose, reduce, or eliminate them [27].

It is now well-established that both financial and non-financial conflicts of interest have negative effects on biomedical research and clinical practice [28, 29]. Therefore, a lack of conflict of interest guidelines or regulations around patient advocacy group involvement in research is of concern. For example, could research participants recruited to clinical trials through advocacy groups be subject to undue pressure to enroll? Could the U.S. Food and Drug Administration (FDA) be pressured by advocacy groups to approve drugs with weak evidence of effectiveness, and are those advocacy groups beholden to industry funders? [30].

Conclusion

Over the last few decades, a number of broad evolutionary shifts have been occurring that have ethical implications for biomedical research generally, and for neuroinnovation specifically. Some of these changes are technological, such as the availability of digitized data relevant to biology, health, and behavior. Others are social, such as the rise of patient advocacy and citizen science. Together, the net effect of these converging phenomena is to displace professional scientists and clinicians from their central role in biomedical innovation.

One potential positive implication of these contextual changes for neuroinnovation is that they make the research more ethical because input from diverse stakeholders on research questions redirects it to solve problems that are more aligned with needs of patients and the public. To be ethical, research must be socially as well as scientifically valuable [31], and with benefits and risks distributed in a way that promotes equity [32]. Second, the advent of the learning health care system could accelerate the timeline between discovery and application to clinical practice and hasten progress towards precision medicine that improves health outcomes through tailored diagnosis and treatments. Third, democratization of access to neurotechnologies could address problems of access to health care, especially for mental health.

There are, however, also several potential pitfalls of these contextual changes. Moving neuroinnovation out of traditional environments means that ethical guidelines and laws to protect research participants and patients from not only physical harms but risks to autonomy, consent, and privacy may no longer be in play. Traditional research and medicine rely on the self-regulatory structures of professionalization, such as accreditation, licensure, and peer review to assure rigor, competency, and ethical behavior and thus minimize harms to research participants and patients. Furthermore, oversight such as by Institutional Review Boards and the FDA may not apply to innovation conducted either in the context of health care, commercial activity, DIY, or citizen science.

Combined with the increasing influence of the private sector in neuroinnovation [33], the evolution of the social context could lead to greater predominance of the market in driving development of and access to technologies and the products of research. In the absence of broad public discussion about whether new neurotechnologies are desirable and whether their potential benefits are worth the potential harms, their emergence and adoption would be determined largely by commercial forces, with few, if any processes to ensure fair access and protection of the vulnerable from exploitation.

To realize fair public benefit from novel neurotechnologies, the innovation process must be trustworthy, which may require transparency, stakeholder involvement, and broad public debate. While these are ambitious goals, societal trends suggest public interest in engaging in the discussion and conduct of innovation.

Key Points

1. Societal shifts over the last few decades have changed how biomedical and translational research is conducted, including the role of the private sector, involvement of patients, patient advocacy organizations and citizen scientists, and the blurring of lines between clinical research and clinical care.
2. Ethical frameworks for biomedical and neuroinnovation have largely been constructed with the main goal of protection of research subjects, and thus are in essence paternalistic and based on professional and fiduciary obligations. The locus of power to determine biomedical research agendas is shifting away from academic scientists but becoming more diffuse.
3. Empowerment of patients and the public in biomedical research has the potential to drive neuroinnovation to serve more just purposes. However, the disruption of established ethical and regulatory frameworks could lead to diminished safety and efficacy of new neurotechnologies.

Questions to Consider

1. What are potential harms and benefits to patients from greater availability of medical and non-medical data for use in neuroinnovation?
2. What are potential effects of increased influence and empowerment of patient advocacy groups on neuroinnovation?
3. How might non-scientists be involved in the neuroinnovation process and how might the process be affected?

References

1. Hall M. Fiduciary principles in health care. Oxford: Oxford University Press; 2019.
2. Thompson D. Understanding financial conflicts of interest. N Engl J Med. 1993;329:573–6.
3. Davis M, Stark A, editors. Conflict of interest in the professions. Oxford: Oxford University Press; 2001.
4. Morley J, Machado C, Burr C, et al. The ethics of AI in health care: a mapping review. Soc Sci Med. 2020;260:113172.
5. Rocher L, Hendrickx J, de Montjoye Y. Estimating the success of re-identifications in incomplete datasets using generative models. Nat Commun. 2019;10:3069.
6. Schwarz CG, Kremers WK, Therneau TM, et al. Identification of anonymous MRI research participants with face-recognition software. N Engl J Med. 2019;381(17):1684–6.
7. Malin B, Sweeney L. Re-identification of DNA through an automated linkage process. J Am Med Informat Assoc. 2001:423–7.
8. Gordon A, Zouk H, Venner E. Frequency of genomic secondary findings among 21,915 eMERGE network participants. Genet Med. 2020;22:1470–7.
9. Lumbreras B, Donat L, Hernández-Aguado I. Incidental findings in imaging diagnostic tests: a systematic review. Br J Radiol. 2010;83:276–89.
10. Martinez-Martin N, Insel T, Dagum P, Greely H, Cho M. Data mining for health: staking out the ethical territory of digital phenotyping. NPJ Digital Med. 2018;1:68.
11. Insel TR. Digital phenotyping: a global tool for psychiatry. World Psychiatry. 2018;17:276–7.
12. Cohen I. Informed consent and medical artificial intelligence: what to tell the patient? Georgetown Law J. 2020;108:1425–69.
13. Hell F, Palleis C, Mehrkens JH, Koeglsperger T, Bötzel K. Deep brain stimulation programming 2.0: future perspectives for target identification and adaptive closed loop stimulation. Front Neurol. 2019;10:314.
14. Kass NE, Faden RR. Ethics and Learning Health Care: The Essential roles of engagement, transparency, and accountability. Learn Health Syst. 2018;2(4):e10066. https://doi.org/10.1002/lrh2.10066.
15. Chang D, Nalls MA, Hallgrímsdóttir IB, et al. A meta-analysis of genome-wide association studies identifies 17 new Parkinson's disease risk loci. Nat Genet. 2017;49(10):1511–6.
16. Brodwin E. DNA-testing companies like 23andMe sell your genetic data to drugmakers and other Silicon Valley startups. Business Insider; 2018.
17. Ekekezie C, Perler BK, Wexler A, Duff C, Lillis CJ, Kelly CR. Understanding the scope of do-it-yourself fecal microbiota transplant. Am J Gastroenterol. 2020;115(4):603–7.
18. Kent C. From grinders to biohackers: where medical technology meets body modification. Medical Device Network; 2020.
19. Wexler A. The social context of "do-it-yourself" brain stimulation: Neurohackers, biohackers, and Lifehackers. Front Hum Neurosci. 2017;11:224.
20. Wexler A. The practices of do-it-yourself brain stimulation: implications for ethical considerations and regulatory proposals. J Med Ethics. 2016;42(4):211–5.
21. Jwa A. Early adopters of the magical thinking cap: a study on do-it-yourself (DIY) transcranial direct current stimulation (tDCS) user community. J Law Biosci. 2015;2(2):292–335.
22. Osuch JR, Silk K, Price C, et al. A historical perspective on breast cancer activism in the United States: from education and support to partnership in scientific research. J Womens Health (Larchmt). 2012;21(3):355–62.
23. Epstein S. Impure science: AIDS, activism, and the politics of knowledge. Berkeley: University of California Press; 1996.
24. Sigal E, Stewart M, Merino D. Advocacy and patient involvement in clinical trials. In: Piantadosi S, Meinert CL, editors. Principles and practice of clinical trials. Cham: Springer International Publishing; 2020. p. 1–13.
25. Mattingly T, Simoni-Wastila L. Patient-centered drug approval: the role of patient advocacy in the drug approval process. J Manag Care Spec Pharm. 2017;23:1078–82.

26. Kesselheim A, Avorn J. Approving a problematic muscular dystrophy drug. Implications for FDA policy. JAMA. 2016;316:2357–8.
27. Rose SL, Highland J, Karafa MT, Joffe S. Patient advocacy organizations, industry funding, and conflicts of interest. JAMA Intern Med. 2017;177(3):344–50.
28. Cherla D, Viso C, Holihan J, et al. The effect of financial conflict of interest, disclosure status, and relevance on medical research from the United States. J Gen Intern Med. 2019;34:429–34.
29. Wood S, Podrasky J, McMonagle M, Raveendran J, Bysshe T, Hogenmiller A. Influence of pharmaceutical marketing on Medicare prescriptions in the District of Columbia. PLoS One. 2017;12:e0186060.
30. Kesselheim AS, Avorn J. Approving a problematic muscular dystrophy drug: implications for FDA policy. JAMA. 2016;316(22):2357–8.
31. Emanuel E, Wendler D, Grady C. What makes clinical research ethical? JAMA. 2000;283:2701–11.
32. Pratt B. Research for health justice: an ethical framework linking global health research to health equity. BMJ Glob Health. 2020;6:e002921.
33. Ienca M, Vayena E. Direct-to-consumer Neurotechnology: what is it and what is it for? AJOB Neurosci. 2019;0:149–51.

Further Reading

Faden RR, Kass NE, Goodman SN, Pronovost P, Tunis S, Beauchamp TL. An ethics framework for a learning health care system: a departure from traditional research ethics and clinical ethics. Hast Center Rep. 2013;43(s1):S16–27.

Institute of Medicine Committee on Conflict of Interest in Medical Research, Education, and Practice; Lo B, Field MJ, editors. Washington (DC): National Academies Press; 2009. ISBN-13:978-0-309-13,188-9.

Wexler A. The social context of "do-it-yourself" brain stimulation: neurohackers, biohackers, and lifehackers. Front Hum Neurosci. 2017;11:224.

Chapter 5
Biomedical Advances: Neuroinnovation and Technology

Nicole Martinez-Martin

Ethical Frameworks for Innovative Biomedical Technologies

"Innovation" can refer to broadly a new idea or method or device. Biomedical research and clinical practice are oriented around producing, translating, and implementing new methods and technologies to apply to health towards innovation. In the U.S., innovation in medicine and surgery generally proceeds under a regulatory ethics framework. This means that novel practices and methods follow a set of procedures to assure the protection of the rights and well-being of research participants [1]. Research for new potential clinical applications of neurotechnologies, as with other potential new medical interventions, generally follows this process for ethical research. Research ethics require that proposed research must be reviewed by an Institutional Review Board (IRB), assessing the evidence and methodology supporting the study protocol, and reviewing other procedures meant to protect the welfare of participants [2]. Informed consent is a critical aspect of research ethics [3]. Research participants must be given information that encompasses issues such as the details of the proposed intervention, the anticipated risks and benefits, and possible alternatives, in order to allow for meaningful consent [4]. Different neurotechnologies may present specific challenges regarding risks and benefits, but some considerations—such as the need to establish safety—are not necessarily that distinct from the considerations necessary for medical innovation generally. The discussion of novel issues in the context of clinical research into neurotechnologies has tended to focus around the role of the brain as the site of consciousness and personality and thus the potential of a neurotechnology to disrupt identity, agency, and other characteristics associated with personhood, as will be discussed more below [5].

N. Martinez-Martin (✉)
Stanford Center for Biomedical Ethics, Stanford, CA, USA
e-mail: nicolemz@stanford.edu

L. W. Roberts (ed.), *Ethics and Clinical Neuroinnovation*,
https://doi.org/10.1007/978-3-031-14339-7_5

Clinical innovation refers to the "use of treatments in a clinical setting that have not been well-proven in a research setting" [6]. Neurosurgery and psychiatry have long histories of clinical innovation. For example, pediatric pharmacology often involved innovative practice because of the relative lack of scientific evidence base regarding children's mental health and pharmacology. In each instance, clinical innovation requires careful consideration of the risks, benefits, and justifications for proceeding. The first item to determine whether clinical innovation is ethically justified is establishing that there is a demonstrable need for the treatment. Factors that can justify the need for the innovative tool or practice include the severity of the condition, lack of adequate treatments, or lack of resources [7]. In addition to establishing the need for the innovation, the ethical acceptability of the specific innovation must be analyzed [8]. One must consider the risks and benefits of the proposed innovation, such as weighing the risks of the underlying condition being treated against the risks posed by the innovation. The evaluation of risk should include consideration of extreme and rare risk. Then, if the innovation is found to be ethically acceptable, attention should turn to what information and practices will be needed to maintain a high standard for informed consent. While informed consent may not be strictly required for clinical innovation, ethical practice suggests that patients receiving innovative treatments should be informed of the purpose, benefits, and risks of the proposed treatment [6].

The use of predictive analytics in mental health care presents a current example of potential clinical innovation. Predictive algorithms apply techniques such as data mining, statistical analysis, and/or machine learning to a range of data to forecast the occurrence of events or states. In medicine, predictive analytics have been applied to a range of clinical applications such as forecasting patient outcomes after surgery [9]. In psychiatry, predictive algorithms could be applied to forecast psychotic episodes or assist with selection of medication [10, 11]. In evaluating clinical innovation involving these algorithms, one would then consider the necessity of the innovation [12]. There is certainly a need for tools that can improve the process of identifying appropriate psychiatric medications for a patient or evaluating the future risk of psychotic episodes [13]. Then one would move to evaluate the risks and benefits of the specific application of technology. Considerations of accuracy, bias or application to local contexts are challenges for algorithms employed in most medical contexts [14, 15]. Careful attention is needed, though, to how these issues may apply to the specific context of a mental health diagnosis or psychiatric population, such as how an app that evaluates risk of suicide interacts with the specific at-risk populations or impacts factors related to suicide risk [16].

The American Medical Association provides guidance regarding ethical innovation in medicine, which includes anticipated risks and burdens of the innovative practice (AMA Code of Medical Ethics Opinion 1.2.11). Clinicians must consider what information will be needed for clear communication of the risks and benefits, including context-specific issues, for informed consent. In the context of the brain, informed consent requires attention to how the innovation may impact personality or cognitive function, as well as the specific vulnerabilities that may influence how people with neurological or psychiatric conditions understand the information [17].

For example, when research began into the use of deep brain stimulation (DBS) for treatment-resistant depression, concerns regarding the decision-making capacity of people with depression prompted efforts to more systematically investigate ethical issues raised by informed consent for DBS in people with depression [18].

It is important to note that currently there are innovations in mental health and direct-to-consumer (DTC) technology taking place outside of these medical and healthcare frameworks. Thus, the ethical and regulatory frameworks that protect the well-being of participants in clinical settings do not necessarily apply to address issues such as safety or privacy. A prime example of such innovation is the proliferation of DTC mental health apps that are available. The U.S. Food and Drug Administration (FDA) regulates the safety of medical devices, but many such apps may not meet the criteria for medical device or are in a category that is considered of sufficiently low-risk that the FDA uses its discretion to not regulate them. Studies have found that most mental health apps are not supported by scientific evidence of safety or effectiveness [19]. Furthermore, in the consumer context, there are not the same rules to protect privacy and confidentiality of sensitive information shared with mental health apps as there are in healthcare settings [20]. The American Psychiatric Association (APA) is one of the organizations that has provided guidance on how to assess mental health apps for safety, data protection, and other key issues [21]. Given the high value of personal data in the current landscape, there remain challenges for the protection of brain and behavioral data that is collected and generated through DTC mental health apps [22].

When innovative technologies emerge, efforts to establish ethical frameworks for the development and use of that technology help provide critical engagement with the risks and benefits of the technology, as well as standards for when and how that technology may be brought into clinical practice. The Precautionary Principle that action must be taken to prevent serious potential harm, even where there is still only suggestive evidence of the threat, is a guiding principle which has supported efforts to assess emerging biotechnologies [23]. The field of synthetic biology, which is an interdisciplinary approach to the design of chemically synthesized DNA, provides an example of how stakeholders came together early in the development of the technology to address ethical issues. The Presidential Commission for the Study of Bioethical Issues produced guiding principles for assessing emerging technologies and used those to assess ethical and social implications of synthetic biology and its potential impact on society [24]. Those guiding principles are: (1) public beneficence, (2) responsible stewardship, (3) intellectual freedom and responsibility, (4) democratic deliberation, and (5) justice and fairness [25]. Principles of stewardship, beneficence, justice and fairness, and public deliberation assist in allowing for examination of how emerging technologies may impact society, as well as setting up key points for considerations of risks and benefits, equity and access, for when the technology transitions to clinical use [26].

In the area of brain research, the BRAIN Initiative included as part of its mission the establishment of ethical principles to guide the advances in neuroimaging and other neurotechnologies. The BRAIN initiative produced guidance for the development of neurotechnologies, such as anticipating special issues, public education, attending to possible malign uses of neurotechnologies, addressing public concerns about the brain and concerns regarding sharing research and benefits of the technology. Their neuroethical principles also address the need for caution as neurotechnologies are moved into medical or non-medical uses, as well as public education and protecting privacy and confidentiality of neural data, [27]. In particular, the principle of caution specifically discourages "the premature widespread use or inappropriate adoption of new technologies, especially those that may be offered directly to consumers or in non-health-care settings, such as in the legal system." [27]. For example, efforts to identify neural correlates for lying or pain might be of interest to a range of organizations, such as law enforcement or insurance companies, and thus lend themselves to premature adoption or misuse. Such innovations could have broader societal impact, through providing new means of surveillance, modifying or even controlling people. While principles such as stewardship or the need for public education are common to emerging technologies with a variety of clinical applications, there are two main areas that neuroethicists tend to identify specifically in the context of neuroinnovations. First, the brain's role in consciousness and identity, and the potential for neuroinnovations to have an impact on personality, agency, and personhood, is viewed as a special and pressing reason for caution and concern in assessing risks and benefits in this context [28]. Additionally, the protection of brain privacy and data is often viewed as an area of particular concern regarding innovative practices and approaches in the context of the brain.

Consciousness, Agency, Identity

In a review of the ethics of innovative brain technology, an issue that is pivotal to many of the specific challenges seen in these innovations is the role of the brain in consciousness and the general perception that the brain is where one's sense of self resides. Interventions in the brain have the potential to impact personality, agency, and other characteristics that make up one's sense of identity and self. Thus, health interventions using such technologies are expected to move cautiously in assessing the potential impact on these aspects of the brain, and explanations of these risks are essential to thorough informed consent for these procedures.

Broadly speaking, medical conditions, even those that are not focused on the brain, can have a significant impact on a person's sense of agency, self, or mood [29], and likewise some medical treatments can produce effects on a person's sense of identity. Studies of the illness narratives of people with cardiac or gastrointestinal conditions, as well as people with chronic pain problems, have demonstrated how these conditions can change a person's sense of self, identity, and personality [30]. There have been reports of more commonly prescribed drugs, such as statins, sometimes producing changes in a person's personality, such as triggering violent mood swings [31]. Nonetheless, when an innovative treatment is specifically meant to exert an effect on the brain or mental health, ethical implementation of that technology or treatment acknowledges the key role of the brain in making a person "who they are" by requiring a careful analysis of the potential impact on agency, identity, or personhood as part of understanding risks and benefits and providing appropriate informed consent. For example, with brain–computer interface (BCI) technologies, there is the potential for individuals to have difficulties with how they experience the boundaries between the device and their sense of self, such as wondering whether the motivation to perform a certain action came from themselves or from the device [32]. Because interventions using BCI may impact the way that a person experiences their actions and motivations or affect their sense of agency, innovations utilizing these technologies should be accompanied by research agendas that work towards better understanding these impacts and how to communicate these risks to patients.

DBS is a technique that has been clinically tested for treatment of Parkinson's disease and has been found to produce changes in personality and mood in some subjects [33]. These types of changes have raised concerns in neuroethics regarding the impact of DBS on identity and agency as well [34, 35]. Research to support ethical innovation in DBS has included study of the extent of these risks, as well as patient perceptions of this impact. Recent approaches to DBS include a "closed loop" design in which a sensor implanted as part of the DBS device monitors brain activity and modulates the device's stimulation effects accordingly, which raised additional concerns regarding how the patient experience of agency is impacted by the device modulation being under its own control [36]. For informed consent purposes, researchers must grapple with how best to explain and present risks for change in personality, especially when a patient may understandably have trouble weighing what may seem like more abstract concerns regarding agency against the debilitating nature of a condition like Parkinson's. As DBS is applied to different conditions, such as dystonia or eating disorders [37, 38], it remains important to evaluate how risks regarding impact on personality or mood compare to or interact with the risks of the specific condition. Empirical research into the experience of those receiving these types of interventions is important for better understanding their impact on personhood issues, as well as assisting with formulating ethical guidance, such as appropriate informed consent. There are also precedents in psychiatry and neurosurgery regarding innovative treatments and impact on personality, agency, or identity that can be examined for useful examples and guidance. For example, a number of psychiatric medications have side effects.

Brain Privacy/Neural Data

Brain privacy and protection of neural data have also been identified as a priority area of concern when innovative neurotechnologies are implemented [39]; emerging technologies utilizing BCI, biomarker projects, computational behavioral analysis, and many consumer neurotechnologies involve the collection of large amounts of fine-grained data about the brain and behavior [40]. Neural data may eventually include information regarding a person's perceptions, emotions, memories, or even thoughts. Neurotechnology initiatives are working on the development of improved methods to measure and analyze brain activity. As technologies that can "decode" neural data are developed, it will likely become possible to decode a person [41]. Similarly, as neuroimaging techniques and data analysis approaches improve, it may be possible to reveal more information about a person's behavioral health from brain scans [42]. Projects to develop "deep phenotyping" for the purposes of precision medicine bring together different layers of data, such as genomic data, behavioral tracking, brain scans, and environmental analysis in order to examine not just the risk of developing a specific brain condition but predict the possibility of specific cognitive states or mental health symptoms arising in the short-term future of an individual [43]. There are different types of brain and behavioral data that will be produced through these neurotechnologies and innovations. The underlying ethical concern is that the production of neural and behavioral data through innovative brain research and its real-world applications will place people at risk of having fine-grained and highly sensitive information about their behavior, mental state, and brain health revealed to others in ways that make them vulnerable to unwanted and unforeseen uses, or simply misuse, of that information.

While some of the neurotechnologies mentioned above are not yet a reality, the protection of brain and behavioral data has become a more pressing concern overall due to the health and behavioral inferences that can be drawn from that type of data through computational analytics [44]. There are diverse forms of data related to the brain and behavior that can be used together with descriptive data, such as clinical interviews, to generate information about mental state or even predictive information regarding behaviors [45]. The availability of information related to biomarkers, such as genetic risk of neurological and behavioral disorders also adds to concerns regarding brain privacy as the combination of different types of information can be used to generate risk prediction and other information about brain health, such as indications of pre-symptomatic and prodromal stages of mental illness [46]. Many people may be unaware of these types of risks when it comes to their neural and behavioral data. Research participants, patients, and consumers must be given clear information about these risks regarding their data and the opportunity to provide consent [47].

In medical settings, the Health Insurance Portability and Accountability Act (HIPAA) privacy protections generally apply to protect personal health information (Pub. L. No. 104–191) [48]. At the same time, patients are often insufficiently aware of the potential for some of their personal health information to be shared with third parties through electronic health records practices [49]. One approach to sharing data for research purposes is to de-identify the data. More and more, though, because of increased technological capabilities and the massive amount of personal information on public databases, it is becoming unfeasible to claim that the risk of re-identification of patient data can be eliminated [50]. For that reason, clinicians and researchers need to consider how to inform participants and patients of these risks and consider how to minimize negative repercussions of re-identified patient data.

In the consumer domain, there are fewer privacy protections for personal information [51]. Consumer data that pertains to a person's brain health or behavior can potentially be sold or shared in ways that the user is not expecting or would not foresee [52]. There have been steps towards increased protection of consumers' personal information in Europe, with the passage of the General Data Protection Regulation and California's Consumer Privacy Act. These regulations provide models for giving consumers better information regarding how their data may be used and opportunity to consent. At the same time, appropriate protections for neural and behavioral data in the consumer domain remain a concern, particularly in light of the value that access to personal data holds for many corporations and organizations.

Risks to an individual of having sensitive mental health and behavioral information being revealed include potential repercussions in areas such as employment, insurance, or education [22]. Neural and behavioral information could also be applied for consumer purposes, such as assisting with targeted marketing. Although marketing is sometimes viewed as a practice that is annoying but benign, mental and behavioral information could be used for purposes that are less benign such as decisions regarding what price points for merchandise are available to certain groups, which employment openings are shown to job searchers, or what type of housing options appear in an internet search [53]. When people are in a vulnerable state, such as a major depressive episode or mania, marketing practices could target them for treatment or purchases that could financially or negatively impact their mental health [54]. Furthermore, data that seemingly would not be labeled as behavioral or even health data, such as shopping purposes or location data, can now be analyzed to produce inferences regarding mental health and behavior [44, 55]. This type of inferential data blurs the lines or defining which data are still in the "context of the brain" – but also underlines the need for paying careful attention to privacy and protection of data when moving forward with neurotechnologies.

Conclusion

Innovation in brain research and technology offers great potential for improving care of people dealing with neurological and mental health disorders. As new methods and approaches for brain interventions are developed, the rapid pace of progress must be accompanied by careful consideration of the ethical challenges faced in the adoption of these innovate technologies. Established ethical frameworks for protecting the safety and welfare of vulnerable people, as well as addressing societal challenges of public engagement and access to innovation, provide useful guidance. At the same time, the context of the brain requires consideration of specific and novel ethical issues, such as the impact of innovative technologies on identity and agency or providing protections for neural and behavioral data.

Key Points

1. Ethical guidance for emerging technologies generally provides frameworks for addressing stewardship, accountability, public education and deliberation, and fairness. In the context of innovative brain technologies, such frameworks include addressing how the public perceives brain technologies and research, as well as anticipating potential areas for premature adoption or misuse of the technology in non-healthcare settings.
2. Clinical innovation may be undertaken when there is a demonstrated need for the innovation and when a careful consideration of risks and benefits shows that the innovation can be undertaken in an ethically acceptable manner. With clinical innovation in psychiatry or neurology, there is often a need to consider special issues such as the potential effects on agency or sense of self and how the underlying condition may affect informed consent.
3. Ethical frameworks that address innovation in the context of the brain generally highlight two special issues that merit concern: (1) the role of the brain in one's sense of agency and personhood, and (2) protection of neural and behavioral data.

Questions to Consider

1. What are principles that have been identified for the ethical development and implementation of emerging brain technologies?
2. How does the brain's role in consciousness and personhood impact the evaluation risks and benefits, as well as informed consent, for clinical innovation in the context of the brain?
3. What are reasons, if any, to consider neural data to be have different qualities or need for heightened protections in comparison to other kinds of health or personal information?

References

1. Agich GJ. Ethics and innovation in medicine. J Med Ethics. 2001;27(5):295–6. https://doi.org/10.1136/jme.27.5.295.
2. Grady C. Institutional review boards. Chest. 2015;148(5):1148–55. https://doi.org/10.1378/chest.15-0706.
3. Grady C, Cummings SR, Rowbotham MC, McConnell MV, Ashley EA, Kang G. Informed consent. N Engl J Med. 2017;376(9):856–67. https://doi.org/10.1056/NEJMra1603773.
4. American Medical Association. Code of medical ethics: current opinions with annotations. Chicago: American Medical Association, Council on Ethical and Judicial Affairs; 2019.
5. Christen M, Müller S. The clinical and ethical practice of neuromodulation – deep brain stimulation and beyond. Frontiers Media SA; 2018.
6. Ghaemi SN, Goodwin FK. The ethics of clinical innovation in psychopharmacology: challenging traditional bioethics. Phil Ethics Human Med. 2007;2:26. https://doi.org/10.1186/1747-5341-2-26.
7. Roberts LW, Jain S. Ethical issues in psychopharmacology: considerations for clinical practice; 2011. https://pro.psychcentral.com/ethical-issues-in-psychopharmacology-considerations-for-clinical-practice/

8. Siegler M, Lantos JD. Commentary: ethical justification for living liver donation. Camb Q Healthc Ethics. 1992;4:320–5.
9. Bellazzi R, Zupan B. Predictive data mining in clinical medicine: current issues and guidelines. Int J Med Inform. 2008;77(2):81–97. https://doi.org/10.1016/j.ijmedinf.2006.11.006.
10. Bzdok D, Meyer-Lindenberg A. Machine learning for precision psychiatry. ArXiv:1705.10553 [stat]; 2017. http://arxiv.org/abs/1705.10553.
11. Koutsouleris N, Kahn RS, Chekroud AM, Leucht S, Falkai P, Wobrock T, Derks EM, Fleischhacker WW, Hasan A. Multisite prediction of 4-week and 52-week treatment outcomes in patients with first-episode psychosis: a machine learning approach. Lancet Psychiatry. 2016;3(10):935–46. https://doi.org/10.1016/S2215-0366(16)30171-7.
12. Martinez-Martin N, Dunn LB, Roberts LW. Is it ethical to use prognostic estimates from machine learning to treat psychosis? AMA J Ethics. 2018;20(9):E804–11. https://doi.org/10.1001/amajethics.2018.804.
13. Hahn T, Nierenberg AA, Whitfield-Gabrieli S. Predictive analytics in mental health: applications, guidelines, challenges and perspectives. Mol Psychiatry. 2017;22(1):37–43. https://doi.org/10.1038/mp.2016.201.
14. Challen R, Denny J, Pitt M, Gompels L, Edwards T, Tsaneva-Atanasova K. Artificial intelligence, bias and clinical safety. BMJ Qual Saf. 2019;28(3):231–7. https://doi.org/10.1136/bmjqs-2018-008370.
15. Char DS, Shah NH, Magnus D. Implementing machine learning in health care—addressing ethical challenges. N Engl J Med. 2018;378(11):981–3. https://doi.org/10.1056/NEJMp1714229.
16. Roberts LW, Berk MS, Lane-McKinley K. Ethical considerations in research on suicide prediction: necessity as the mother of invention. JAMA Psychiat. 2019;76(9):883–4. https://doi.org/10.1001/jamapsychiatry.2019.1228.
17. Dunn LB, Candilis PJ, Roberts LW. Emerging empirical evidence on the ethics of schizophrenia research. Schizophr Bull. 2006;32(1):47–68. https://doi.org/10.1093/schbul/sbj012.
18. Dunn LB, Holtzheimer PE, Hoop JG, Mayberg HS, Appelbaum PS. Ethical issues in deep brain stimulation research for treatment-resistant depression: focus on risk and consent. AJOB Neurosci. 2011;2(1):29–36.
19. Weisel KK, Fuhrmann LM, Berking M, Baumeister H, Cuijpers P, Ebert DD. Standalone smartphone apps for mental health—a systematic review and meta-analysis. NPJ Digital Med. 2019;2(1):1–10. https://doi.org/10.1038/s41746-019-0188-8.
20. Yuste R, Goering S, Arcas BAY, Bi G, Carmena JM, Carter A, Fins JJ, Friesen P, Gallant J, Huggins JE, Illes J, Kellmeyer P, Klein E, Marblestone A, Mitchell C, Parens E, Pham M, Rubel A, Sadato N, et al. Four ethical priorities for neurotechnologies and AI. Nature News. 2017;551(7679):159. https://doi.org/10.1038/551159a.
21. Chan S, Torous J, Hinton L, Yellowlees P. Towards a framework for evaluating mobile mental health apps. Telemed J E-Health. 2015;21(12):1038–41. https://doi.org/10.1089/tmj.2015.0002.
22. Martinez-Martin N, Insel TR, Dagum P, Greely HT, Cho MK. Data mining for health: staking out the ethical territory of digital phenotyping. NPJ Digital Med. 2018b;1(1):68. https://doi.org/10.1038/s41746-018-0075-8.
23. Grandjean P. Implications of the precautionary principle for primary prevention and research. Annu Rev Public Health. 2004;25(1):199–223. https://doi.org/10.1146/annurev.publhealth.25.050503.153941.
24. Gutmann A. The ethics of synthetic biology: guiding principles for emerging technologies. Hast Cent Rep. 2011;41(4):17–22. https://doi.org/10.1002/j.1552-146x.2011.tb00118.x.
25. Häyry M. Synthetic biology and ethics: past, present, and future. Camb Q Healthc Ethics. 2017;26(2):186–205. https://doi.org/10.1017/S0963180116000803.
26. Lipworth W, Axler R. Towards a bioethics of innovation. J Med Ethics. 2016;42(7):445–9. https://doi.org/10.1136/medethics-2015-103048.
27. Greely HT, Grady C, Ramos KM, Chiong W, Eberwine J, Farahany NA, Johnson LSM, Hyman BT, Hyman SE, Rommelfanger KS, Serrano EE. Neuroethics guiding principles for the NIH BRAIN initiative. J Neurosci. 2018;38(50):10586–8. https://doi.org/10.1523/JNEUROSCI.2077-18.2018.

28. Ramos KM, Grady C, Greely HT, Chiong W, Eberwine J, Farahany NA, Johnson LSM, Hyman BT, Hyman SE, Rommelfanger KS, Serrano EE, Churchill JD, Gordon JA, Koroshetz WJ. The NIH BRAIN initiative: integrating Neuroethics and neuroscience. Neuron. 2019;101(3):394–8. https://doi.org/10.1016/j.neuron.2019.01.024.
29. Oris L, Luyckx K, Rassart J, Goubert L, Goossens E, Apers S, Arat S, Vandenberghe J, Westhovens R, Moons P. Illness identity in adults with a chronic illness. J Clin Psychol Med Settings. 2018;25(4):429–40. https://doi.org/10.1007/s10880-018-9552-0.
30. Barg FK, Kellom K, Ziv T, Hull SC, Suhail-Sindhu S, Kirkpatrick JN. LVAD-DT: culture of rescue and liminal experience in the treatment of heart failure. Am J Bioethics. 2017;17(2):3–11. https://doi.org/10.1080/15265161.2016.1265162.
31. Gorvett Z. The medications that change who we are. BBC; 2020. https://www.bbc.com/future/article/20200108-the-medications-that-change-who-we-are.
32. Drew L. The ethics of brain–computer interfaces. Nature. 2019;571(7766):S19–21. https://doi.org/10.1038/d41586-019-02214-2.
33. Lewis CJ, Maier F, Horstkötter N, Zywczok A, Witt K, Eggers C, Meyer TD, Dembek TA, Maarouf M, Moro E, Zurowski M, Woopen C, Kuhn J, Timmermann L. Subjectively perceived personality and mood changes associated with subthalamic stimulation in patients with Parkinson's disease. Psychol Med. 2015;45(1):73–85. https://doi.org/10.1017/S0033291714001081.
34. Gibert SH. Closed-loop deep brain stimulation and its compatibility with autonomous agency. AJOB Neurosci. 2017;8(2):88–90. https://doi.org/10.1080/21507740.2017.1320326.
35. Goering S, Klein E, Dougherty DD, Widge AS. Staying in the loop: relational agency and identity in next-generation DBS for psychiatry. AJOB Neurosci. 2017;8(2):59–70. https://doi.org/10.1080/21507740.2017.1320320.
36. Meynen, Gerben, Widdershoven G. The Impact of closed-loop DBS on agency: an open question. AJOB Neuroscience 2017;8(2):79–80. https://doi.org/10.1080/21507740.2017.1320325.
37. Gardner J. A history of deep brain stimulation: technological innovation and the role of clinical assessment tools. Soc Stud Sci. 2013;43(5):707–28. https://doi.org/10.1177/0306312713483678.
38. Park RJ, Singh I, Pike AC, Tan JOA. Deep brain stimulation in anorexia nervosa: Hope for the hopeless or exploitation of the vulnerable? The Oxford Neuroethics gold standard framework. Front Psych. 2017;8:44. https://doi.org/10.3389/fpsyt.2017.00044.
39. Ienca M, Haselager P, Emanuel EJ. Brain leaks and consumer neurotechnology. Nat Biotechnol. 2018;36(9):805–10. https://doi.org/10.1038/nbt.4240.
40. Emerging Issues Task Force, International Neuroethics Society. Neuroethics at 15: the current and future environment for Neuroethics. AJOB Neurosci. 2019;10(3):104–10. https://doi.org/10.1080/21507740.2019.1632958.
41. Glaser JI, Benjamin AS, Chowdhury RH, Perich MG, Miller LE, Kording KP. Machine learning for neural decoding. ArXiv:1708.00909 [Cs, q-Bio, Stat]; 2019. http://arxiv.org/abs/1708.00909
42. Underwood. Brain scans could help personalize treatment for people who are depressed or suicidal. Science | AAAS; 2019. https://www.sciencemag.org/news/2019/08/brain-scans-could-help-personalize-treatment-people-who-are-depressed-or-suicidal
43. Delude CM. Deep phenotyping: the details of disease. Nature. 2015;527(7576):S14–5. https://doi.org/10.1038/527S14a.
44. Wachter S, Mittelstadt B. A right to reasonable inferences: re-thinking data protection law in the age of big data and AI (SSRN scholarly paper ID 3248829). Social Science Research Network; 2019. https://papers.ssrn.com/abstract=3248829.
45. Harari GM, Lane ND, Wang R, Crosier BS, Campbell AT, Gosling SD. Using smartphones to collect behavioral data in psychological science: opportunities, practical considerations, and challenges. Perspect Psychol Sci. 2016;11(6):838–54. https://doi.org/10.1177/1745691616650285.
46. Wang L, Liu Z-P. Detecting diagnostic biomarkers of Alzheimer's disease by integrating gene expression data in six brain regions. Front Genet. 2019;10:157. https://doi.org/10.3389/fgene.2019.00157.
47. American Medical Association. Code of Medical Ethics Opinion 1.2.11.

48. Health Insurance Portability and Accountability Act. Pub. L. No. 104–191, § 264, 110 Stat.1936.
49. Rothstein MA. Health privacy in the electronic age. J Leg Med. 2007;28(4):487–501. https://doi.org/10.1080/01947640701732148.
50. Rocher L, Hendrickx JM, de Montjoye Y-A. Estimating the success of re-identifications in incomplete datasets using generative models. Nat Commun. 2019;10(1):1–9. https://doi.org/10.1038/s41467-019-10933-3.
51. Glenn T, Monteith S. Privacy in the digital world: medical and health data outside of HIPAA protections. Curr Psychiatry Rep. 2014;16(11):494. https://doi.org/10.1007/s11920-014-0494-4.
52. Huckvale K, Torous J, Larsen ME. Assessment of the data sharing and privacy practices of smartphone apps for depression and smoking cessation. JAMA Netw Open. 2019;2(4):e192542. https://doi.org/10.1001/jamanetworkopen.2019.2542.
53. Allen M. Health insurers are vacuuming up consumer data that could be used to raise rates | HealthLeaders media. Kaiser Health News; 2019. https://www.healthleadersmedia.com/finance/health-insurers-are-vacuuming-consumer-data-could-be-used-raise-rates
54. Van Bulck L, Luyckx K, Goossens E, Oris L, Moons P. Illness identity: capturing the influence of illness on the person's sense of self. Eur J Cardiovasc Nurs. 2019;18(1):4–6. https://doi.org/10.1177/1474515118811960.
55. Insel TR. Digital phenotyping: technology for a new science of behavior. JAMA. 2017;318(13):1215–6. https://doi.org/10.1001/jama.2017.11295.
56. Kühn AA, Volkmann J. Innovations in deep brain stimulation methodology. Mov Disord. 2017;32(1):11–9. https://doi.org/10.1002/mds.26703.
57. Martinez-Martin N, Kreitmair K. Ethical issues for direct-to-consumer digital psychotherapy apps: addressing accountability, data protection, and consent. JMIR Mental Health. 2018;5(2):e32. https://doi.org/10.2196/mental.9423.
58. National Academies of Sciences, Engineering, and Medicine. Framework for addressing ethical dimensions of emerging and innovative biomedical technologies: a synthesis of relevant National Academies Reports. Washington, DC: The National Academies Press; 2019. https://doi.org/10.17226/25491.
59. Savage N. How AI and neuroscience drive each other forwards. Nature. 2019;571:S15–7. https://doi.org/10.1038/d41586-019-02212-4.
60. Torous J, Keshavan M. A new window into psychosis: the rise digital phenotyping, smartphone assessment, and mobile monitoring. Schizophr Res. 2018;197:67–8. https://doi.org/10.1016/j.schres.2018.01.005.
61. Torous J, Roberts LW. The ethical use of Mobile Health Technology in Clinical Psychiatry. J Nerv Ment Dis. 2017;205(1):4–8. https://doi.org/10.1097/NMD.0000000000000596.

Further Reading

Emerging Issues Task Force, International Neuroethics Society. Neuroethics at 15: the current and future environment for neuroethics. AJOB Neurosci. 2019;10(3):104–10. https://doi.org/10.1080/21507740.2019.1632958.

Ghaemi SN, Goodwin FK. The ethics of clinical innovation in psychopharmacology: challenging traditional bioethics. Phil Ethics Human Med. 2007;2:26. https://doi.org/10.1186/1747-5341-2-26.

Martinez-Martin N, Dunn LB, Roberts LW. Is it ethical to use prognostic estimates from machine learning to treat psychosis? AMA J Ethics. 2018;20(9):E804–11. https://doi.org/10.1001/amajethics.2018.804.

Roberts LW, Berk MS, Lane-McKinley K. Ethical considerations in research on suicide prediction: necessity as the mother of invention. JAMA Psychiatry. 2019;76(9):883–4. https://doi.org/10.1001/jamapsychiatry.2019.1228.

Rothstein MA. Health privacy in the electronic age. J Legal Med. 2007;28(4):487–501. https://doi.org/10.1080/01947640701732148.

Chapter 6
The NIH's BRAIN 2025 Agenda: Attention to Related Ethical Considerations

Tenzin Tsungmey, Jodi Paik, Laura Turner-Essel, and Laura Weiss Roberts

Introduction

The brain is the most complex part of the human body and is responsible for creating human intelligence, interpreting sensory inputs, driving body movement, and controlling human behavior. For years, doctors, scientists, and philosophers have been captivated by the brain. In the last 15 years, the accelerating pace of research in neuroscience and the development of innovative research methods and practices have allowed scientists to learn more about the human brain than in all previous centuries. In recognition of this advancement in understanding, Congress named the 1990s the Decade of the Brain [1].

BRAIN Initiative Background

Understanding how the brain works is possibly one of the greatest scientific challenges of our time. Experts have put a lot of effort into understanding and explaining how different regions of the brain operate, but no general theory of brain function is accepted universally. Neuroscientists are still in the infant stages of understanding how neurons are connected throughout the brain and how these connections change behaviors. Understanding brain microcircuitry is essential for neuroscience to move forward [2]. On April 2, 2013, President Barack Obama launched the Brain Research

T. Tsungmey · J. Paik · L. Turner-Essel · L. W. Roberts (✉)
Department of Psychiatry and Behavioral Sciences, Stanford University School of Medicine, Stanford, CA, USA
e-mail: jodi.paik@stanford.edu; lte@stanford.edu; robertsl@stanford.edu

L. W. Roberts (ed.), *Ethics and Clinical Neuroinnovation*,
https://doi.org/10.1007/978-3-031-14339-7_6

Through Advancing Innovative Neurotechnologies (BRAIN) Initiative® to "accelerate the development and application of new technologies that will enable researchers to produce dynamic pictures of the brain that show how individual brain cells and complex neural circuits interact at the speed of thought" [3]. An initiative of this size requires ideas and partnership from scientists, doctors, and engineers across many different disciplines and divisions. The National Institutes of Health (NIH) has collaborated with other government agencies such as the Defense Advanced Research Projects Agency (DARPA), National Science Foundation (NSF), the U.S. Food and Drug Administration (FDA), and Intelligence Advanced Research Projects Activity (IARPA) [4].

The BRAIN Initiative has made remarkable progress toward understanding the brain, facilitating the development of new technologies and tools that appear to match the challenge of understanding the brain. New and unprecedented methods are revolutionizing exploration from the nanoscale to brain-and organism-wide investigation [5]. Through 2019, the NIH has made over 700 awards to investigators, totaling nearly $1.3 billion. The NIH is expected to ramp up its funding in the second half of the BRAIN Initiative project with more than half of the lifetime funding (estimated $4.9 billion) still to be spent. An allocation of funds via the Omnibus Appropriation Bill for fiscal year 2021 included support for two new transformative projects: $40 million for the Human Brain Cell Atlas and $20 million for the Armamentarium for Brain Cell Access. With the positive progress and outcomes from the early part of the initiative and the continued investment in the program to come, it is predicted that the most productive period of the BRAIN Initiative lies ahead [5].

Goals of the BRAIN Initiative

In response to the grand challenge issued by President Obama's proclamation, the NIH assembled the BRAIN Multi-Council Working Group (MCWG) to provide ongoing oversight of the vision and roadmap of the BRAIN Initiative. After assessing the challenges and opportunities in the field of neuroscience, seven areas were classified as high significance for the BRAIN Initiative. The text of the June 5, 2014 working group report, *BRAIN 2025: A Scientific Vision*, specifically identifies the following high priority subjects (a more detailed explanation of the seven goals can be found in Section II and Section III of the *BRAIN 2025* Report [3]) (Table 6.1).

Neuroethics and the BRAIN Initiative

Neuroethics is a field that studies the ethical, legal, and societal consequences of neuroscience. As an area of study, it is unique in that it involves experts from many different disciplines, including philosophy, psychology, law, theology, and

Table 6.1 Seven goals of the BRAIN Initiative

Seven goals of the BRAIN Initiative	
"Discovering diversity: Identify and provide experimental access to the different brain cell types to determine their roles in health and disease" (pg. 6)	"It is within reach to characterize all cell types in the nervous system, and to develop tools to record, mark, and manipulate these precisely defined neurons in the living brain. We envision an integrated, systematic census of neuronal and glial cell types, and new genetic and non-genetic tools to deliver genes, proteins, and chemicals to cells of interest in non-human animals and in humans." (pg. 6)
"Maps at multiple scales: Generate circuit diagrams that vary in resolution from synapses to the whole brain." (pg. 6)	"It is increasingly possible to map connected neurons in local circuits and distributed brain systems, enabling an understanding of the relationship between neuronal structure and function. We envision improved technologies—faster, less expensive, scalable—for anatomic reconstruction of neural circuits at all scales, from non-invasive whole human brain imaging to dense reconstruction of synaptic inputs and outputs at the subcellular level." (pg. 6)
"The brain in action: Produce a dynamic picture of the functioning brain by developing and applying improved methods for large-scale monitoring of neural activity." (pg. 6)	"We should seize the challenge of recording dynamic neuronal activity from complete neural networks, over long periods, in all areas of the brain. There are promising opportunities both for improving existing technologies and for developing entirely new technologies for neuronal recording, including methods based on electrodes, optics, molecular genetics, and nanoscience, and encompassing different facets of brain activity." (pg. 6)
"Demonstrating causality: Link brain activity to behavior with precise interventional tools that change neural circuit dynamics." (pg. 6)	"By directly activating and inhibiting populations of neurons, neuroscience is progressing from observation to causation, and much more is possible. To enable the immense potential of circuit manipulation, a new generation of tools for optogenetics, chemogenetics, and biochemical and electromagnetic modulation should be developed for use in animals and eventually in human patients." (pg. 6)
"Identifying fundamental principles: Produce conceptual foundations for understanding the biological basis of mental processes through development of new theoretical and data analysis tools." (pg. 6)	"Rigorous theory, modeling, and statistics are advancing our understanding of complex, nonlinear brain functions where human intuition fails. New kinds of data are accruing at increasing rates, mandating new methods of data analysis and interpretation. To enable progress in theory and data analysis, we must foster collaborations between experimentalists and scientists from statistics, physics, mathematics, engineering, and computer science." (pg. 6)
"Advancing human neuroscience: Develop innovative technologies to understand the human brain and treat its disorders; create and support integrated human brain research networks." (pg. 6)	"Consenting humans who are undergoing diagnostic brain monitoring, or receiving neurotechnology for clinical applications, provide an extraordinary opportunity for scientific research. This setting enables research on human brain function, the mechanisms of human brain disorders, the effect of therapy, and the value of diagnostics. Meeting this opportunity requires closely integrated research teams performing according to the highest ethical standards of clinical care and research. New mechanisms are needed to maximize the collection of this priceless information and ensure that its benefits people with brain disorders." (pgs. 6–7)

(continued)

Table 6.1 (continued)

Seven goals of the BRAIN Initiative	
"From BRAIN Initiative to the brain: Integrate new technological and conceptual approaches produced in Goals #1–6 to discover how dynamic patterns of neural activity are transformed into cognition, emotion, perception, and action in health and disease." (pg. 7)	"The most important outcome of the BRAIN Initiative will be a comprehensive, mechanistic understanding of mental function that emerges from synergistic application of the new technologies and conceptual structures developed under the BRAIN Initiative." (pg. 7)

Brain Research through Advancing Innovative Neurotechnologies Working Group. BRAIN 2025 Report [Internet]. National Institutes of Health; 2014. Available from: https://braininitiative.nih.gov/strategic-planning/brain-2025-report

sociology, among others. Although many of the ethical concerns in brain research are common with those found in biomedical research in general, there are additional, special ethical considerations that impact neuroscience research because the brain is responsible for our consciousness, our innermost thoughts, and our most basic human needs [6]. Additionally, as increasingly innovative neurotechnologies are developed as research tools and as therapies to better understand the brain and treat brain disorders, neuroethics plays a crucial role in the advancement of neuroscience as a field; any technology that informs our understanding of the brain and its functions, including higher-order activities like consciousness and thought, requires ethical guidance for the development, application, and consequences of its use [7].

Discussions about the need for an Ethical, Legal, and Social Implications (ELSI) Program for neuroscience began in the early twenty-first century, inspired by the model of the ELSI Program of the NIH's Human Genome Project. Cognizant that neuroinnovation was happening *now,* and that the nature of neuroscience research necessitated consideration of complex issues involving decision-making capacity, autonomy, and consent, neuroethicists were attuned to the need for ethical guidance in the neurosciences. From its beginning, members of the BRAIN Initiative community were concerned with the ethical implications of innovative neuroscience research and application and advocated to include support for ethical inquiry within its structure. Initial inquiry and identification of ethical issues began in 2013, when, as part of his additional BRAIN Initiative announcement, President Obama tasked the Presidential Commission for the Study of Bioethical Issues to consider and create a set of proactive ethical standards to guide neuroscience research. Three public stakeholder meetings were held in August 2013, December 2013, and February 2014, and the Commission issued its first set of recommendations in 2014. *Gray Matters: Integrative Approaches for Neuroscience, Ethics, and Society (Gray Matters*, Volume 1), emphasized the need to integrate ethics throughout the research endeavor and explicitly called for funding for ethics. A second, further report, *Gray Matters: Topics and the Intersection of Neuroscience, Ethics, and Society (Gray Matters*, Volume 2*)* was released in March of 2015 and focused attention on three topics of heightened concern arising from neuroscience research: cognitive enhancement, capacity for consent, and neuroscience and the law. The reception to the *Gray Matters* reports was mixed, with some researchers critical of its lack of specific advocacy for an ELSI-style initiative.

Attention to ethics was included as an important component of the BRAIN Initiative's strategic plan. The initial *BRAIN 2025* report highlighted four specific goals for neuroethics: (1) the establishment of a shared vision for the ethical conduct of neuroscience research; (2) the development of resources for collecting and disseminating best practices in the conduct of ethical scientific research, particularly clinical research; (3) the support of data-driven research to inform ethical issues arising from BRAIN 2025 projects; and (4) the creation of outreach activities for diverse stakeholders to promote the discussion and examination of the social and ethical implications of neuroscience research [3]. As part of this strategy, a specific Neuroethics Working Group (NEWG) was created to provide guidance for the Initiative in 2015.

With the scientific goals of the BRAIN Initiative in mind, the NEWG developed a list of Neuroethics Guiding Principles to frame and address neuroethical issues of concern for research projects funded by the BRAIN Initiative, with the goal to facilitate progress in neuroscience while helping to ensure that neuroscientific advancements support human well-being. The Working Group's Guiding Principles were outlined specifically in a December 2018 article in the Journal of Neuroscience [8] (Table 6.2).

Most recently, informed by the BRAIN Initiative's interim review (*BRAIN 2.0*), the NIH convened an additional advisory group, the BRAIN 2.0 Neuroethics Subgroup (BNS), reporting directly to the Advisory Committee to the Director (ACD). The creation of the BNS again highlighted the value of integrating neuroscience and ethics to confront the challenges and emerging ethical questions arising from the study of the brain, anticipating ethical concerns and guiding navigation through them and influencing how neuroscience research should be designed, conducted, interpreted, and applied. The BNS was tasked with reviewing the BRAIN 2025 goals, evaluating their progress in their current context, and developing a specific "Neuroethics Roadmap" supporting the integration of neuroethics and

Table 6.2 Neuroethics guiding principles

1.	Make assessing safety paramount.
2.	Anticipate special issues related to capacity, autonomy, and agency.
3.	Protect the privacy and confidentiality of neural data.
4.	Attend to possible malign uses of neuroscience tools and neurotechnologies.
5.	Use caution when moving neuroscience tools and neurotechnologies into medical or non-medical uses.
6.	Identify and address specific concerns of the public about the brain.
7.	Encourage public education and dialogue.
8.	Behave justly and share the benefits of neuroscience research and resulting technologies.

neuroscience, releasing their findings independently but in concert with the *BRAIN 2.0* report [7]. Five major findings, highlighting distinct areas for enhanced BRAIN Initiative attention and support, were described [7]:

1. The BRAIN Initiative should enhance integration of neuroscience and neuroethics.
2. The BRAIN Initiative should provide additional tools and resources for neuroethics research and for neuroscientists to appreciate neuroethics issues.
3. The BRAIN Initiative should monitor the development and use of animal and other biological models aimed to more closely approximate human brain function.
4. The BRAIN Initiative should establish guidelines for the neuroscience data ecosystem that address data capture, storage, sharing, and translation to humans and society.
5. The BRAIN Initiative should initiate conversations and collaborations to address neuroscience applications beyond biomedical and clinical contexts.

Following the model established by the ELSI Program of the NIH Human Genome Project, which required that "not less than" 5% of funding be dedicated to research on the ethical, legal, and social implications of genetic research [9], the BNS recommended that the ELSI standard for the BRAIN Initiative also work toward a 5% goal of ethics-related spending. Current spending on neuroethics via BRAIN Initiative funds hovers around 1% to date. ELSI funding under the Human Genome Project was 3% of total budget in the first fiscal year (1990) and quickly reached its 5% goal in 1992. By the end of its first 5 years, ELSI-HGP had funded 126 unique research and education projects, and funding for genome-related ELSI projects continues to this day via the NIH's National Human Genome Research Institute. If the BRAIN Initiative hopes to reach its ethics funding goal of 5%, there is room for growth.

An internal neuroethics project team, composed of NIH program staff, ensures that neuroethics are considered throughout the grant cycle, from their inclusion in the development of funding plans for review, to approval of the funding by the NIH, and throughout the duration of completing the funded project. The Neuroethics Guiding Principles may evolve over time as the BRAIN Initiative continues to grow and increase our understanding of the human brain and how it functions.

BRAIN Initiative: Research on the Ethical Implications of Advancements in Neurotechnology and Brain Science

The BRAIN Initiative supports research and study on neuroethics in multiple ways, via a funded neuroethics research portfolio, supported neuroethics training and postdoctoral fellowships, and administrative supplements that include neuroethics inquiry into existing BRAIN Initiative projects. Many current neuroethics projects are grounded in and informed by the BRAIN Initiative's research portfolio, and BRAIN Initiative-sponsored research efforts toward understanding brain function and treating brain disorders in human subjects have informed neuroethics research that both addresses classical bioethics and also requires new, specific analyses of ethical concerns that arise due to the uniqueness of the brain, namely its role as the organ of the mind. Likewise, emerging neuroscience tools and technologies powerfully inform our understanding of the brain and provide potentially unprecedented ways to modify its function.

The initial funding opportunity announcement (FOA) specifically for neuroethics study, RFA-MH-17-260, was posted on October 21, 2016 and opened for submission on December 30 of that year. Informed by the goals of the BRAIN Initiative, the FOA provided support for research addressing core ethical issues arising from innovative neuroscience research and technology development funded by the Initiative. To date, 15 projects have been funded through three ethics-specific RFAs (Table 6.3), representing an investment of over $15 million since 2017.

Table 6.3 BRAIN Initiative neuroethics projects funded in three initial ethics-specific RFAs

Year	Title	Institution
2017	Achieving ethical integration in the development of novel neurotechnologies	UCSF
	Enabling ethical participation in innovative neuroscience on mental illness and addiction: towards a new screening tool enhancing informed consent for transformative research on the human brain	Stanford University
	Ethics of patients and care partners perspectives on personality change in Parkinson's disease and deep brain stimulation	Cleveland Clinic
	Neuroethics of a DBS system targeting neuropsychiatric and movement disorders	Baylor College of Medicine
2018	Assessing the effects of deep brain stimulation on agency	Dartmouth College
	Human agency and brain–computer interfaces: understanding users' experiences and developing a tool for improved consent	University of Washington
	Informing choice for neurotechnological innovation in pediatric epilepsy surgery	University of British Columbia
	Is the treatment perceived to be worse than the disease? Ethical concerns and attitudes towards psychiatric electroceutical interventions	Michigan State University
	The brainstorm project: a collaborative approach to facilitating the neuroethics of bioengineered brain modeling research	Case Western Reserve University
2019	Cognitive restoration: neuroethics and disability rights	Weill Medical College of Cornell University
	Leveraging ethical dissension among capacity, beneficence and justice in clinical trials of neurotherapeutics in the severely disabled: lessons from schizophrenia	University of Colorado
	Pediatric deep brain stimulation: neuroethics and decision-making	Baylor College of Medicine
2020	Neuroethics of non-therapeutic invasive human neurophysiologic research	UCLA
	Highly portable and cloud-enabled neuroimaging research: confronting ethics challenges in field research with new populations	University of Minnesota
	Ethics of the choice of invasive versus non-invasive neurosurgery: different stakeholders' perspectives, surgical decision-making, and impact on patient sense of control	Cleveland Clinic/ Case Western Reserve University

Supported neuroethics research projects often, although not exclusively, complement BRAIN Initiative-funded research involving human subjects. Awareness of potential downstream effects, on both an individual and societal basis, inform neuroethical research design, with research addressing issues of societal reentry and cognitive restoration [10] and the meaning of unintended effects in aspects of cognition, behavior, and emotion that potentially arise from novel neurointervention [11]. Funded projects also anticipate ethical questions that are likely to arise from still-developing research-specific technology; as an example, researchers are proactively examining ethical questions that spring from increasingly more complex brain organoid research [12].

As the BRAIN Initiative research portfolio expands to include study of implanted neuromodulation devices, research into the ethical concerns related to the use of these innovative devices has grown. Several projects have explicitly explored the ethical implications of deep brain stimulation (DBS), addressing concerns about autonomy and identity [13]; querying perspectives on valued personality characteristics and perceived changes after DBS [14, 15]; and developing assessment tools to better catalogue changes in agency in patients post-DBS [16].

Research into the different values and ethical concerns of different stakeholder groups have also been an important part of the BRAIN Initiative neuroethics research portfolio, with the hope that recognition, understanding, and integration of the unique perspectives of patients and researchers will aid in the ethical design and implementation of innovative neuroscience. These research endeavors have been both technology-specific (incorporating stakeholder perspectives in DBS [17]), technology- and group-specific (developing a clinical decision aid for the use of DBS in a pediatric population [18, 19]), and investigative of ethical concerns, and their similarities and their differences, across neuroinnovative research types and stakeholders [20, 21].

Genesis of the "Enabling Ethical Participation in Innovative Neuroscience on Mental Illness and Addiction" Project at Stanford

Great discoveries in neuroscience hold promise for reducing the burden of many of the most disabling conditions that threaten human health worldwide, including mental illnesses and addictions. Thanks to the BRAIN Initiative, this challenge is now being surmounted with new technologies and methods emerging at an unprecedented level of innovation. Such efforts include brain–machine interfaces, circuit-based neuromodulation trials, gene editing, optogenetic manipulations, behavioral vaccine development, in vitro cellular models from induced pluripotent stem cells, big data/machine learning and artificial intelligence, among others. Increasingly, exceptionally innovative science inspires hope that devastating brain-based disorders may be prevented, treated, and even cured.

As the *BRAIN 2025* scientific vision notes, however, a suite of novel ethical challenges confronts those engaged in innovative neuroscience. These concerns include the deepest questions about what defines humanity and personhood, what forms of novel inquiry may exceed ethically acceptable limits in society, and how to perform ethically sound studies with volunteers who may be vulnerable to exploitation in the research situation. Such issues are particularly salient in mental illness and addiction research because these conditions affect cognition, emotion, motivation, behavior, and self-governance of potential participants. Importantly, some of these ethical issues are amenable to empirical study, which can yield valuable insights and evidence-informed practices that strengthen and enable ethically sound human brain investigation.

In January 2018, a group of researchers at Stanford School of Medicine proposed a study to do just that. Led by co-investigators Dr. Laura Weiss Roberts (Chair of the Department of Psychiatry and Behavioral Sciences) and Dr. Laura B. Dunn (at the time, faculty member and clinician in the same department) the study's overarching goal was to accelerate neuroscience toward lessening the burden of mental illness and addiction through a hypothesis-driven empirical ethics inquiry in three parts. Both investigators had previously led research teams examining the ethical research participation of individuals with serious mental illness and/or addiction, and the evolution of the BRAIN Neuroethics grant provided an opportunity for further collaboration in this important topic area. Biostatistician Dr. Jane Paik Kim was brought on as a co-investigator and proved to be instrumental in developing measures for the study, as well as analyzing and interpreting statistical results. A scientific advisory board comprised of psychiatrists, bioethicists, and research experts was assembled to support both goals of the study and along with a team of diverse research personnel.

The first aim of the study was to determine the distinct ethical issues and problems encountered in innovative neuroscience related to mental illness and addiction through semi-structured interviews with neuroscientists, neuroethicists, and institutional review board members. Informed by the research team's past work and grounded in a rigorous conceptual model, the second aim of the study was to examine factors (both negative and positive) that influence research decision-making by people with mental illness and addiction in the context of innovative neuroscience research.

Maximizing an established record of expertise in empirical ethics investigations and neuroethics, the project leveraged access to the exceptional neuroscience research conducted at Stanford University, including work by BRAIN initiative investigators; provided extensive, systematically collected data on influences on decision-making about innovative neuroscience research participation by individuals with mental or physical illness and healthy controls; and held promise for the development of new evidence-informed best practices in safeguarding human volunteers in cutting-edge neuroscience. It continues to be our hope that such work will ultimately mitigate fears and biases that continue to adversely affect scientific, regulatory, and public practices related to research involving people with neuropsychiatric conditions.

The following section of this book, *Special Topics in Clinical Neuroinnovation*, provides "'deep dives" into unique issues in the field, exploring the ethics of neuro-innovative research in the context of neurosurgery, disorders of consciousness, psychedelic drugs, in the courts, and in concert with industry. The final section of this book, *Neuroethics and Innovation: Inquiry informed by the Roberts Valence Model*, discusses our particular project in finer detail and presents qualitative results from the first aim as well as some quantitative data generated and analyzed in relation to the second aim. For further description of the grant which funded the work described in this book, and a rationale for the chosen focus of the study, please see Chap. 12.

References

1. Brain Basics: Know Your Brain | National Institute of Neurological Disorders and Stroke [Internet]. [cited 2021 Nov 9]. https://www.ninds.nih.gov/Disorders/Patient-Caregiver-Education/Know-Your-Brain
2. Alivisatos AP, Chun M, Church GM, Greenspan RJ, Roukes ML, Yuste R. The brain activity map project and the challenge of functional connectomics. Neuron. 2012;74(6):970–4.
3. Brain Research through Advancing Innovative Neurotechnologies Working Group. BRAIN 2025 Report [Internet]. National Institutes of Health; 2014. https://braininitiative.nih.gov/strategic-planning/brain-2025-report
4. Overview | Brain Initiative [Internet]. [cited 2021 Nov 9]. https://braininitiative.nih.gov/about/overview
5. Advisory Committee to the NIH Director BRAIN Initiative Working Group 2.0. The BRAIN Initiative® 2.0: from cells to circuits, toward cures [Internet]. National Institutes of Health; 2019. https://braininitiative.nih.gov/strategic-planning/acd-working-groups/brain-initiative%C2%AE-20-cells-circuits-toward-cures
6. Neuroethics and the BRAIN Initiative [Internet]. [cited 2020 Mar 23]. http://www.theneuroethicsblog.com/2016/04/neuroethics-and-brain-initiative.html
7. ACD Working Group on BRAIN 2.0 Neuroethics Subgroup. The BRAIN Initiative® and neuroethics: enabling and enhancing neuroscience advances for society [Internet]. National Institutes of Health; 2019. https://braininitiative.nih.gov/strategic-planning/acd-working-groups/brain-initiative%C2%AE-and-neuroethics-enabling-and-enhancing
8. Bianchi DW, Cooper JA, Gordon JA, Heemskerk J, Hodes R, Koob GF, et al. Neuroethics for the National Institutes of Health BRAIN initiative. J Neurosci. 2018;38(50):10583–5.
9. ELSI Planning and Evaluation History [Internet]. Genome.gov. [cited 2021 Nov 9]. https://www.genome.gov/10001754/elsi-planning-and-evaluation-history
10. Hendriks S, Grady C, Ramos KM, Chiong W, Fins JJ, Ford P, et al. Ethical challenges of risk, informed consent, and posttrial responsibilities in human research with neural devices: a review. JAMA Neurol. 2019;76(12):1506–14.
11. Cabrera LY, Gilbert MMC, McCright AM, Achtyes ED, Bluhm R. Beyond the Cuckoo's nest: patient and public attitudes about psychiatric electroceutical interventions. Psychiatry Q. 2021;92(4):1425–38.
12. Hyun I. Moonshots and other metaphors: the BRAIN initiative. AJOB Neurosci. 2020;11(3):198–200.
13. Ramirez-Zamora A, Giordano JJ, Gunduz A, Brown P, Sanchez JC, Foote KD, et al. Evolving applications, technological challenges and future opportunities in neuromodulation: proceedings of the fifth annual deep brain stimulation think tank. Front Neurosci. 2018;11:734.
14. Kubu CS, Frazier T, Cooper SE, Machado A, Vitek J, Ford PJ. Patients' shifting goals for deep brain stimulation and informed consent. Neurology. 2018;91(5):e472–8.

15. Kubu CS, Ford PJ, Wilt JA, Merner AR, Montpetite M, Zeigler J, et al. Pragmatism and the importance of interdisciplinary teams in investigating personality changes following DBS. Neuroethics. 2021;14(Suppl. 1):95–105.
16. Assessing the Effects of Deep Brain Stimulation on Agency [Internet]. [cited 2021 Nov 10]. http://www.theneuroethicsblog.com/2020/02/assessing-effects-of-deep-brain_11.html
17. Chiong W, Leonard MK, Chang EF. Neurosurgical patients as human research subjects: ethical considerations in intracranial electrophysiology research. Neurosurgery. 2018;83(1):29–37.
18. MuÑoz KA, Blumenthal-Barby J, Storch EA, Torgerson L, LÁzaro-MuÑoz G. Pediatric deep brain stimulation for dystonia: current state and ethical considerations. Camb Q Healthc. 2020;29(4):557–73.
19. Muñoz KA, Kostick K, Torgerson L, Zuk P, Kalwani L, Sanchez C, et al. Pressing ethical issues in considering pediatric deep brain stimulation for obsessive-compulsive disorder. Brain Stimulat. 2021;14(6):1566–72.
20. Roberts LW, Kim JP. Healthy individuals' perspectives on clinical research protocols and influences on enrollment decisions. AJOB Empir Bioeth. 2017;8(2):1–10.
21. Tsungmey T, Kim JP, Dunn LB, Ryan K, Lane-McKinley K, Roberts LW. Negative association of perceived risk and willingness to participate in innovative psychiatric research protocols. J Psychiatr Res. 2020;122:9–16.

Further Reading

Bianchi DW, Cooper JA, Gordon JA, Heemskerk J, Hodes R, Koob GF, et al. Neuroethics for the National Institutes of Health BRAIN initiative. J Neurosci. 2018;38(50):10583–5.

Roberts LW, Kim JP. Healthy individuals' perspectives on clinical research protocols and influences on enrollment decisions. AJOB Empir Bioeth. 2017;8(2):1–10.

Tsungmey T, Kim JP, Dunn LB, Ryan K, Lane-McKinley K, Roberts LW. Negative association of perceived risk and willingness to participate in innovative psychiatric research protocols. J Psychiatr Res. 2020;122:9–16.

Part II
Special Topics in Clinical Neuroinnovation

Chapter 7
Neurosurgery and Neuroinnovation in the Surgical Suite: The Ethics of Neurostimulation for Severe Obesity

Disep I. Ojukwu, Daniel A. N. Barbosa, Arthur L. Caplan, and Casey H. Halpern

Introduction

In the words of Goel and Kothari, "neurosurgery has the magical touch of restoring the might to the paralyzed. Equally, the opposite can as well happen [1]." Herein lies the juxtaposition of innovation and ethics, which has been intertwined with neurosurgery since its infancy. Indeed, the history of neurosurgery and the journey of its development have been paved with courage, ingenuity, and the desire to attenuate patients' suffering. This history was hinged upon the understanding of anatomy and advancement in surgical techniques during the Renaissance period in Europe [2]. Jacopo Berengario da Carpi (1460–1530), an Italian surgeon and anatomist, was the first to publish "an anatomical text supplemented with illustrations [and] a monograph dedicated to head injury" [2]. Over the next centuries, tremendous accomplishments were achieved in anatomical knowledge, with American neurosurgeon Albert L. Rhoton (1932–2016) rendering modern contributions to the understanding of microsurgical neuroanatomy. Rhoton notoriously published numerous illustrations and stepwise dissections "to permit the neurosurgeon to navigate accurately, gently, and safely around and through the cerebrum and intracranial space" [3].

D. I. Ojukwu
Department of Neurosurgery, Stanford University School of Medicine, Stanford, CA, USA
e-mail: dojukwu@stanford.edu

D. A. N. Barbosa · C. H. Halpern (✉)
Department of Neurosurgery, Perelman School of Medicine, Richards Medical Research Laboratories, Pennsylvania Hospital, University of Pennsylvania, Philadelphia, PA, USA
e-mail: Dbarbosa@pennmedicine.upenn.edu; Casey.Halpern@pennmedicine.upenn.edu

A. L. Caplan
Division of Medical Ethics, New York University Grossman School of Medicine, New York, NY, USA
e-mail: Arthur.Caplan@nyulangone.org

L. W. Roberts (ed.), *Ethics and Clinical Neuroinnovation*,
https://doi.org/10.1007/978-3-031-14339-7_7

Although cranial surgeries for traumatic injuries had been conducted throughout history, the advent of modern neurosurgical techniques did not arise until the 19th century. In 1879, Scottish surgeon Sir William Macewen (1848–1924) performed his first tumor resection [4]. The surgery, which marked "the first elective surgery specifically targeting a lesion deep to the dura," was successfully conducted on a 14-year-old patient who likely had a tuberculoma [5]. In 1884, English surgeon Sir Rickman J. Godlee (1849–1925) excised a tumor in a patient with a history of severe headaches and worsening paralysis of one side of the body. The patient survived for a few months after the surgery, at which point postmortem examination revealed the presence of a malignant tumor [4, 5]. Although Godlee never undertook another intracranial operation, he is considered important to the historical development of neurosurgery as a discipline [5]. These monumental surgeries performed by Macewen and Godlee were made possible by advancements in anesthesia to induce insensitivity to pain, antiseptic concepts developed by British surgeon Joseph Lister to minimize infection, hemostasis for the prevention of blood loss intraoperatively, and accurate cerebral localization borne from clinical observation and examination [2, 4, 5]. Still, strides were yet to be made.

In 1886, Victor A.H. Horsley (1857–1916) was considered the first "neurosurgeon" in history following his appointment as a surgeon of cranial surgery, albeit he continued to practice as a general surgeon [4]. Primarily practicing in London, Horsley became renowned for performing the largest volume of cranial surgeries worldwide. Within a year, he performed 11 operations with one death in a boy with a posterior fossa tumor [4]. In addition, Horsley is credited as being the first surgeon to perform operations for epilepsy, pituitary tumors, subarachnoid hemorrhage, and resection of a spinal tumor [4]. While many regarded him as dangerous and rough in his surgical approaches, others classified him as a masterful technical surgeon for his excellent survival rates without the use of medications or imaging modalities [4]. In addition to his anatomical and surgical skills, his contributions to surgical innovation also shaped neurosurgery for more than a century. For instance, his conception of the stereotactic frame has been revolutionary for precise brain localization, although it was only used in a non-human primate [6]. Horsley also implemented staging for difficult procedures, in which cranial decompression was performed to reduce brain swelling before attempting a second procedure on the same patient [4].

Amidst these pivotal advancements in Europe, the field of neurosurgery remained infantile during the late 19th and 20th centuries [5]. Conversely, in the United States, "the rigid application of techniques, new insights into the physiology of the brain and cerebral circulation, and innovative techniques and devices that helped to achieve hemostasis in the brain were developed" [5]. Most notably, American neurosurgeon Harvey W. Cushing (1869–1939) is credited as founding neurosurgery as an independent surgical specialty and writing the first substantial text on neurosurgery [7]. In addition, Cushing was the first to use the new technology of X-rays in surgical practice, in which he used the technology to localize and remove a bullet in the cervical spinal canal [7]. Among other accomplishments, Cushing was also the first to add blood pressure measurement to the ether management chart of surgical patients in the United States, although this practice was contemptuously viewed by most surgeons.

As one surgeon remarked, "there is nothing to be learned from the measurement of blood pressure that cannot be learned by the skilled palpating finger on the pulse" [7].

The ingenuities from prominent individuals truly shaped the discipline of neurosurgery. Moreover, neurosurgical innovation continued to be a cornerstone for the advancement of the specialty. From Neolithic human skeletal trepanations (i.e., surgically created holes in the skull) to NeuroMate surgical robots, innovation has been indispensable for pioneering clinical practice [8–10]. Using cerebral localization performed by Alexander H. Bennett (a neurologist), Godlee's excision of the tumor from the Rolandic area of the brain marked the first time localization was used to successfully localize and remove a tumor from a human brain [2, 11]. Another example of historical neurosurgical innovation occurred in 1957, when American neurosurgeon Theodore Kurze (1922–2002) became the first neurosurgeon to use a binocular microscope in the operating room. After a year of practicing middle ear dissections, Kurze translated those surgical skills to neurosurgery to remove a Schwannoma of the seventh cranial nerve in a 5-year-old patient [2].

Embedded within the historical tapestry of neurosurgical ingenuity lies an undercurrent of ethical principles, which arguably are more relevant today. According to Wall et al., "an ethical problem occurs when an agent must choose between mutually exclusive options, both of which either have equal elements of right and wrong or are perceived as equally obligatory. The essential element that [defines] an ethical problem [...] is the element of choice" [12]. Although the choices and intrinsic nature of moral dilemmas in neurosurgery evolved over time, the implications of moral decisions remained significant. To this point, inherent in the surgeon's scalpel is the contradiction to one of the most valued principles of clinical ethics —*primum non nocere*, that is "first, do no harm." By using the scalpel, however, surgeons must first "harm" before cure [13]. This necessary incision must be performed with sympathy and care for the patient. Eighteenth century Scottish physician and moralist John Gregory implored physicians to utilize "that sensibility of heart which makes us feel for the distresses of our fellow-creatures ... [and] ... incites us in the most powerful manner to relieve them" [13]. This abiding compassion is one of the many points of the moral compass that continues to guide the field of neurosurgery and its innovative advancements.

Deep Brain Stimulation (DBS)

Neurostimulation, such as deep brain stimulation (DBS), is one advancement in neurosurgery that has helped to relieve distress in patients with movement or psychiatric disorders [14]. The surgical procedure involves placement of electrodes within designated targets of the brain that are involved in the pathophysiology of the disorder [15]. The history of neurostimulation, or electrostimulation of the brain, dates back to prehistory. While much has been documented, this narrative will briefly highlight the historical background of modern DBS therapies in the United States [16–19].

In the 1930s, neurosurgeons used ablative therapy as definitive treatment for epilepsy. The patient was kept awake while the neurosurgeon used an electrical probe to deliberately destroy a pre-identified region of the cerebral cortex that was presumed malfunctioned [20]. In 1947, the stereotactic apparatus was used on humans for the first time, enabling surgeons to ablate deeper areas within the brain [6, 20]. The emergence of this apparatus also ushered in the establishment of a new neurosurgical specialty—stereotactic neurosurgery—which ultimately developed the knowledge, skills, and equipment that positioned DBS as a viable therapy [20]. During this "era of experimentation," neurosurgeons continued to identify brain targets for movement disorders and patients typically received chronic stimulation as therapy for those disorders. "Electrodes were left in situ and protruding from the skull for several weeks, enabling the surgeons to identify and lesion the optimal areas for ablation in an incremental fashion" [20]. Subsequent to the introduction of levodopa as medical management for Parkinson's disease (PD) in 1968, however, stereotactic neurosurgery waned. Because neurologists were reluctant to refer PD patients for neurosurgical intervention, only a few academic centers continued to offer stereotactic neurosurgery to "patients with chronic, untreatable pain and those with movement disorders that would not respond to the new levodopa medications" [20]. These academic centers would eventually serve as the epicenter of DBS evolution and expansion.

At that time, neurostimulation still consisted of external electrodes which protruded "from the [patient's] head in order to link with a power source, which at the time [was] large, cumbersome and certainly not implantable" [20]. In the 1960s, the medical device manufacturer Medtronic introduced a groundbreaking product—the first commercially available cardiac pacemaker, which heralded the advent of a small, mobile power source [20]. With the financial success and advancement garnered from this new technology, "the development of the neurostimulator piggybacked on the success of the cardiac pacemaker" [20]. A milestone was reached when American neurosurgeon C. Norman Shealy (b. 1932) adapted the technology of the cardiac pacemaker for treatment of patients with chronic, intractable pain [20]. Utilizing implantable electrodes within the spinal cord, Shealy delivered modified radio frequency energy through the skin to an implanted receiver [20].

After much success of this modified technology for treating chronic pain, the medical device industry developed a neurostimulator device specific for such purpose. In 1968, Medtronic released the first device, then manufacturer Avery Laboratories released their device in 1972, and finally manufacturer Cordis released their neurostimulator device [20]. Soon thereafter, these neurostimulators were used to stimulate deep areas of the brain as treatment for various motor disorders, cerebral palsy, epilepsy, schizophrenia, severe depression, and dystonia. Over time, DBS devices incorporated lithium batteries that extended the lifespan of implanted neurostimulators to several years, improved neurostimulator leads, and included wireless programming consoles [20]. However, widespread adoption of DBS remained hindered due to the lack of device approval from the United States Food and Drug Administration (FDA).

Beginning in 1987, French neurosurgeon Alim Louis Benabid (b. 1942) conducted clinical trials among patients with PD or essential tremor [20, 21]. These trials utilized high frequency stimulation from Medtronic neurostimulator electrodes placed within the thalamus of the brain, although Benabid later targeted the subthalamic nucleus in the brain as a more effective treatment for PD [20, 22]. Serendipitously, Benabid's trials occurred when physicians began seeking alternative therapeutic approaches to PD management. Although medications, such as levodopa, were initially successful in managing the motor symptoms of PD, their effectiveness diminished over time. By the mid-1980s, there was an emergence of severely affected PD patients whose symptoms had been inadequately managed with medications [20]. In conjunction with the need for more effective PD therapies, an objective tool for assessing the efficacy of PD treatments had been developed—namely, the Unified Parkinson's Disease Rating Scale (UPDRS) [20]. The combination of these factors, among others, created the ideal environment for the advancement of DBS. Medtronic subsequently employed Benabid to design international (Europe and United States) multi-site clinical trials with the aim of assessing the efficacy of DBS. All of the clinical trial sites used the five-part UPDRS as a standardized, objective outcome measure of the severity of PD symptoms before and after treatment. Findings from the multi-year Medtronic clinical trials demonstrated that DBS therapy resulted in statistically significant reductions in tremor and global disability. Moreover, "the efficacy of [DBS] treatment appeared to be greater than that of available medications" [20].

Upon presenting these findings to the FDA's Neurological Devices Panel Advisory Committee in 1997, the FDA formally approved the use of unilateral DBS for the treatment of tremor related to essential tremor and severe PD [20, 23]. Subsequently, the FDA approved DBS for more general cases of PD in 2002, dystonia in 2003, and obsessive-compulsive disorder in 2009 [20, 24–26]. Due to the efficacious use of DBS for psychiatric and movement disorders, neurosurgeons have begun to re-explore its potential benefits in treating other medical "conditions inadequately managed with medications" or standard therapies [20]. One such emerging indication is neurostimulation for the treatment of severe obesity.

Obesity continues to be an increasing challenge, with nearly 25% of U.S. adults projected to have severe obesity (i.e., body mass index [BMI] of 35 kg/m^2 or greater) by the year 2030 [27]. Additionally, severe obesity is predicted to be the most common BMI category among women, non-Hispanic Black adults, and low-income adults in the U.S [27]. These trends will ultimately exert an increased burden on the health care system. A BMI of 35–40 kg/m^2 has been associated with a 50% increase in health care expenditures above persons with normal weight, while a BMI of over 40 kg/m^2 has been associated with a 100% increase in health care costs [28].

The societal challenges associated with obesity are also compounded by the realities of ineffective long-term treatments. Laparoscopic Roux-en-Y gastric bypass surgery is the gold standard surgical treatment for severe obesity; however, the long-term failure rate of this intervention is reported as 20–35% [29]. Hence, novel surgical interventions are being implemented to address this growing and unmet need. In particular, closed-loop DBS of the nucleus accumbens (NAc) has shown promise to provide long-term benefits for treatment-refractory obesity, without altering one's social behavior [30]. Instead of dispensing chronic and continuous electrical stimulation (i.e., open-loop DBS), closed-loop DBS systems administer "brain-responsive neurostimulation therapy" [30]. Similar to the treatment modalities for epilepsy, "in which epileptiform activity can precede electrographic and clinical seizures," temporal activity specific to the anticipatory period preceding feeding provides a brief window of opportunity for intervention [30]. Ultimately, the aim of brain-responsive neurostimulation is to restore inhibitory control over loss of control eating during vulnerable moments [30]. In light of this novel surgical approach for the treatment of severe obesity, this chapter focuses on the classical ethical concerns of this indication—a treatment modality that is one of the most innovative approaches in neurosurgery—beginning with the surgeon–patient relationship [31–36].

The Surgeon–Patient Relationship

The foundation of the surgeon–patient relationship is trust. Although neurosurgeons typically have a short amount of time to develop a meaningful physician–patient relationship "from the time of referral to surgery, the degree of trust that patients must place in their surgeons creates an intricate covenant that begins in the office and carries over to the operating theater" [13]. This level of trust differs from that in general medicine, simply because of the difference in the temporality of the patient–physician relationship. In primary care medicine, patients are perceived as continuous shared decision makers in their medical care. For instance, if a patient is not tolerating a particular medication or refuses to continue treatment, the patient maintains control over his or her choice in real time [37]. However, the surgical environment differs. Although "awake DBS" procedures are performed in which the patient undergoes conscious sedation, most of the time "in the operating room, consciousness and thus autonomy are suspended, leaving the patient's life in the hands of the surgeon and the anesthesiologist. It is not feasible to discuss ahead of time *all* [emphasis added] the possible intraoperative outcomes and decisions required during the course of a surgical procedure. So after a reasonable discussion with the surgeon, the patient relies on the surgeon's judgment [and experience] to do 'the right thing' during the operation" [37]. Hence, a high level of trust is inherent in the surgeon–patient relationship. Patients with severe obesity who choose to undergo neurostimulation are especially vulnerable. Because their medical condition is not usually acute (e.g., ruptured appendix, strangulated hernia, intracranial hemorrhage), they electively enter the operating room in a fairly stable state of health. The patient trusts the neurosurgeon to use correct judgment if and when complications arise.

Essential to any surgical procedure is also the understanding of risks associated with the procedure. Additionally, neurosurgeons are tasked with the responsibility of balancing patients' expectations with the realities of what the surgical procedure can offer, with the aim of achieving a mutually agreed upon satisfactory outcome for both parties. In order for patients to have adequate understanding of expected outcomes and risks associated with the procedure, effective communication is vital to the surgeon–patient relationship [38]. Patients with obesity, however, often experience a disconnect in communication from their medical providers due to obesity or weight bias. In most modern societies, obesity is a highly stigmatized condition [39–42]. "In many cases, people with obesity are blamed for irresponsible overeating or inactivity, or both" [39]. Unfortunately, this stigma is not limited to societal interactions. There is increasing evidence that weight bias exists within health care settings [39, 40]. Some physicians view patients with obesity as "annoying" or a "greater waste of their time" [39]. Additionally, physicians have been shown to have less emotional rapport with their patients who have obesity [43]. These "negative attitudes and biases place the patient–physician relationship at risk by reducing patient satisfaction and the quality of the patient encounter, which can lead to negative patient outcomes" [39, 40]. Therefore, in the setting of utilizing neurostimulation as treatment for severe obesity, neurosurgeons must be cognizant of their own biases and sensitive to not only the physical, but also the emotional needs of their patients.

Autonomy, Decisional Capacity, and Surgical Informed Consent

Respect for autonomy, meaning respect for the decisions of patients regarding their medical care, is necessary in all patient interactions [12, 44]. In situations when the patient has decisional capacity, he or she is able to make his or her own medical decisions. However, there are circumstances in which patients may not have decisional capacity. In those circumstances, the patient's rights must still be honored. For example, if a patient with intellectual disability is under the care of a caregiver and that caregiver desires neurostimulation for the patient's refractory obesity, the onus rests on the health care team to ascertain, to the degree possible, the patient's true wishes and viewpoints regarding neurostimulation before the procedure is performed. Another example is a patient with diminished mental capacity resulting from a long-standing history of seizures. If this patient had refractory obesity and his or her caregiver desired neurostimulation as treatment for obesity, an ethical construct must be implemented to ensure that the patient's rights were not being infringed upon. Just as this patient relies on his or her caregiver for decisions pertaining to management of diabetes mellitus, cancer treatment, or cardiac pacemaker placement, the patient would rely on the caregiver for decisions regarding neurostimulation.

Inherent in the notion of honoring the patient's rights is the act of obtaining informed consent before any surgical procedure is performed. The five components of the informed consent process are disclosure, decisional capacity, patient understanding of the information, voluntariness, and consent [12]. In order to adequately obtain consent for the surgical intervention, a neurosurgeon "must provide the patient with information about the nature of the surgery, the expected benefits, material risks and adverse effects, alternate treatments and the consequences of not having the surgery. Material risks include [potential] risks common to all [surgeries] and [potential] risks specific for the proposed surgery, even if they are rare" [45]. Neurosurgeons are also encouraged to discuss the postoperative course that is expected, including any additional interventions that might be needed if complications arise [12]. As stated by the World Federation of Neurosurgical Societies, "unless [patients] are unable to read, competent patients participating in research must be fully informed, in writing, about the purpose and methods of the research and must give their voluntary, fully informed, and explicit consent to participate" [46]. Patients should also have the freedom to make their decisions without coercion or intimidation [47]. These guidelines are relevant to research as well as surgical procedures.

The use of informed consent is especially critical before novel surgical operations are performed, such as neurostimulation as treatment for severe obesity. Patients must receive full disclosure regarding the novelty of the procedure and the associated risks. "The informed consent process should include specific discussion of the innovative aspect of the procedure. Any omission of such discussion arguably involves deception and violates patient autonomy-based rights to submit to care, and could create potential liability for surgeons and their institutions" [48]. In addition, the use of "off-label" modalities must be disclosed to the patient. When a drug or device is used "off-label," it is used in a patient population, route of administration, or preparation that is different from its approved usage. It is, therefore, the responsibility of the physician to disclose the relative risk to the patient as part of the informed consent process [48]. Informed consent is a process, and as such, physicians need to secure both a signed consent form and document that comprehension was attained through the use of questions or quizzes soliciting patient responses, which also ought to be documented.

Since limited historical data may exist regarding outcomes associated with neurostimulation for severe obesity, a patient must also be informed that he or she may be the first patient to experience any type of complication. In essence, "a greater understanding by patients of the many uncertainties associated with new surgical techniques will serve to improve the informed consent process and lead to more reasonable expectations by patients of what new procedures might offer" [49]. This level of transparency helps to clarify a patient's expectations regarding potential weight loss, weight gain, or adverse events related to neurostimulation.

Nonmaleficence

Nonmaleficence is the bioethical construct of "to do no harm" [12, 50]. As such, patient safety is vitally important in neurosurgical procedures. In order to enhance patient safety and treatment efficacy, preclinical research provides in-depth knowledge regarding the scientific underpinnings of pathological states. This valuable information aides the thought process of intended surgical interventions before translation to the clinical setting. Specific to neurostimulation as treatment for severe obesity, research has afforded understanding of neurobiological factors involved in food reward and excessive food intake paradigms, the role of neural dopamine D2 receptors in loss of control behavior, and the notion that food intake regulation and addiction share similar neurobiological circuitry [51–53]. Animal and human models also demonstrate weight loss and decreased caloric intake following neurostimulation of the NAc, hence providing insights regarding auspicious neural targets for surgical intervention [54–57].

Although neurosurgeons utilize preclinical research to acquire knowledge and attenuate nonmaleficence, there usually exists a learning curve during the early period after introduction of a novel surgical technique. In order to ensure patient safety while surgeons gain the necessary experience to execute novel procedures, surgeons typically begin by learning new techniques on inanimate models, animal models, or human cadavers before attempting to use the technique on a human being [36, 49]. In addition to these modalities, experienced proctors and visiting surgeons are often invited to share their knowledge and expertise as learner surgeons observe their techniques. Neurostimulation has been no exception to this practice.

Furthermore, when neurostimulation targets the NAc in efforts of treating severe obesity, nonmaleficence must be considered when modulating this component of the limbic system in the brain. Normally, the NAc contributes to reward processes, emotional and behavioral components of feelings, motivation, and pleasure [15, 58]. Disruption or elimination of these neural pathways may have negative effects on one's normal brain processes [15, 33]. From mirroring the smile of a laughing baby to feeling a sense of accomplishment after completing a simple household chore, a patient's sense of reward, satisfaction, and wellbeing may be substantially altered. On the other hand, the reversible aspect and less invasive nature of DBS to the brain tissue offer patients the option of device modulation, deactivation, or possible extraction if undesired behavioral changes develop [15, 59, 60]. Additionally, the use of closed-loop NAc DBS devices is thought to cause less alteration to one's normal behavior, as electrical stimulation is "delivered in response to a behaviorally specific fluctuation in NAc physiology" [30].

Notwithstanding the potential for physical "harm" caused by neurostimulation, other constructs of psychological "harm" must also be addressed. Variability of beliefs exists from one religion to another regarding implantation of medical devices. For example, patients who are Jewish or Muslim may be willing to accept medical devices (e.g., porcine heart valves) that are contrary to their moral reasoning and dietary beliefs, given the imperative to save or preserve human life [61]. In addition, patients who are Protestant Christian may accept medical technologies such as "filled or false teeth, glasses or contact lenses, hearing aids, pacemakers, dialysis, hairpieces, and... vaccinations" as a means of restoring one's "organic functioning or processes," while remaining skeptical of other technologies or devices that appear to "[enhance] capacities or [...] alter conscious identity [61]." On the other hand, patients who are Buddhists or Hindus may be somewhat ambivalent regarding implantation of medical devices. While the implantation of medical devices may not violate Buddhist or Hindu teachings related to their "authentic Self (which cannot be improved or damaged), patients may be concerned that implanted medical devices will hinder "the need to advance spiritual welfare detached from the [physical] body" [61].

Culture may also impact how patients perceive the potential harms and benefits of innovative technologies. For instance, South Asian and East Asian traditions, although very different, include a shared "ambivalence toward the body" that does not exist in Western Cartesian attitudes [61]. Individuals from Asian traditions may believe that "improved psychophysical integration is to be attained through spiritual practice. This view does not deny the value of technological interventions if they can reduce suffering, such as correcting disabilities. However, mechanical devices cannot address [one's] deepest suffering (dukkha); rather, [one's] efforts to transform [his or her] ways of thinking, feeling, and acting are required" [61].

In addition to religious and cultural perspectives, one's ethical viewpoint of "human-machine interactions" should be considered [62]. "Many people experience a psychological difference between a device attached externally to the body and a device that penetrates the skin. The skin has always been a boundary for the self, and penetrating this boundary can constitute a profound invasive violation of not only the body but the person" [62]. There are also individuals who share similar perspectives as American technologist and science fiction author Ramez Naam, who notes that "as we learn how to repair damaged brains, we'll discover an immense amount about how the brain works. That in turn will lead to devices that can improve our mental abilities .... These abilities will pose serious questions to our sense of identity and individuality. They will blur the line between man and machine" [62]. While these viewpoints may seem extreme to some people, they may factor into a patient's moral persuasions regarding implantable devices, such as neurostimulation.

Justice and Beneficence

When considering the use of neurostimulation for the treatment of severe obesity, one must invariably ponder the notion of justice. Justice is a moral principle that encompasses fairness, equality, and impartiality—the obligation to be fair to all people [50]. In medicine, distributive justice is most commonly used to determine the fair distribution of medical goods and services [12]. Essentially, "individuals have the right to be treated equally regardless of ethnic group or race, gender, culture, age, marital status, medical diagnosis, social standing, economic level, political or religious beliefs, or any other individual characteristics" [50]. This ensures fair access to medical resources and equity.

DBS systems typically cost $35,000 to $50,000, and up to $100,000 for bilateral procedures [63]. These costs include surgical expenses, devices, anesthesia, hospital fees, and physician fees [63]. Data suggest that obesity prevalence rates among adults in the United States tend to be higher in lower-income groups [64]. In addition, U.S. adults ages 20–64 years with public health insurance (e.g., Medicaid) are more likely to be extremely obese (i.e., BMI of 40 kg/m^2 or greater) as compared to individuals with private health insurance or those who are uninsured (10.4%, 4.0%, 4.7%, respectively) [65]. In light of these disparities, patients with severe obesity may desire neurostimulation as treatment for their condition, yet not have the financial means to pay for this novel procedure. This financial conundrum is further exacerbated by the reality that most health insurance plans (public or private) offer coverage to only FDA-approved indications for neurostimulation (e.g., essential tremor, Parkinson's disease). Because neurostimulation as treatment for severe obesity is not currently FDA-approved, eligible individuals may have the option of enrolling in clinical trials as a means of having the expenses covered. Ideally, lower-income groups would have an equal chance of participating in such studies. Institutional review boards (IRBs) are responsible for ensuring fair access as a condition of conducting clinical trials. However, lack of access to adequate health care services may impede these opportunities for low-income groups. This financial limitation and disparity necessitate that the distribution of medical treatments be fairly appropriated to those most in need in order to exercise not only justice, but also beneficence—that is, the notion of "doing good," maximizing benefits, and minimizing harm to patients [12, 50].

One could argue that this notion of "doing good" for patients with obesity entails mitigating long-term complications from the disease. These complications include, but are not limited to, diabetes mellitus, hypertension, hypercholesterolemia, stroke, heart disease, certain types of cancers, and surgical complications such as wound infections and venous thromboembolism [39]. In addition to curtailing these patient-level sequelae, there are also ramifications to the health care system as a whole. "Direct and indirect costs arising from the medical care, increased morbidity and mortality, and decreased productivity related to obesity create a significant economic effect on the U.S. health care system. For example, the number of sick days and medical claims increase as a person's BMI increases, and adults with severe or morbid obesity (BMI more than 40 kg/m^2) have per capita health care costs that are 81% higher than those of healthy-weight adults" [39]. Furthermore, the care of patients with obesity may require physicians to expend more time and medical resources, such as "the availability of specialized equipment […] that can accommodate a higher maximum weight, and specially designed instruments for use in the operating room. Additionally, surgical procedures that often are performed in more cost-effective outpatient surgical centers may need to be undertaken in hospitals because of increased anesthesia risks to patients with obesity, along with other medical considerations. These surgical procedures may be more complex, and they may be of longer duration" [39]. Hence, novel treatments for obesity are not simply cosmetic procedures. Equity needs to be employed to ensure that patients with the greatest need have access to these treatments and are also selected to receive these treatments. Sponsors of clinical trials must be strongly encouraged to help cover the costs for those in greatest need who cannot pay.

On the other hand, some may question the morality of investing resources into implantable medical devices when approximately 8.5% of the U.S. population does not have health insurance coverage and "a good share of the world's population lacks access to vaccinations and basic public health measures of sanitation" [61,

66]. As Campbell et al. asserts, "not to ask such questions is a moral failing and evades the central issues of contemporary biomedical ethics" [61]. However, theological ethicist Gene Outka suggested several criteria to provide guidance for competing health care priorities. These criteria include (1) frequency of the condition or illness within a population, (2) the level of risks of communicability to a broader population, (3) costs of treatment, (4) extent of pain and suffering experienced by those afflicted with the condition, and (5) prospects for rehabilitation with treatment [61]. Although many patients with severe obesity have undergone conventional treatments such as gastric bypass surgery, there is doubt regarding their long-term efficacy and capacity for rehabilitation. Magro et al. reported that approximately 50% of patients with obesity experienced weight regain within 24 months after gastric bypass surgery, while patients with severe obesity regained up to 18.8% at 48 months after surgery [67]. These findings reveal the need for cheaper and more effective innovative approaches to address this life-impacting disease.

To this end, others may also question the fairness of lower-income individuals using government assistance to finance innovative health care treatments, claiming such practice to be a misuse of taxpayer dollars. However, similar to costs associated with diabetes mellitus or hypertension management among patients with obesity, taxpayer dollars are already being allocated to fund many obesity-related medical expenses. Hence, with the prevalence rates of obesity (BMI of greater than or equal to 30 kg/m^2) and severe obesity (BMI of greater than or equal to 40 kg/m^2) among U.S. adults being the highest they have been in the past 10 years, novel treatment modalities for obesity should not be considered lavish or cosmetic, but necessary [68].

Conflict of Interest

According to the World Federation of Neurosurgical Societies, "a conflict of interest exists when an investigator, author, reviewer, or editor has a financial or personal relationship that inappropriately influences or biases his or her actions. Financial relationships, such as employment, consultancies, stock ownership, honoraria, and paid expert testimony are the most easily identifiable conflicts and have the greatest potential to undermine the credibility of academic institutions, investigators, authors, journals, and science itself. However, conflicts can occur for other reasons, such as personal relationships, academic competition, and intellectual passion" [46]. While conflict of interest in medicine has been of concern, the relationships that surgeons have with industry differ from industry's relationships with other medical professionals [12]. Because surgeons are the only persons "who are licensed to use surgical devices on living humans," surgeons are essential to industry for the development and testing of such instruments and devices [12]. Likewise, surgeons need private industry in order to design, test, and market surgical devices [12]. This mutually beneficial relationship is indispensable to the success of surgical advancements. Therefore, equipoise must exist in this synergistic relationship.

Conflicts of interest may also arise when financial and other rewards bias physicians regarding their primary duty to care for their patients. In order to circumvent such bias, "the purposes, applications, consequences, and sponsorship of research projects should be clearly disclosed to all individuals who are materially affected, including patients participating in the research project, subjects, collaborators, and funders" [46]. Although this recommendation is specific to research projects, it can equally be applied to the clinical setting in which any potential conflicts of interest with industry may adversely influence the informed consent process or surgical outcomes. The use of neurostimulation for severe obesity highlights a clinical scenario in which these potential conflicts of interest may exist. This indication is currently considered experimental because it is not FDA-approved. Therefore, sponsorship for clinical trials stems from grant funders, with most study aims focused primarily on assessing feasibility and effectiveness of the procedure. In order to maintain transparency, uphold moral principles, and minimize distrust from society, physician researchers must inform study participants, IRBs, their universities, and other stakeholders regarding potential conflicts of interest with grant sponsorships and industry involvement.

Clinical Trials

In order to maintain high moral integrity of clinical trials associated with neurostimulation for severe obesity, the safety of trial participants must remain of utmost importance at all times throughout any clinical trial. Appropriate safeguards and periodic safety evaluations are critical for early detection of adverse effects and tolerance of stimulation. In addition, clinical trials should be conducted in full accordance with the ICH E6 Guideline for Good Clinical Practice, the principles of the Declaration of Helsinki, as well as relevant regulatory entities (e.g., U.S. FDA regulations, IRB guidelines) [69, 70]. Before the commencement of any study, the principal investigator should have written and dated approval from the IRB for the trial protocol, informed consent and competence assessment documents, subject recruitment procedures (e.g., advertisements), alternative treatments, registry participation, and any other written information to be provided to trial participants. To further maintain transparency and adequate dissemination of acquired scientific knowledge, results and accomplishments from the study should be made available by registry on ClinicalTrials.gov and publication to study participants, health care professionals, other relevant groups, and the public. De-identified study findings should be freely shared through publications in peer-reviewed academic journals and presentations at peer-reviewed scientific meetings.

Protected Populations

Our understanding of vulnerable populations has changed over time and continues to be refined. As such, regulations to protect these groups have evolved accordingly. When evaluating the appropriateness of neurostimulation, certain vulnerable populations warrant careful attention—including children, prisoners, and pregnant women. First, because obesity among children is of increasing concern, there may be an inclination to offer neurostimulation to a pediatric patient as a treatment option [71]. Indeed, there is increased encouragement to include children in the decision-making process of their medical care. "When minor children are patients, agreement on the course of treatment should be reached with the patient's parents or with the person legally responsible when there are no parents. Children who are younger than the age of consent but able to understand what is proposed should be informed and consulted regarding their treatment" [46]. Children (including those who are wards of the State, but excluding emancipated minors) are a protected group [72]; therefore, care should be taken when considering them for surgically implantable devices, especially if the device is considered elective and not medically required. Consultation in advance with an IRB is essential.

Another protected group to consider are prisoners. The United States National Institutes of Health (NIH) advises that "because prisoners may not be free to make a truly voluntary and uncoerced decision regarding whether or not to participate in research, the regulations require additional safeguards for the protection of prisoners" [72]. Although research is specified in the NIH guideline, novel treatments requiring informed consent are also applicable.

Third, pregnant women are also viewed as a protected group. The NIH advises that "because research may pose additional and/or unknown risks to pregnant women, human fetuses and neonates, the regulations require additional safeguards in research" [72]. Again, the safeguards pertinent to research are also relevant to novel treatments. Working in advance with both legal counsel and IRBs is indispensable before commencing a trial involving pregnant women.

Preoperative Assessments and Postoperative Tracking

Similar to bariatric surgery, the decision to implant a neurostimulation device as treatment for severe obesity necessitates the involvement of a multidisciplinary team with medical, surgical, psychiatric, and nutritional expertise [33, 73, 74]. A psychological assessment of each patient is inherent to this decision-making process. Although there is no formal standard regarding the psychological assessment, there are components that should be incorporated to determine a patient's appropriateness for surgery [73, 75]. First, a clinical interview is conducted in which the patient meets with a psychologist or psychiatrist to assess behavior and psychiatric symptoms. In addition, "the core parts of the clinical interview include reasons for seeking surgery, weight and diet history, current eating behaviors, understanding of the surgery and its [possible] associated lifestyle changes, social supports and history, and psychiatric symptoms (current and past)" [73]. The patient then completes psychological testing which provides an objective measure of his or her presentation style, psychological adjustment, and readiness for surgery [73]. This thorough evaluation before surgery can help to ascertain the patient's support system, realistically frame the patient's expectations of surgical outcomes, and contribute to long-term success postoperatively [76]. Additionally, ongoing psychological assessment after implantation of the neuromodulatory device would serve to monitor the patient's adaptation to stimulation and intercept deleterious outcomes.

In addition to a psychological assessment, all patients must demonstrate prior failure of conventional therapies for weight loss. These therapies may include diet, exercise, behavioral therapy, pharmacotherapy, and surgical approaches [77]. This assessment of past therapy must occur before neurostimulation is offered as a novel treatment option for severe obesity. Similar to bariatric surgery, a detailed weight loss history must be obtained and documented [78]. Although there is no specific guideline for this criteria, patients must have attempted behavioral and medical interventions for obesity that were not successful over time [78, 79].

Furthermore, the health, safety, and privacy of patients continue to be of utmost importance after implantation of the neuromodulatory device. As such, postoperative monitoring of patients is essential. Long-term tracking of protected health information can best be achieved using a secure registry database, with findings reported in aggregate and in compliance with the Health Insurance Portability and Accountability Act of 1996 (HIPAA) [80]. Routine data collection and analysis will ensure that patient outcomes are monitored for weight loss trends and overall health. This registry would also serve as an additional mechanism to ascertain device complications and assist in mitigating deleterious outcomes. Ultimately, findings from registry data will help to guide the informed consent process, as well as amend future clinical approaches for long-term success and optimal patient care.

Conclusion

Although the use of neurostimulation in movement disorders and psychiatric conditions has been well-established, its utilization as treatment for severe obesity highlights an example of innovation in modern neurosurgery. Several ethical constructs accompany this novel advancement, including the need for informed consent, the avoidance of "harm" associated with the implantation of a neuromodulatory device within the reward center of the brain, and equitable distribution of medical resources to those most in need of this surgical intervention. In addition, protected populations who may not be able to make voluntary decisions about their medical care must be rigorously safeguarded. In the future, as non-invasive neuromodulatory therapies advance, the ethical construct of nonmaleficence will be of greater importance. One will invariably question the utility of invasive treatments if non-invasive alternatives provide equal or greater therapeutic benefit. This introspective process of continuously evaluating therapeutic value highlights not only the inherent science, but also the art of innovation in neurosurgery.

Key Points

1. The use of neurostimulation as treatment for severe obesity is novel and implores several ethical considerations.
2. As part of the trust-building process between neurosurgeons and patients, informed consent must be obtained from the patient before any surgical procedure is performed.
3. The notion of avoiding "harm" extends beyond physical harm to include religious and cultural considerations that may contribute to a patient's perception of "harm."
4. Neurosurgeons should maintain key relationships, particularly with industry partners, without compromising the patient's care or instigating conflicts of interest.
5. Vulnerable populations (e.g., children, pregnant women, prisoners) require additional safeguards and careful attention when evaluating their appropriateness for neurostimulation.

Questions to Consider

1. Why is trust critical to the surgeon–patient relationship?
2. What are the risks associated with neurostimulation of the nucleus accumbens as treatment for severe obesity?
3. What role does justice play when using neurostimulation to treat severe obesity?
4. What steps can be taken to uphold high moral integrity of clinical trials associated with neurostimulation for severe obesity?
5. When a patient desires neurostimulation as treatment for severe obesity, who should be involved in the decision-making process to determine the patient's appropriateness for surgery?
6. What assessments are important before and after surgery when a patient receives neurostimulation for severe obesity?

References

1. Goel A, Kothari M. Ethics and neurosurgery. Neurol India. 2006;54(1):11.
2. Ormond DR, Hadjipanayis CG. The history of neurosurgery and its relation to the development and refinement of the frontotemporal craniotomy. Neurosurg Focus. 2014;36(4):E12.
3. Rhoton AL Jr. The cerebrum. Anatomy. Neurosurgery. 2007;61(1 Suppl):37–118; discussion 118–9.
4. Powell MP. Sir Victor Horsley at the birth of neurosurgery. Brain. 2016;139(Pt 2):631–4.
5. Prestigiacomo CJ. The history of the Journal of Neurosurgery: How the "White Journal" helped in the growth of neurological surgery. J Neurosurg. 2019;131(5):1347–55.
6. Khan FR, Henderson JM. Deep brain stimulation surgical techniques. Handb Clin Neurol. 2013;116:27–37.
7. Long DM. Harvey Cushing at Johns Hopkins. Neurosurgery. 1999;45(5):983–9.
8. Alt KW, Jeunesse C, Buitrago-Téllez CH, Wächter R, Boës E, Pichler SL. Evidence for stone age cranial surgery. Nature. 1997;387(6631):360.

9. Barbosa DAN, De Oliveira-Souza R, Monte Santo F, De Oliveira Faria AC, Gorgulho AA, De Salles AAF. The hypothalamus at the crossroads of psychopathology and neurosurgery. Neurosurg Focus. 2017;43(3):E15.
10. Muzumdar D. Neurosurgery in the past and future: An appraisal. Ann Med Surg. 2012;1:13–5.
11. Alexander Hughes Bennett (1848–1901): Rickman John Godlee (1849–1925). CA Cancer J Clin. 1974;24(3):169–70.
12. Wall A, Angelos P, Brown D, Kodner IJ, Keune JD. Ethics in surgery. Curr Probl Surg. 2013;50(3):99–134.
13. Namm JP, Siegler M, Brander C, Kim TY, Lowe C, Angelos P. History and evolution of surgical ethics: John Gregory to the twenty-first century. World J Surg. 2014;38(7):1568–73.
14. Marcus HJ, Hughes-Hallett A, Kwasnicki RM, Darzi A, Yang G-Z, Nandi D. Technological innovation in neurosurgery: A quantitative study. J Neurosurg. 2015;123(1):174–81.
15. Grant RA, Halpern CH, Baltuch GH, O'Reardon JP, Caplan A. Ethical considerations in deep brain stimulation for psychiatric illness. J Clin Neurosci. 2014;21(1):1–5.
16. Schwalb JM, Hamani C. The history and future of deep brain stimulation. Neurotherapeutics. 2008;5(1):3–13.
17. Sironi VA. Origin and evolution of deep brain stimulation. Front Integr Neurosci. 2011;5:42.
18. Pycroft L, Stein J, Aziz T. Deep brain stimulation: An overview of history, methods, and future developments. Brain Neurosci Adv. 2018;2:2398212818816017.
19. Coffey RJ. Deep brain stimulation devices: A brief technical history and review. Artif Organs. 2009;33(3):208–20.
20. Gardner J. A history of deep brain stimulation: Technological innovation and the role of clinical assessment tools. Soc Stud Sci. 2013;43(5):707–28.
21. C.V. de Alim-Louis Benabid (in French). France: Membre de l'Académie des sciences; 2008. https://web.archive.org/web/20131103213212/http://www.academie-sciences.fr/academie/membre/BenabidAL_bio1208.pdf.
22. Williams R. Alim-Louis Benabid: stimulation and serendipity. Lancet Neurol. 2010;9(12):1152.
23. FDA. Premarket Approval (PMA) Database: Medtronic Activa Tremor Control System. Silver Spring, MD: U.S. Food and Drug Administration; 2020 [updated April 13, 2020]. https://www.accessdata.fda.gov/scripts/cdrh/cfdocs/cfpma/pma.cfm?id=P960009.
24. FDA. Humanitarian Device Exemption (HDE) Database: Medtronic Activa Deep Brain Stimulation (DBS) for OCD therapy. Silver Spring, MD: U.S. Food and Drug Administration; 2020 [updated April 13, 2020]. https://www.accessdata.fda.gov/scripts/cdrh/cfdocs/cfhde/hde.cfm?id=375533.
25. FDA. Premarket Approval (PMA) Database: Medtronic Activa Parkinson's Control System. Silver Spring, MD: U.S. Food and Drug Administration; 2020 [updated April 13, 2020]. https://www.accessdata.fda.gov/scripts/cdrh/cfdocs/cfpma/pma.cfm?id=P960009S007.
26. FDA. Humanitarian Device Exemption (HDE) Database: Medtronic Activa Deep Brain Stimulation (DBS) System for dystonia therapy. Silver Spring, MD: U.S. Food and Drug Administration; 2020 [updated April 13, 2020]. https://www.accessdata.fda.gov/scripts/cdrh/cfdocs/cfhde/hde.cfm?id=375511.
27. Ward ZJ, Bleich SN, Cradock AL, Barrett JL, Giles CM, Flax C, et al. Projected U.S. state-level prevalence of adult obesity and severe obesity. N Engl J Med. 2019;381(25):2440–50.
28. Andreyeva T, Sturm R, Ringel JS. Moderate and severe obesity have large differences in health care costs. Obes Res. 2004;12(12):1936–43.
29. Elnahas AI, Jackson TD, Hong D. Management of failed laparoscopic Roux-en-Y gastric bypass. Bariatr Surg Pract Patient Care. 2014;9(1):36–40.
30. Wu H, Adler S, Azagury DE, Bohon C, Safer DL, Barbosa DAN, et al. Brain-responsive neurostimulation for loss of control eating: Early feasibility study. Neurosurgery. 2020;87(6):1277–88.
31. Pisapia JM, Halpern CH, Muller UJ, Vinai P, Wolf JA, Whiting DM, et al. Ethical considerations in deep brain stimulation for the treatment of addiction and overeating associated with obesity. AJOB Neurosci. 2013;4(2):35–46.

32. Halpern CH, Wolf JA, Bale TL, Stunkard AJ, Danish SF, Grossman M, et al. Deep brain stimulation in the treatment of obesity. J Neurosurg. 2008;109(4):625–34.
33. Ho AL, Sussman ES, Pendharkar AV, Azagury DE, Bohon C, Halpern CH. Deep brain stimulation for obesity: Rationale and approach to trial design. Neurosurg Focus. 2015;38(6):E8.
34. Formolo DA, Gaspar JM, Melo HM, Eichwald T, Zepeda RJ, Latini A, et al. Deep brain stimulation for obesity: A review and future directions. Front Neurosci. 2019;13:323.
35. Kumar R, Simpson CV, Froelich CA, Baughman BC, Gienapp AJ, Sillay KA. Obesity and deep brain stimulation: An overview. Ann Neurosci. 2015;22(3):181–8.
36. Nangunoori RK, Tomycz ND, Oh MY, Whiting DM. Deep brain stimulation for obesity: From a theoretical framework to practical application. Neural Plast. 2016;2016:1–8.
37. Ferreres AR, Angelos P, Singer EA, Blair PG. Ethical issues in surgical care. Chicago, IL: American College of Surgeons; 2017.
38. Pellegrini CA, Ferreres A. Surgical Ethics Symposium: Ethical dilemmas in surgical practice. World J Surg. 2014;38(7):1565–6.
39. ACOG. Ethical considerations for the care of patients with obesity: American College of Obstetricians and Gynecologists (ACOG) Committee Opinion No. 763. Obstet Gynecol. 2019;133(1):e90–e96.
40. Phelan SM, Burgess DJ, Yeazel MW, Hellerstedt WL, Griffin JM, Van Ryn M. Impact of weight bias and stigma on quality of care and outcomes for patients with obesity. Obes Rev. 2015;16(4):319–26.
41. Spahlholz J, Baer N, König H-H, Riedel-Heller SG, Luck-Sikorski C. Obesity and discrimination: A systematic review and meta-analysis of observational studies. Obes Rev. 2016;17(1):43–55.
42. Wee CC, Davis RB, Huskey KW, Jones DB, Hamel MB. Quality of life among obese patients seeking weight loss surgery: The importance of obesity-related social stigma and functional status. J Gen Intern Med. 2013;28(2):231–8.
43. Gudzune KA, Beach MC, Roter DL, Cooper LA. Physicians build less rapport with obese patients. Obesity. 2013;21(10):2146–52.
44. Jaworska A. Advance directives and substitute decision-making: The Stanford Encyclopedia of Philosophy. Stanford, CA: Stanford University; 2009. https://plato.stanford.edu/entries/advance-directives/.
45. Hanson M, Pitt D. Informed consent for surgery: Risk discussion and documentation. Can J Surg. 2017;60(1):69–70.
46. Umansky F, Black PL, DiRocco C, Ferrer E, Goel A, Malik GM, et al. Statement of ethics in neurosurgery of the World Federation of Neurosurgical Societies. World Neurosurg. 2011;76(3):239–47.
47. Hotopf M. The assessment of mental capacity. Clin Med. 2005;5(6):580–4.
48. Biffl WL, Spain DA, Reitsma AM, Minter RM, Upperman J, Wilson M, et al. Responsible development and application of surgical innovations: A position statement of the Society of University Surgeons. J Am Coll Surg. 2008;206(6):1204–9.
49. Angelos P. The ethical challenges of surgical innovation for patient care. Lancet. 2010;376(9746):1046–7.
50. OU. Health management, ethics and research module: 7. Principles of healthcare ethics.: The Open University (OU); n.d. https://www.open.edu/openlearncreate/mod/oucontent/view.php?id=225&printable=1.
51. Lennerz B, Lennerz JK. Food addiction, high-glycemic-index carbohydrates, and obesity. Clin Chem. 2018;64(1):64–71.
52. Beaver JD, Lawrence AD, van Ditzhuijzen J, Davis MH, Woods A, Calder AJ. Individual differences in reward drive predict neural responses to images of food. J Neurosci. 2006;26(19):5160–6.
53. Volkow ND, Fowler JS, Wang GJ, Baler R, Telang F. Imaging dopamine's role in drug abuse and addiction. Neuropharmacology. 2009;56:3–8.
54. Harat M, Rudaś M, Zieliński P, Birska J, Sokal P. Nucleus accumbens stimulation in pathological obesity. Neurol Neurochir Pol. 2016;50(3):207–10.

55. Mantione M, van de Brink W, Schuurman PR, Denys D. Smoking cessation and weight loss after chronic deep brain stimulation of the nucleus accumbens: Therapeutic and research implications: Case report. Neurosurgery. 2010;66(1):E218.
56. Tronnier VM, Rasche D, Thorns V, Alvarez-Fischer D, Münte TF, Zurowski B. Massive weight loss following deep brain stimulation of the nucleus accumbens in a depressed woman. Neurocase. 2018;24(1):49–53.
57. Halpern CH, Tekriwal A, Santollo J, Keating JG, Wolf JA, Daniels D, et al. Amelioration of binge eating by nucleus accumbens shell deep brain stimulation in mice involves D2 receptor modulation. J Neurosci. 2013;33(17):7122–9.
58. Mavridis I. The role of the nucleus accumbens in psychiatric disorders. Psychiatriki. 2015;25(4):282–94.
59. Qu L, Ge S, Li N, Wang W, Yang K, Wu P, et al. Clinical evaluation of deep brain stimulation of nucleus accumbens/anterior limb of internal capsule for opioid relapse prevention: Protocol of a multicentre, prospective and double-blinded study. BMJ Open. 2019;9(2):e023516.
60. Real E, Plans G, Alonso P, Aparicio MA, Segalàs C, Cardoner N, et al. Removing and reimplanting deep brain stimulation therapy devices in resistant OCD (when the patient does not respond): Case report. BMC Psychiatry. 2016;16(1).
61. Campbell CS, Clark LA, Loy D, Keenan JF, Matthews K, Winograd T, et al. The bodily incorporation of mechanical devices: Ethical and religious issues (part 2). Camb Q Healthc Ethics. 2007;16(3):268–80.
62. Campbell CS, Clark LA, Loy D, Keenan JF, Matthews K, Winograd T, et al. The bodily incorporation of mechanical devices: Ethical and religious issues (part 1). Camb Q Healthc Ethics. 2007;16(2):229–39.
63. Okun MS, Zeilman PR, Foundation NP. Parkinson's disease: Deep brain stimulation a practical guide for patients and families. 2 ed; 2014. https://www.med.upenn.edu/pdmdc/assets/user-content/deep-brain-stimulation-guide-parkinsons.pdf.
64. Ogden CL, Fakhouri TH, Carroll MD, Hales CM, Fryar CD, Li X, et al. Prevalence of obesity among adults, by household income and education: United States, 2011-2014. MMWR Morb Mortal Wkly Rep. 2017;66(50):1369–73.
65. Carroll W, Rhoades J. Obesity in America: Estimates for the U.S. civilian noninstitutionalized population age 20 and older, 2009. Rockville, MD: Agency for Healthcare Research and Quality (AHRQ); 2012. https://www.meps.ahrq.gov/data_files/publications/st364/stat364.shtml.
66. Berchick ER, Barnett JC, Upton RD. Health insurance coverage in the United States: 2018, report number P60–267 (RV). U.S. Department of Commerce. U.S. Census Bureau; 2019. https://www.census.gov/library/publications/2019/demo/p60-267.html.
67. Magro DO, Geloneze B, Delfini R, Pareja BC, Callejas F, Pareja JC. Long-term weight regain after gastric bypass: A 5-year prospective study. Obes Surg. 2008;18(6):648–51.
68. Hales CM, Carroll MD, Fryar CD, Ogden CL. Prevalence of obesity and severe obesity among adults: United States, 2017–2018. Hyattsville, MD: National Center for Health Statistics (NCHS); 2020. https://www.cdc.gov/nchs/products/databriefs/db360.htm.
69. HHS. E6(R2) Good clinical practice: Integrated addendum to ICH E6(R1): Guidance for industry. Silver Spring, MD: U.S. Department of Health & Human Services, U.S. Food and Drug Administration, Center for Drug Evaluation and Research (CDER), Center for Biologics Evaluation and Research (CBER); 2018 [updated March 2018]. https://www.fda.gov/media/93884/download.
70. WMA. WMA Declaration of Helsinki – Ethical principles for medical research involving human subjects. Helsinki, Finland: World Medical Association (WMA); 1964 [updated October 2013 in Fortaleza, Brazil]. https://www.wma.net/policies-post/wma-declaration-of-helsinki-ethical-principles-for-medical-research-involving-human-subjects/.
71. Hales CM, Carroll MD, Fryar CD, Ogden CL. Prevalence of obesity among adults and youth: United States, 2015–2016. NCHS data brief, no 288. Hyattsville, MD: National Center for Health Statistics (NCHS); 2017. https://www.cdc.gov/nchs/data/databriefs/db288.pdf.

72. NIH. Vulnerable and other populations requiring additional protections. Bethesda, MD: National Institutes of Health; 2019 [updated January 7, 2019]. https://grants.nih.gov/policy/humansubjects/policies-and-regulations/vulnerable-populations.htm.
73. Snyder AG. Psychological assessment of the patient undergoing bariatric surgery. Ochsner J. 2009;9(3):144–8.
74. Hubbard VS, Hall WH. Gastrointestinal surgery for severe obesity. Obes Surg. 1991;1(3):257–65.
75. Flores CA. Psychological assessment for bariatric surgery: Current practices. ABCD Arquivos Brasileiros de Cirurgia Digestiva (São Paulo). 2014;27(suppl 1):59–62.
76. Engstrom D. Why is it required to have a psychological evaluation prior to weight-loss surgery? Obesity Action Coalition (OAC); 2006. https://www.obesityaction.org/community/article-library/why-is-it-required-to-have-a-psychological-evaluation-prior-to-weight-loss-surgery/.
77. Wyatt HR. Update on treatment strategies for obesity. J Clin Endocrinol Metab. 2013;98(4):1299–306.
78. Karmali S, Johnson Stoklossa C, Sharma A, Stadnyk J, Christiansen S, Cottreau D, et al. Bariatric surgery: A primer. Can Fam Physician. 2010;56(9):873–9.
79. Wolfe BM, Kvach E, Eckel RH. Treatment of obesity: Weight loss and bariatric surgery. Circ Res. 2016;118(11):1844–55.
80. HHS-OCR. Summary of the HIPAA privacy rule. Washington, D.C.: U.S. Department of Health & Human Services; 2013 [updated July 26, 2013]. https://www.hhs.gov/hipaa/for-professionals/privacy/laws-regulations/index.html.

Further Reading

Halpern CH, Wolf JA, Bale TL, Stunkard AJ, Danish SF, Grossman M, et al. Deep brain stimulation in the treatment of obesity. J Neurosurg. 2008;109(4):625–34.
Ho AL, Sussman ES, Pendharkar AV, Azagury DE, Bohon C, Halpern CH. Deep brain stimulation for obesity: Rationale and approach to trial design. Neurosurg Focus. 2015;38(6):E8.
Khan FR, Henderson JM. Deep brain stimulation surgical techniques. Handb Clin Neurol. 2013;116:27–37.
Lennerz B, Lennerz JK. Food addiction, high-glycemic-index carbohydrates, and obesity. Clin Chem. 2018;64(1):64–71.
Wu H, Adler S, Azagury DE, Bohon C, Safer DL, Barbosa DAN, et al. Brain-responsive neurostimulation for loss of control eating: Early feasibility study. Neurosurgery. 2020;87(6):1277–88.

Chapter 8
In the Midst of Uncertainty: Neuroinnovation at the Edge of Consciousness

Laura P. Dresser and Christos Lazaridis

Brain–Computer Interfaces and Related Ethical Dilemmas

Multiple areas of medicine and research have worked alongside to establish and advance the field of brain–computer interfaces—the foundations of this area of neuroscience date back to several decades of ongoing developments. The conception of electroencephalography (EEG) by Hans Berger in 1929 [1] is probably the first breakthrough in the field, with subsequent efforts to use EEG technology for neurofeedback therapy starting early on. These efforts have since evolved into sophisticated ideas on how to utilize computer interfaces, and the past 20 years have seen the initial translational applications of prior neuroscientific discoveries. Innovation is fostered by the collective work of different fields, including engineering, neurophysiology, psychology, medicine, and computer science.

The definition of brain–computer interfaces (BCI) is not collectively agreed upon in the neuroscientific community, driven mainly by the fact that it is an area of rapid development, and there is considerable variability in its application. The basic notion propelled by many in the community is that of a neurotechnological device able to detect brain activity and translate into a command executed by a machine [2, 3]. Some authors and, in many instances, the media and public include brain stimulation techniques in the definition of BCI [4, 5]. The crucial aspects of BCI functioning include its ability to detect and classify brain activity and to immediately, or almost immediately, provide feedback to the user about its intended goal [6]. BCIs

L. P. Dresser (✉)
Department of Neurology, University of Chicago, Chicago, IL, USA

MS & Neuromuscular Center of Excellence, Clearwater, Chicago, IL, USA

C. Lazaridis
Department of Neurology and Neurosurgery, University of Chicago Medical Center, Chicago, IL, USA
e-mail: Lazaridis@uchicago.edu

L. W. Roberts (ed.), *Ethics and Clinical Neuroinnovation*,
https://doi.org/10.1007/978-3-031-14339-7_8

are usually classified according to the type of signal they detect into active, reactive, and passive. In active BCI, the user performs a mental task that is then translated into an executable command by the machine. The most common application of this type is motor imagery, by which patients can, for example, control a robotic arm. Reactive or evoked BCI systems utilize an external stimulus and measure selective attention. Passive BCI measures baseline or background brain activity [3, 7]. The way signals are acquired also plays a role in the classification of BCIs, with both invasive (for example, implanted cortical grids, which measure activity in the outer brain layer) and non-invasive technologies (such as non-invasive surface EEG and functional magnetic resonance imagine (fMRI)) available.

The potential applications of this technology are many and varied but fall mostly into two main categories: (1) exclusively for medical and therapeutic application and (2) consumer enhancing technologies. So far, both areas have been developed on a relatively separate playing field, and many aspects of each largely remain on a research-only stage. In recent years, there has been a growing interest in utilizing developments in informatics and computer science to advance the technology and its potential applications, leading to the belief that the field will rapidly expand with input from emerging companies. Neuralink, created by Elon Musk, recently published data about their new research platform, described as a "scalable, high-bandwidth brain-machine interface," which aims at providing more reliable, functional signals, recorded from a larger cortical source [8]. The tremendous innovation and possibility for advancement, as well as increasing interest in non-medical applications, have bolstered the awareness of hidden controversies and possible obstacles to its responsible development.

Ethical Concerns

Multiple ethical concerns have emerged in the context of brain–computer interfaces (Table 8.1). The debate that arises takes place in different arenas, some of which are more of a philosophical and abstract nature with open-ended questions, such as the idea of enacted or embodied mind. On the other hand, other aspects of the debate have a more concrete structure, such as the analysis of potential benefits vs. risks. As the technology continues to advance, these questions become ever more relevant as only addressing them in a timely fashion will allow us to create a regulated pathway to the development of new technologies, as well as to manage their societal impact [9].

Table 8.1 Potential applications and ethical considerations of brain–machine interfaces

	Applications	Social and ethical concerns
Medical and therapeutics	**Neurorehabilitation:** retraining neuromuscular function after brain or spinal injury **Movement and communication control:** direct control of assistive device for motor and language performance **Environmental control:** control of domestic devices and other factors in home or office environment	**Risk/benefit ratio:** risks from surgical procedures, implanted devices or long-term use of technology **Privacy loss:** risk of misuse of individual's data **Questions about identity, personhood and agency** **Ethical concerns about utilization in research**: includes risks/benefit ratio of interventions, obtaining consent
Direct-to-consumer and enterprises	**Performance enhancement:** improving performance in healthy adults **Everyday tasks:** facilitating and improving efficiency of a multitude of tasks, including typing and voice commands **Entertainment and gaming**	**Social perception:** changes in how society perceives the impact of technology in values and day-to-day life **Trends in social norms**: changes in culture and behavior based on expectations of technology consumers **Enterprise ethics:** challenges in regulation of behavior of companies and consumers

Risk and Benefit Ratio

Medical safety issues include those derived from the device itself, such as immediate surgical risks associated with invasive technologies, as well as longer-term risks arising from reactive scarring or gliosis if they are implanted in or near the brain cortex. Non-invasive technologies do not carry these direct effects but can be associated with the potential for changes in electrochemical signaling or brain plasticity. Even then, it is still unknown if changes could be reversible [10].

Questions about Identity, Personhood, and Autonomy

One of the most controversial debates around BCI is whether those who use the technology, patients, or healthy individuals experience any sort of changes to their identity or personality. Some argue that these questions lack any validity, as technology has already become an intricate part of human life. The concept of body schema has already been altered by technology and medical procedures, with the everyday and widely accepted use of artificial devices for improvement of health and functionality. In addition, some argue that personality changes can occur as part of the regular course of life, as a result of chronic illnesses, medications for neurological or psychological disorders, or due to invasive therapeutic procedures. Even if BCI was associated with changes in personality or identity, deterring its development or application could prevent us from providing real and tangible benefits to society. Inasmuch as these arguments sound enough to propel the debate forward, many others believe that the risks are real and just too high. For example, Demetriades et al. argue that BCI makes individuals more robotic and "less human" [11].

These concepts can then be used to understand the potential effects of BCI on an individual's autonomy, perception of self, and societal acceptance or stigma. One of the main applications for BCI is assistive technology for patients with disabilities, with both clinicians and researchers putting forward their goal as that of restoring function to "normal levels" and enhanced quality of life [3]. On the other hand, many believe that this goal can be conducive to further stigmatization of disability and to classify disability as a deficit or a burden to society [12, 13]. In the context of disorders of consciousness, one of the main applications of this technology in the field of neurocritical care, this rationale will likely lose its traction, as this technology will provide patients with the potential for communication.

Finally, many have addressed the question of autonomy in the setting of BCI. One aspect of the discussion focuses on the effects of this technology on an individual's ability for self-determination. Is BCI able to alter an individual's capacity to make decisions? If so, how does it affect our current view of agency and responsibility for one's actions? For example, some researchers believe that BCI may be more responsive to brain activity than one's own body, so that it may be more difficult for an individual to censor actions taken by the machine interface [14]. Moral and legal responsibility and liability stemming from actions aided or taken by BCI technology are also a concern. Some scholars have created parallels between this situation and that of operating dangerous tools, but there is a general agreement that the BCI user is responsible for its actions [15, 16]. On the other hand, BCIs can improve an individual's autonomy by providing a renewed sense of independence through the enhancement of motor abilities, speech, and language. Boosting independence and self-reliance can be instrumental to the perception of human dignity and the key to personhood.

Privacy Concerns

The questions around privacy issues in BCIs are brought up both by researchers and potential users. Two main categories arise: (1) risk of extraction of brain signals that can be used as markers for behavior, perceptions, or disease states and that in turn could be further utilized to classify individuals into categories and for potential discrimination and stigmatization [17, 18]; and (2) security concerns over the hacking of computer platforms and introduction of unwanted behaviors, or extraction of relevant data [13, 14].

Research Ethics

Many issues arise in this field, including (1) informed consent and voluntariness, (2) disclosure and approach to incidental findings, and (3) justice. The questions about informed consent are primarily centered around voluntariness in patients with diminished cognitive capacity or those with depressed motor abilities, which impair their expression and interaction with society, such as in locked-in syndrome or minimally conscious state. As indicated in the Belmont Report, based on the principle of respect for persons, subjects have the right to make decisions about their participation in research to the extent of their abilities. Provision of the necessary information, adequate comprehension, and voluntariness are necessary requisites to informed consent. Ascertaining adequate comprehension in vulnerable populations is challenging, but still necessary and paramount, independent of the potential benefits yielded by the technology [15, 19, 20]. Surrogate decision-makers are sometimes needed in making the decision, but the subject has the right to object to their participation in any research venture, unless it constitutes the only alternative to treatment that is unavailable elsewhere. Participation of third parties in the decision-making process adds an extra layer of complexity. The state of vulnerability in some populations raises concerns about the voluntariness to participate in research, as at times the decision may be driven by desperation to obtain therapeutic benefits through specific interventions, despite high risks and potentially minimal benefits.

Justice is a concern for both the research and clinical arena. The basic ethical principle of justice refers to the fairness of distribution of the benefits and burdens of research. Once technologies are developed, will patients be able to access them fairly? [17, 21] According to the principle of justice, vulnerable populations should only participate in research for the development or application of BCI if this technology will then be accessible to them for therapeutic purposes. More so, these new devices and technologies should not be developed uniquely for the enhancement of cognitive or physical abilities in healthy individuals, as this would further contribute to social disparity and stratification [18]. Finally, should research participants be entitled to keep the technology once the study is completed? If BCIs become widely available, are individuals entitled to updating them with the advancement of technology? And, are companies obligated to provide technical and medical support after that? These, and potentially many more questions, are complex as they have financial and potential political implications for the society at large. Use of novel technologies, such as BCI, can add significant costs to the care of vulnerable patients and may be perceived as a financial burden on already strained healthcare systems. Nonetheless, this should not be a limitation to their development and application. These questions will continue to play a critical role in the landscape of BCI technology and companies should use them as an incentive to create more accessible and cost-efficient technologies.

Table 8.2 Summary of disorders of consciousness

Diagnosis	Description	Characteristics and behavioral response	Response to verbal commands	Communication
Coma	Absence of wakefulness and awareness. Acute and transient condition, lasting no more than 4 weeks	Eyes closed. No reproducible and consistent behavioral responses	None	None
Vegetative state/ unresponsive wakefulness state (VS/ UWS)	Return of wakefulness without evidence of awareness (lack of purposeful behavior)	Spontaneous eye-opening. Sleep-wake cycle is present. Reflexive behavior present, including startle response, withdrawal, localization to sound, other reflexive motor and oral behaviors	No coherent or purposeful response to verbal command. Reflexive responses, such as localization to sound are present	None
Minimally conscious state (MCS)	Return of wakefulness and awareness, characterized by ability to generate variable, but reproducible behavior	Spontaneous eye-opening. Sleep-wake cycle is present. Reproducible and non-reflexive behavior	Behavioral responses to verbal commands are variable	Variable. If present, can include reproducible motor response to command or verbalizations (intelligible or non-functional)

Ethical Considerations in the Discovery of Covert Conscience

Severe brain injuries can lead to the development of disorders of consciousness, which are characterized by reduced or lack of awareness, and include coma, the minimally conscious state (MCS), and the unresponsive wakefulness syndrome (UWS), previously referred to as vegetative state (VS) (Table 8.2). Coma is an acute and transient condition, characterized by the absence of both wakefulness and awareness, which occurs immediately after brain injury. MCS and UWS/VS are chronic disorders of consciousness, characterized by the return of wakefulness or eye-opening, and differentiated by the presence of reproducible behavioral responses (motor, verbal, visual tracking) in the first, but absence in the later [22, 23]. Patients admitted to the intensive care unit (ICU) with severe brain injuries undergo frequent bedside assessments to determine their level of consciousness; these evaluations are then used to predict their recovery potential. Bedside behavioral assessments have significant limitations, mainly related to the patient's underlying motor, speech, and sensory deficits as a result of the initial injury. Previous studies have shown that up to 40% of patients admitted to the ICU due to brain injuries can be misdiagnosed with a lack of awareness by bedside examinations [24]. A recent study conducted by Claassen et al. found that 15% of unresponsive patients due to acute brain injuries (median time from injury 4 days) exhibited brain activation in response to the command of hand movement, measured by qEEG [25]. Accurate bedside diagnosis is, therefore, challenging, even for astute and seasoned clinicians.

The decision to proceed with or withhold further treatment after severe brain injury is highly influenced by the treating clinician and the family's perception of the potential for recovery. Diagnostic and prognostic uncertainty in this context, therefore, has significant consequences; it is well known that the proximate leading cause of death in the neurological ICU is withdrawal of life-sustaining measures [26]. Claasen et al., reported that behaviorally unresponsive patients in a neurological ICU that were found to have appropriate brain activation in response to motor commands, recovered faster than those who didn't, and also had better long-term prognosis [25]. So, what if a patient is perceived to be comatose, or in UWS/VS through bedside assessments, when they are actually aware of their surroundings, but unable to express it through motor or verbal behaviors? What if a decision is made to withdraw life-sustaining measures based on this perception? Furthermore, what are the clinical, financial, legal, and ethical ramifications of this decision? These and many more questions are fundamental in the care of patients with severe brain injuries, as well as in the development of ethical frameworks for neuroinnovative technologies. We believe that as these technologies become more efficient and widely available in clinical practice, physicians caring for patients with disorders of consciousness will have better tools at their disposal to be able to answer these questions and avoid dire repercussions. In the interim, it is essential for healthcare providers to acknowledge the limitations of current assessments and diagnostic tools, and to learn how to adequately relay this information to family members as they make decisions regarding goals of care. Diagnostic and prognostic uncertainty are a reality of clinical care for this patient population, and neuroinnovative technologies will not eliminate them completely, but they may be useful in augmenting confidence among physicians and surrogate decision-makers.

Owen et al., presented the first case of a woman diagnosed as being in VS who was able to produce changes in brain activity in response to verbal commands, measured through functional magnetic resonance imaging (fMRI). Ensuing research studies around this technique have consistently reproduced these initial findings, showing that a significant proportion of patients diagnosed with chronic disorders of consciousness retain the ability to covertly follow verbal commands, answer simple yes/no questions, retrieve memories, and even process complex logical questions [27–30]. Based on analysis of biological and epidemiological data, the American Academy of Neurology, in conjunction with the American College of Rehabilitation Medicine (ACRM), and the National Institute on Disability, Independent Living and Rehabilitation Research (NIDILRR) published an updated practice guideline addressing the diagnosis, treatment, and prognosis of chronic disorders of consciousness [31]. One of the main contributions of this document to the neuroethics field was the redefinition of persistent VS/UWS to chronic VS/UWS, based on the fact that up to 20% of patients with this condition could regain awareness; these are probably patients mis-classified as UWS when in fact they are MCS or exhibit cognitive-motor dissociation (CMD). In addition, it included fMRI and task-based electroencephalography (EEG) as diagnostic techniques to be considered in these scenarios. The significance of the use of these innovative techniques goes beyond clinical practice in neurocritical care and touches on important ethical, cultural, and legal aspects. The debate includes questions about the moral significance of consciousness, implications on quality of life, fiduciary duty, as well as legal implications on the right-to-die notion, and societal expectations about the irreversibility of the condition [32–35].

Definition of Consciousness and Moral Implications

As previously noted, the main appeal of innovative neurodiagnostic techniques in disorders of consciousness is that they provide an alternative, non-behavioral means of detecting awareness. The fallible nature of bedside behavioral assessments has led to growing distrust in our ability to diagnose and predict outcomes in brain injury accurately; as best explained by Wilkison et al., we should not define a phenomenon based on our ability to find out about it [34]. Neuroimaging techniques, both fMRI and PET imaging, as well as task-based quantitative EEG (qEEG), have flourished in this field. The absence of conscious awareness has been a critical determinant of the ethical discourse around patients in UWS/VS, which highlights the importance of these advances. As encompassed in its definition, UWS/VS is characterized by a lack of awareness of oneself, others, and the environment, while MCS presupposes some degree of awareness, albeit limited by severe cognitive impairments. Therefore, the lack of conscious awareness in UWS/VS, in its basic moral premise, depicts a separation from MCS. This distinction has broad implications on the legal and ethical aspects of decisions about removal of life-sustaining measures in patients in UWS/VS, as consciousness has long been regarded as the essential and valuable aspect of human life. Thus, it is of paramount importance to accurately differentiate between lack of consciousness and its presence, even if minimal.

The next big question then becomes what consciousness is, and thus far, this continues to be a source of contention. The concept is rooted not only in the medical tradition but also in the legal and philosophical arena, each with different views and definitions. In medicine, the determination of consciousness is essential in the diagnosis, management, and prognosis of brain injury. Jennet and Plum were the first to coin the term vegetative state in 1972, and they described life in such a state as "merely a physical life, devoid of intellectual activity and social intercourse" [36]. In legal terms, the permanent absence of consciousness has been the driving force behind the "right-to-die" movement, dating back to Quinlan's case in 1975. From a philosophical standpoint, the debate regarding the definition and moral implications of consciousness has been alive for thousands of years and continues to today. Perhaps, an easy way to approach the significance of consciousness, or lack thereof, within this population is by understanding it through the lens of "qualia" or phenomenological consciousness. Qualia is a term coined by philosophers to explain the phenomenal aspects of a mental state, meaning what "it is like" to undergo any state. Hence, a state would meet the status of phenomenological consciousness if being in it feels like something.

Neuroinnovative technologies could provide an answer to the epistemological problem of differentiating between the conscious and the unconscious. Consciousness is sometimes thought of as a dichotomous outcome but creating a discrete line to separate UWS from MCS is challenging, both with traditional and novel assessments. Defining consciousness as a continuum is supported both from an experiential and neurobiological point of view. The experience of awareness is characterized by degrees, from the basic awareness of internal or external stimuli to the more complex integration of self and environment. The neurobiology of consciousness is complex and relies on the integration of multiple systems, some more important than others. Accordingly, if we conceptualize it as a continuum, is there a degree of consciousness that permeates into clinical, ethical, or moral significance? Is being conscious better or worse for patients with no behavioral manifestation of their awareness? These questions are tricky and, by and large, remain unanswered. The ethical and social debate will continue on the background of neuroinnovation in disorders of consciousness but developing a framework for research and clinical practice will be crucial for these technologies to become relevant in the lives of patients in UWS and MCS.

Quality of Life in Chronic Disorders of Consciousness

Assessment of quality of life is subjective and, therefore, we are unable to address it directly in patients with impaired consciousness. Historically, three main philosophical traditions have approached the question of what a good life entails. Perfectionism weighs quality on the base of accomplishing objective and significant human potentials, but it does not take into account the individual's system of values. Hedonism focuses on the procurement of pleasure and avoidance of pain, such that a life worth living is that in which there are pleasant mental states, but no painful ones. Finally, the preference theory highlights the individual's interests and values; in its basic premise, it states that quality of life is measured by whether individuals can get what they want [37].

Previous efforts to evaluate quality of life in patients suffering from disorders of consciousness have relied on the public's opinion on this matter, with many respondents agreeing that life in UWS/VS and MCS is not worth living [38]. One could reach a similar conclusion using the perfectionism and hedonism theories, as based on their physical limitation, these patients will not be objectively able to fulfill what others consider important human potentials. It is also possible that in the course of their disease, they will be subjected to painful experiences, therefore defeating the hedonistic goal. Nonetheless, it would be impossible, based on our current way of measuring behavior, to ascertain an answer from the perspective of the preference theory. Innovative techniques such as task-based fMRI, PET imaging, and qEEG could play a role in enabling communication of desires or pain. Unfortunately, these technologies have yet to evolve to provide reliable communication with patients in either UWS/VS or MCS. Only a small proportion of patients diagnosed as UWS/VS exhibit measurable changes in brain activity in response to external stimuli, so considerable efforts are still required to advance the field.

In an effort to develop an instrument to define and measure quality of life in this population, Tung et al. conducted several surveys with five expert groups, including healthcare workers and patient advocates. They found that the highest-ranked domain by all groups was "bodily pain and discomfort," but patient advocates prioritized "social functioning." At the same time, professionals (healthcare workers, bioethicists, neuroscientists, and quality of life methodologists) highlighted the importance of "cognitive functioning" and "communication ability" [39]. Once again, this highlights the need to develop tools that can provide direct communication with those patients that are behaviorally unresponsive but covertly aware.

Assistive technologies, such as BCIs, could help improve quality of life for patients in UWS or MCS by providing better diagnostic accuracy of their underlying level of consciousness, and by establishing a means for communication. BCIs could detect measurable and reliable neural signals that can then be used to measure awareness. The first challenge is to reach a consensus on what kind of signals are most indicative of consciousness; for example, a commonly used method is measuring objective responses to command-following tasks and comparing them to baseline brain activity. Examples include changes in qEEG amplitude in specific brain regions or specific event-related potentials (ERP) [40].

Justice and Rights

Patients with disorders of consciousness have been marginalized from society, mostly segregated to chronic care institutions, separated from the advances of science and medicine. After the acute care finishes, families then feel banished to the chronic care system, where many times they must fight a long battle to obtain appropriate rehabilitation services. New technologies may allow families to prove covert consciousness and have more tools to advocate for the provision of necessary assistance. Unfortunately, neuroscience development has had little impact on the experiences of patients diagnosed as UWS/VS or MCS and their families [41].

Distributive justice is a debated topic for many chronic illnesses, including severe brain injury. Neurotechnologies investigating covert consciousness could renew debates about justice and the distribution of resources. Once these technologies become part of standard clinical practice in the management of brain injuries, they will further add to the significant financial expenses incurred by this population [35].

Deep Brain Stimulation for Disorders of Consciousness

Neuromodulation techniques include invasive and non-invasive methods, with varied applications in neurology, neurosurgery, and psychiatry. The field is rapidly evolving, and, along with it, new ethical dilemmas have surfaced regarding its application, distribution, and impact. One of these applications is the use of deep brain stimulation (DBS), an invasive neuromodulation technique, in the treatment of disorders of consciousness. Efforts started back in the 1960s, but the path for its development has been riddled with concerns about its safety, efficacy, and ethical implications, delaying the efforts considerably. Still, there is tremendous hope and promise for neurostimulation techniques in restoring some cognitive and behavioral functions in patients with altered states of consciousness after brain injury. In 2006, a collaborative group in the United States presented the first case of therapeutic DBS in a patient in MCS. This 38-year-old individual suffered a severe brain injury after an assault and had remained in MCS for over 6 years. After DBS implantation, the patient progressively exhibited gradual improvement in his level of awareness, language, and motor abilities over 6 months [42, 43]. Similar efforts were carried out in 1990 in Japan for patients in UWS/VS, showing significant physiological activation of the cerebrum with DBS of the thalamus; nonetheless, there was no clinical benefit [44]. Most efforts have since been limited to patients in MCS, further emphasizing the conflation of all patients with disorders of consciousness into a single category without care for their underlying dissimilarities. This assumption could have profound consequences on our ability to propose treatments and evolve the field. Since this technology is limited mainly to the research realm, we will mostly focus this discussion on the ethical implications of its investigation, touching on potential clinical issues.

Consent for Research

Patients with disorders of consciousness lack decisional capacity; therefore, participation in research would require that consent be obtained from a surrogate decision-maker. This population has been designated as vulnerable; as such, preventing exploitation through research is of paramount importance [45]. Although this is certainly a laudable and essential effort, the regulatory complexities that constitute its framework can also halt attempts to move the field forward. For example, patient's surrogates are limited in their ability to consent to procedures with undetermined therapeutic efficacy; therefore, efforts to conduct phase I trials are inevitably curtailed. Most authors agree on the use of multidisciplinary panels as part of the consent process [45, 46], with only one group proposing the removal of informed consent based on the idea of this treatment being essential [47].

Outcomes: The Potential for Benefit and Harm

The first concern arising from neuromodulation interventions in patients with disorders of consciousness is the lack of reliable measures to assess outcomes. Efforts are underway to create clinical and paraclinical tools to assess changes in consciousness, but these are riddled with the same issues discussed about the determination of consciousness itself. Schiff et al. proposed that the primary goal of DBS in this population should be "restoration of consistent communication," as opposed to changes in standardized measurements, such as the coma recovery scale (CRS-R) [42].

Perhaps the most appealing debate is that of the "self-awareness paradox," which worries about possible harms that could result from the resurgence of self-awareness. Patients may become attuned to their disability and deficits but will enjoy little to no benefits from a limited degree of consciousness [48]. Contrary to this belief, some argue that patients in MCS may already have some awareness of their situation independent of their ability to communicate. This notion could potentially be extrapolated to patients with retained awareness who were previously diagnosed as UWS/PVS [49]. Moreso, it is unknown if DBS effects are potentially reversible.

Societal Neglect Syndrome and the Potential Impact of Neuromodulation on the "Right-to-Die"

Many authors have argued that society has neglected to provide adequate care for patients with disorders of consciousness. Once acute care is completed, patients are transferred to chronic care centers for long-term management where they can remain years without neurological or imaging follow up. There are no specific protocols for their management, including development of specific rehabilitation goals and procedures [50]. Some even argue that patients are discharged sooner than what would be considered medically appropriate, further worsening their possibilities for good neurological outcomes [41]. The story is eerily similar when it comes to research, which remains mostly underfunded and with limited access to patients. Societal neglect syndrome or therapeutic nihilism results from society's denial of this population's needs.

Joseph Fins argued that one of the main challenges that emerging technologies face in this field is that they intend to create treatment options for the very same kind of patients represented in the landmark "right-to-die" cases. This intention contradicts society's view that patients in UWS/VS are "hopelessly damaged" due to the irretrievable loss of the "cognitive sapient state" [33]. Hence, the road to development and application of technologies for neuroinnovation are laced with public questions and concerns, and scientists and physicians may have to continue to face many challenges.

This chapter has discussed some of the main concerns that arise in the ethical arena regarding innovative technologies in the diagnosis and potential management of patients with disorders of consciousness. This is a rapidly evolving landscape and ethical questions will continue to arise and morph along with it. Conversations around these topics have been happening for decades, even before neuroinnovation was technologically feasible, but their significance has never been higher. As these technologies continue to mature, it is imperative that scientists and clinicians continue to prioritize responsible development and applications. Our aim was to provide some background to these questions and a potential platform to continue to fuel these conversations.

Key Points

1. Neuroinnovative technologies are advancing at a rapid pace, providing scientists and clinicians with novel tools to diagnose, interact with, and potentially treat patients with disorders of consciousness.
2. Emerging technologies carry ethical and societal concerns. Their implications are ever more relevant now that technologies are becoming more feasible and available.
3. Brain–computer interfaces can be useful in many facets of medicine and life, including discovery of covert consciousness in patients with brain injuries.
4. Ethical concerns around brain–computer interfaces include understanding and explaining risk-benefit ratios, guaranteeing privacy for users, and discerning their impact on patient's autonomy and identity.
5. Up to 40% of patients who suffer brain injuries can be misdiagnosed as lacking awareness by bedside examinations. This has significant prognostic implications as clinicians and surrogates decide on life-sustaining measures and procedures.

Questions to Consider

1. What are the pitfalls of the current diagnostic approach to disorders of consciousness? What are the ethical implications of these pitfalls?
2. What are the ethical and societal implications of detecting covert consciousness in patients with severe motor and language impairments?
3. Do the benefits of using brain–machine interfaces in the diagnosis and management of brain injury outweigh potential ethical risks?

References

1. Stone JL, Hughes JR. Early history of electroencephalography and establishment of the American Clinical Neurophysiology Society. J Clin Neurophysiol. 2013;30:28–44.
2. Mak JN, Wolpaw JR. Clinical applications of brain-computer interfaces: current state and future prospects. IEEE Rev Biomed Eng. 2009;2:187–99.
3. Nijboer F, Clausen J, Allison BZ, Haselager P. The Asilomar survey: stakeholders' opinions on ethical issues related to brain-computer interfacing. Neuroethics. 2013;6:541–78.
4. Racine E, Waldman S, Palmour N, Risse D, Illes J. "Currents of Hope": Neurostimulation techniques in U.S. and U.K. Print Media Camb Q Healthc Ethics [Internet]; 2007;16. [cited 2020 Mar 24]. http://www.journals.cambridge.org/abstract_S0963180107070351
5. Racine E, Waldman S, Rosenberg J, Illes J. Details for Manuscript Number: SSM-D-09-00651 R2 "Contemporary neuroscience in the media." Soc Sci Med 1982. 2010;71:725–733.
6. Burwell S, Sample M, Racine E. Ethical aspects of brain computer interfaces: a scoping review. BMC Med Ethics. 2017;18:60.

7. Zander TO, Krol LR. Team PhyPA: brain-computer interfacing for everyday human-computer interaction. Period Polytech Electr Eng Comput Sci. 2017;61:209–16.
8. Musk E, Neuralink. An integrated brain-machine interface platform with thousands of channels [internet]. Neuroscience; 2019. http://biorxiv.org/lookup/doi/10.1101/703801
9. Royal Society. iHuman: blurring lines between mind and machine [internet]. London: The Royal Society; 2019. [cited 2020 Mar 23]. https://royalsociety.org/-/media/policy/projects/ihuman/report-neural-interfaces.pdf
10. McGie SC, Nagai MK, Artinian-Shaheen T. Clinical ethical concerns in the implantation of brain-machine interfaces: part II: specific clinical and technical issues affecting ethical soundness. IEEE Pulse. 2013;4:32–7.
11. Demetriades AK, Demetriades CK, Watts C, Ashkan K. Brain-machine interface: the challenge of neuroethics. Surgeon. 2010;8:267–9.
12. Aas S, Wasserman D. Brain–computer interfaces and disability: extending embodiment, reducing stigma? J Med Ethics. 2016;42:37–40.
13. Klein E. Informed consent in implantable BCI research: identifying risks and exploring meaning. Sci Eng Ethics. 2016;22:1299–317.
14. Glannon W. Neuromodulation, agency and autonomy. Brain Topogr. 2014;27:46–54.
15. Clausen J. Conceptual and ethical issues with brain–hardware interfaces. Curr Opin Psychiatry. 2011;24:495–501.
16. Tamburrini G. Brain to computer communication: ethical perspectives on interaction models. Neuroethics. 2009;2:137–49.
17. Jebari K, Hansson S-O. European public deliberation on brain machine Interface technology: five convergence seminars. Sci Eng Ethics. 2013;19:1071–86.
18. Vlek RJ, Steines D, Szibbo D, Kübler A, Schneider M-J, Haselager P, et al. Ethical issues in brain–computer interface research, development, and dissemination. J Neurol Phys Ther. 2012;36:94–9.
19. Farisco M, Laureys S, Evers K. Externalization of consciousness. Scientific possibilities and clinical implications. In: Lee G, Illes J, Ohl F, editors. Ethical issues Behav Neurosci [internet]. Berlin, Heidelberg: Springer; 2015. p. 205–22. https://doi.org/10.1007/7854_2014_338. [cited 2020 Mar 24].
20. Farisco M, Salles A, Evers K. Neuroethics: a conceptual approach. Camb Q Healthc Ethics. 2018;27:717–27.
21. Schicktanz S, Amelung T, Rieger JW. Qualitative assessment of patients' attitudes and expectations toward BCIs and implications for future technology development. Front Syst Neurosci [Internet]. 2015;9 [cited 2020 Mar 23]. https://www.ncbi.nlm.nih.gov/pmc/articles/PMC4410612/
22. Machado C. The minimally conscious state: definition and diagnostic criteria; 2020 [cited 2020 Mar 29]. https://n.neurology.org/content/minimally-conscious-state-definition-and-diagnostic-criteria-1
23. The European Task Force on Disorders of Consciousness, Laureys S, Celesia GG, Cohadon F, Lavrijsen J, León-Carrión J, et al. Unresponsive wakefulness syndrome: a new name for the vegetative state or apallic syndrome. BMC Med. 2010;8:68.
24. Schnakers C, Vanhaudenhuyse A, Giacino J, Ventura M, Boly M, Majerus S, et al. Diagnostic accuracy of the vegetative and minimally conscious state: clinical consensus versus standardized neurobehavioral assessment. BMC Neurol. 2009;9:35.
25. Claassen J, Doyle K, Matory A, Couch C, Burger KM, Velazquez A, et al. Detection of brain activation in unresponsive patients with acute brain injury. N Engl J Med. 2019;380:2497–505.
26. Turgeon AF, Lauzier F, Simard J-F, Scales DC, Burns KEA, Moore L, et al. Mortality associated with withdrawal of life-sustaining therapy for patients with severe traumatic brain injury: a Canadian multicentre cohort study. CMAJ Can Med Assoc J. 2011;183:1581–8.
27. Fernández-Espejo D, Owen AM. Detecting awareness after severe brain injury. Nat Rev Neurosci. 2013;14:801–9.

28. Hampshire A, Parkin BL, Cusack R, Espejo DF, Allanson J, Kamau E, et al. Assessing residual reasoning ability in overtly non-communicative patients using fMRI. NeuroImage Clin. 2012;2:174–83.
29. Monti MM, Vanhaudenhuyse A, Coleman MR, Boly M, Pickard JD, Tshibanda L, et al. Willful modulation of brain activity in disorders of consciousness. N Engl J Med. 2010;362:579–89.
30. Naci L, Owen AM. Making every word count for nonresponsive patients. JAMA Neurol. 2013;70:1235–41.
31. Giacino JT, Katz DI, Schiff ND, Whyte J, Ashman EJ, Ashwal S, et al. Comprehensive systematic review update summary: disorders of consciousness. Neurology. 2018;91:461–70.
32. Edlow BL, Fins JJ. Assessment of covert consciousness in the intensive care unit: clinical and ethical considerations. J Head Trauma Rehabil. 2018;33:424–34.
33. Fins JJ. Disorders of consciousness, past, present, and future. Camb Q Healthc Ethics. 2019;28:603–15.
34. Wilkinson D, Kahane G, Savulescu J. "Neglected personhood" and neglected questions: remarks on the moral significance of consciousness. Am J Bioeth. 2008;8:31–3.
35. Wilkinson D, Savulescu J. Is it better to be minimally conscious than vegetative? J Med Ethics. 2013;39:557–8.
36. Jennett B, Plum F. Persistent vegetative state after brain damage. A syndrome in search of a name. Lancet Lond Engl. 1972;1:734–7.
37. Sandøe P. Quality of life - three competing views. Qual Life. 1999;2:11–23.
38. Kuehlmeyer K, Palmour N, Riopelle RJ, Bernat JL, Jox RJ, Racine E. Physicians' attitudes toward medical and ethical challenges for patients in the vegetative state: comparing Canadian and German perspectives in a vignette survey. BMC Neurol. 2014;14:119.
39. Tung J, Speechley KN, Gofton T, Gonzalez-Lara LE, Graham M, Naci L, et al. Towards the assessment of quality of life in patients with disorders of consciousness. Qual Life Res [Internet]. 2019 [cited 2020 Mar 29]. http://link.springer.com/10.1007/s11136-019-02390-8
40. Annen J, Laureys S, Gosseries O. Brain-computer interfaces for consciousness assessment and communication in severely brain-injured patients. Handb Clin Neurol [Internet]. Elsevier; 2020, p. 137–152. [cited 2020 Apr 6]. https://linkinghub.elsevier.com/retrieve/pii/B9780444639349000111
41. Fins JJ. Rights Come to Mind: Brain Injury, Ethics, and the Struggle for Consciousness [Internet]. Camb. Core. Cambridge University Press; 2015. [cited 2020 Mar 30]. https://pubmed.ncbi.nlm.nih.gov/30210658/
42. Schiff ND, Giacino JT, Kalmar K, Victor JD, Baker K, Gerber M, et al. Behavioural improvements with thalamic stimulation after severe traumatic brain injury. Nature. 2007;448:600–3.
43. Shirvalkar P, Seth M, Schiff ND, Herrera DG. Cognitive enhancement with central thalamic electrical stimulation. Proc Natl Acad Sci U S A. 2006;103:17007–12.
44. Tsubokawa T, Yamamoto T, Katayama Y, Hirayama T, Maejima S, Moriya T. Deep-brain stimulation in a persistent vegetative state: follow-up results and criteria for selection of candidates. Brain Inj. 1990;4:315–27.
45. Glannon W. Neurostimulation and the minimally conscious state. Bioethics. 2008;22:337–45.
46. Giacino J, Fins JJ, Machado A, Schiff ND. Central thalamic deep brain stimulation to promote recovery from chronic posttraumatic minimally conscious state: challenges and opportunities. Neuromodulation. 2012;15:339–49.
47. Patuzzo S, Manganotti P. Deep Brain Stimulation in Persistent Vegetative States: Ethical Issues Governing Decision Making [Internet]. Behav. Neurol. Hindawi; 2014, p. e641213. [cited 2020 Mar 31]. https://www.hindawi.com/journals/bn/2014/641213/
48. Yamamoto T, Katayama Y, Kobayashi K, Oshima H, Fukaya C, Tsubokawa T. Deep brain stimulation for the treatment of vegetative state. Eur J Neurosci. 2010;32:1145–51.
49. Schiff ND, Giacino JT, Fins JJ. Deep brain stimulation, neuroethics, and the minimally conscious state: moving beyond proof of principle. Arch Neurol. 2009;66:697–702.
50. Lanoix M. Where angels fear to tread: proxy consent and novel technologies. Brain Inj. 2010;24:1336–42.

Further Reading

Claassen J, Doyle K, Matory A, Couch C, Burger KM, Velazquez A, et al. Detection of brain activation in unresponsive patients with acute brain injury. N Engl J Med. 2019;380:2497–505.

Fins JJ. Disorders of consciousness, past, present, and future. Camb Q Healthc Ethics. 2019;28:603–15.

Graham M, Wallace E, Doherty C, Mccann A, Naci L. From awareness to prognosis: ethical implications of uncovering hidden awareness in behaviorally nonresponsive patients. Camb Q Healthc Ethics. 2019;28:616–31.

Kögel J, Jox RJ, Friedrich O. What is it like to use a BCI? – insights from an interview study with brain-computer interface users. BMC Med Ethics. 2020;21:2.

Peterson A, Owen AM, Karlawish J. Translating the discovery of covert consciousness into clinical practice. JAMA Neurol. 2020;77:541.

Chapter 9
On the Edges: The Ethics of Human Studies with Psychedelic Substances

Sabrina Correa da Costa and Mehmet Sofuoglu

Introduction

What more can a person gain in life than what God-Nature itself reveals to him?

— Goethe

Psychedelic substances are a large and diverse group of natural, synthetic, or semisynthetic compounds with distinct molecular structures, receptor affinity, and pharmacological effects [1]. Examples of classical psychedelics, based on their chemical structure, include ergolines (lysergic acid diethylamide (LSD)), indole ethylamines (psilocybin, *N*, *N*-dimethyltryptamine (DMT)), phenylethylamines (mescaline, 3,4-methylenedioxy-methylamphetamine (MDMA)), and bicyclic diterpenoids (salvia divinorum), as shown in Table 9.1 [1, 2, 31]. Other psychedelics include ibogaine, ketamine, and phencyclidine (PCP).

The term "psychedelic," first coined by Humphrey Osmond in 1957, was used to denote the "mind-revealing" or "mind-manifesting" properties and mystical experiences associated with the use of such substances like LSD or psilocybin [32]. Psychedelic substances differ from other psychoactive substances, like opioids or psychostimulants, mainly by their effects on conscious experience that may include altered sense of time and space, dissociative symptoms, or the separation between one's thoughts, feelings, memories, actions, and the environment, distorted perceptions, and loss of normal boundaries of the self, which is usually described as "ego dissolution." Mystical experiences, near-death experiences, and a renewed sense of

S. C. da Costa (✉)
Department of Psychiatry and Psychology, Mayo Clinic College of Medicine and Science, Rochester, MN, USA

M. Sofuoglu
Yale University School of Medicine, New Haven, CT, USA

VA Connecticut Healthcare System, West Haven, CT, USA

L. W. Roberts (ed.), *Ethics and Clinical Neuroinnovation*,
https://doi.org/10.1007/978-3-031-14339-7_9

Table 9.1 Examples of Hallucinogens and their use in human research

Class	Molecular structure	Examples of use in human research
Ergolines **LSD**[a] **Semisynthetic**	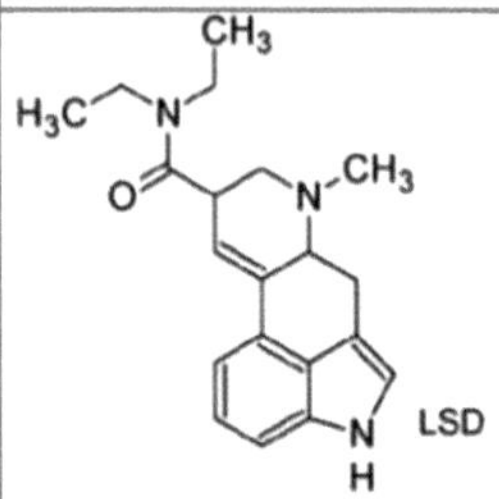	Neural and biological mechanisms [2–4], alcohol use disorder [5–7], cancer-related anxiety [8], “obsessional depression and neurosis [9],” schizophrenia [9], “sociopathic disorder [9],” depression [4], cluster headaches [10]
Indole ethylamines **Psilocybin (“magic mushrooms”)** **DMT (Psychotria viridis, Mimosa hostilis)** **Natural, Plant-based hallucinogens**	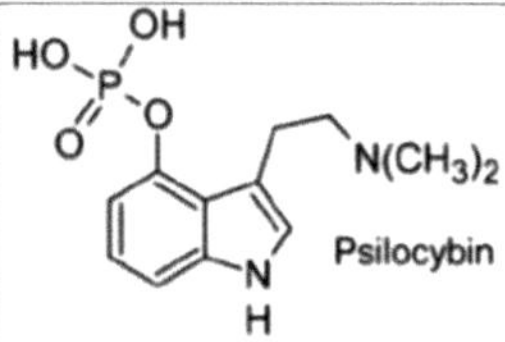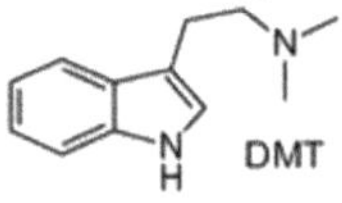	Psilocybin: Neural and biological mechanisms [11–13], treatment-resistant major depressive disorder [14], alcohol dependence [6, 15, 16], anxiety/adjustment disorder secondary to life-threatening conditions [17], obsessive-compulsive disorder [18], tobacco use disorder [19], cannabis, opioid, and stimulant misuse [20], cluster headaches [10], “schizophrenia-like symptoms” [21] DMT/ayahuasca[b]: Recurrent major depressive disorder [4]
Phenylethylamines **Mescaline, Natural, Plant-based hallucinogen** **MDMA, synthetic**[a]	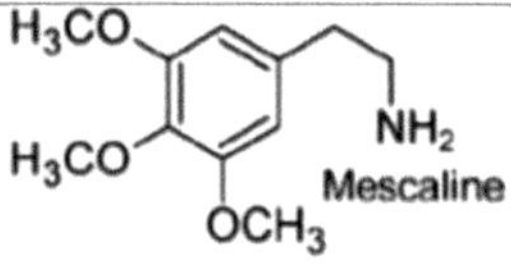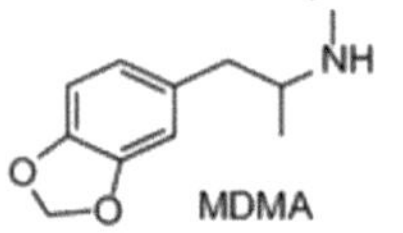	Mescaline: Schizophrenia/psychosis [4], military and intelligence services [22] MDMA: Treatment-resistant PTSD [23, 24], social anxiety in adults with autism spectrum disorder [25], anxiety secondary to life-threatening conditions [6], substance interactions [26–29], military and intelligence services [22]
Bicyclic diterpenoid **Salvia – Salvia divinorum**[b] **Natural, plant-based**	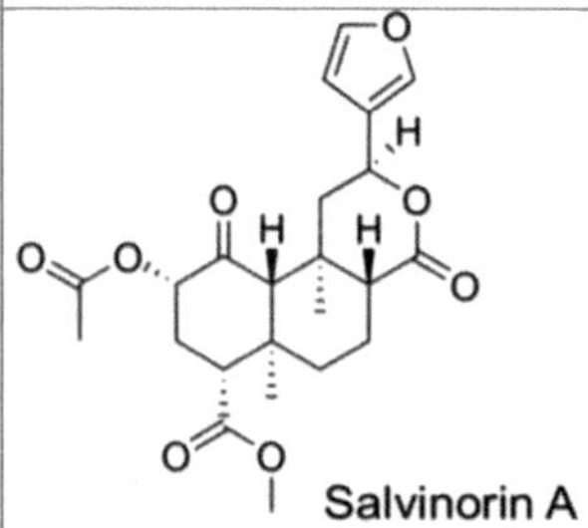	Case reports of addictive potential and persistent psychosis with salvia [30], human research on mechanisms of action, safety, and tolerability in healthy subjects [3, 4]

DMT – *N N*-dimethyltryptamine, *LSD* lysergic acid diethylamide, *MDMA* 3,4-methylenedioxy-methylamphetamine, *PTSD* posttraumatic stress disorder

[a]Empathogens or entactogens (“en” – Greek – “within,” “tactus” Latin “touch,” “gen” – Greek “generate”) are believed to have the potential to enhance closeness and connectedness, ability to decrease anxiety, increase trust and self-acceptance

[b]Entheogen – God within (En – within, Greek Theos – god, gen – produce) – used in spiritual or religious rituals

purpose and meaning are also commonly described with the use of psychedelics and seem to be a consequence of the unique *entheogenic* ("*generating God within*") properties of these substances [32].

The biological mechanisms underlying these complex psychoactive effects remain poorly understood. The basic pharmacology of psychedelics includes agonist and antagonist activity at serotonergic (5HT)-2A, 5HT-2C, 5HT-1A receptors, dopamine (D)-2 receptors, kappa opioid receptors, monoamine transporters (serotonergic, dopaminergic, noradrenergic), and glutamatergic system [1, 31]. This broad range of complex pharmacological effects raise concerns over the potential for acute and chronic toxicity as well as interactions with multiple drugs and disease states associated with psychedelic exposure [3].

Psychedelic substances have a rich history and have been used for many centuries in religious, shamanic, and therapeutic settings [33, 34]. The surge in psychedelic human research in the United States (U.S.) during the 1950s and 60s was followed by over two decades of dormancy as a consequence of substantial regulatory constraints enforced by psychedelic Schedule I classification in 1970. However, since 1990, there has been a resurgence in psychedelic research, with multiple studies testing the potential efficacy of psychedelics as novel treatments for a broad range of psychiatric disorders, besides basic and translational research on their neurobiology and psychopharmacology. The increased interest in psychedelic substances has brought with it several issues regarding the ethics of conducting human studies with psychedelics, which will be the primary goal of this chapter. We first provide a brief background on psychedelics, followed by a summary of the current state of human psychedelic research. We next discuss the ethical issues related to conducting research studies with psychedelics, followed by concluding remarks.

History of Psychedelic Substances

Psychedelic substances, including mescaline, psilocybin, ibogaine, DMT, and salvia divinorum, also known as "hallucinogenic" substances, have been used for millennia by South and Mesoamerican, Asian, and European cultures [33–35]. Evidence suggests that psilocybin and lysergamides were used in religious and shamanic rituals by Aztecs; DMT, and ayahuasca by indigenes in the Amazonian regions of South America, while mescaline, the psychoactive compound of peyote cactus, has been used by tribes in the Northern Mexico for more than 3000 years. Mescaline continues to be used in religious ceremonies by the Native American Church in the U.S. and Canada to this day, similarly to ayahuasca in South America [33–35].

Although LSD was first synthesized in 1938, its psychoactive effects were only serendipitously discovered in 1943 by the Swiss scientist Albert Hofmann [36, 37]. By the 1950s, however, LSD and other psychedelics prompted considerable interest in psychiatric research, leading to more than 1000 scientific publications by the 1970s [38, 39]. Interestingly, psychedelic-assisted psychotherapy became commonly applied not only in research, but also in clinical practice, where it was used to facilitate progress in psychotherapy through self-reflection, ego dissolution (i.e., the loss of boundaries between one's self or identity and the external environment), and access to unconscious material [4]. It is noteworthy that both LSD and psilocybin were marketed under brand names by pharmaceutical companies and made available to physicians in the 1950s and 1960s [40]. During the same period, psychedelics were even used as "truth serums," adjunctive to hypnosis, for military and intelligence purposes [22].

Due to psychedelic-induced perceptual disturbances and dissociative states, some researchers proposed that psychedelic substances could be used as disease models of schizophrenia and other psychotic disorders [21, 41]. However, most of the initial studies focused on the safety and potential clinical utility of these substances for the treatment of psychiatric conditions such as depression, anxiety, "neurotic," and psychosomatic disorders [5, 8, 9]. Early studies deemed psychedelics safe, even in medically complex patients, such as in terminal cancer, although usually devoid of reliable descriptions of adverse events from these compounds. In addition to perceptual and dissociative symptoms following psychedelic use, early observations also described feelings of euphoria, happiness and relaxation, increased energy, sociability, and empathy. Psychedelics, particularly mescaline and LSD, were believed to provoke extraordinary insights into the essence of the creative process, besides mystico-religious experiences. Curiosity and interest outside of the medical field emerged, particularly on the part of artists and other intellectuals. Nonmedical use of these compounds spread with an epidemic-like pace, and careless experimentation led to multiple negative incidents, including prolonged psychosis, erratic behaviors, accidents, and even criminal acts and tragic events [30, 42, 43]. While early research suggested that medically supervised use of LSD, for instance, was rarely associated with negative outcomes, the increasing number of catastrophic events associated with nonmedical use of these compounds became unmanageable, and Sandoz eventually discontinued the medication (Delysid® – LSD 25) in 1965.

In the 1960s, the widespread recreational use of psychedelics and their association with political activism and the counterculture movement resulted in significant stigma against these substances [44]. Research findings on their clinical benefits were eventually viewed as inconsistent, and concerns for safety and the quick escalation of their nonmedical use resulted in the criminalization of psychedelic substances [45]. Many psychedelic substances, including LSD and psilocybin, were classified as Schedule I substances by the Controlled Substance Act of 1970, denoting high potential for abuse, no accepted medical use, and lack of accepted safety for use of these substances under medical supervision [45]. Similar to other substances of its class, MDMA, also known as ecstasy, although first synthesized in 1912, was only introduced as a recreational drug in the 1960s. By the mid-1980s, after achieving widespread recreational use, MDMA was also classified as a Schedule I substance [46]. Following this restrictive scheduling by the U.S. Drug Enforcement Agency (DEA), research on psychedelics dramatically reduced [47].

Current Status

Once a major issue in the 1960s and 1970s, the recreational use of psychedelics has significantly decreased since its criminalization. According to the 2018 National Survey on Drug Use and Health (NSDUH), the lifetime prevalence of hallucinogen use in the U.S. population aged 12 and older is 15.8%, representing about 42 million Americans [48]. It is also estimated that about 2% of the U.S. population, or 5.6 million Americans aged 12 or older were past-year hallucinogen users, whereas 0.6% were past-month users. The age group with the highest rates of hallucinogen use was young adults between ages of 18 and 25 (6.9%) [48]. It is important to note that, in this epidemiological study, the "hallucinogen" category encompasses many commonly used psychedelics, including LSD, PCP, peyote, mescaline, psilocybin mushrooms, MDMA, ketamine, DMT, and *Salvia divinorum*. Interestingly, according to the 2018 NSDUH, higher rates of psychedelic use are typically observed with MDMA (also known as "ecstasy"), which seems to confer a higher risk of compulsive use, physiological dependence, and addiction when compared to other psychedelic substances, particularly due to its amphetamine-like properties leading to reinforcing and habit-forming effects [48].

Following a period of dormancy after the 1970s, research on psychedelics has received a renewed interest, particularly over the past two decades [6, 49, 50]. Promising findings from early research and the urgent need for novel pharmacologic agents in psychiatry resulted in a resurgence of research on psychedelics. For instance, LSD, psilocybin, and MDMA have been under investigation for further elucidation of their biological and neural mechanisms [11–13, 51]. In addition, ongoing clinical trials are testing the safety and efficacy of psychedelic substances for medical and psychiatric conditions. Case reports suggest efficacy of LSD and psilocybin for acute management and prophylaxis of cluster headaches [10]. Moreover, preliminary findings from a small pilot study suggest improvement of symptoms in obsessive-compulsive disorder (OCD) following psilocybin-assisted psychotherapy [18]. Similarly, findings from an open-label pilot study on the effects of a single-dose psilocybin intervention for treatment-resistant depression suggest significant and sustained improvements of depressive symptoms at 1 week and 3 months post-psychedelic treatment [14]. Additionally, MDMA-assisted psychotherapy has shown positive results for the treatment of social anxiety in adults with autism spectrum disorder [25]. Yet, more compelling findings of efficacy and safety of psychedelics for psychiatric disorders have been described in post-traumatic stress disorder (PTSD) [23], depression and anxiety in terminal cancer and life-threatening conditions [17], and substance use disorders [6]. For instance, a Phase 2 randomized controlled trial of MDMA-assisted psychotherapy for PTSD has shown short-term (1 month) and long-term (12 months) improvements of overall PTSD symptoms and no significant adverse events [24]. In addition, early experiments and contemporary studies have demonstrated efficacy and safety of psilocybin for anxiety and depressive symptoms in life-threatening conditions, such as terminal cancer, leading to less distressing reactions to terminal illness, a new sense of purpose and meaning, and improvements in overall quality of life [17]. The use of psychedelics for the treatment of substance use disorders has shown preliminary efficacy of psilocybin for alcohol use disorder [15, 16, 19] and smoking cessation [52], and ibogaine for the treatment of opioid withdrawal symptoms [6]. Interestingly, a recent study has shown that naturalistic psychedelic use that included moderate or high doses of LSD or psilocybin mushrooms was followed by persisting reductions in cannabis, opioid, or stimulant use in individuals who met criteria for severe substance use disorders at baseline, questioning the potential benefits of psychedelic substances as harm reduction strategies [20].

Noteworthy, research on these substances has proven to be particularly challenging. Despite promising results, methodological limitations challenge generalizability of findings and applicability of these interventions into large scale populations. For instance, the majority of early studies were descriptive reports or uncontrolled clinical trials. Besides, most studies involved small sample sizes and study populations that were not representative of the general population. Moreover, heterogeneity of the treatment groups, inconsistent interventions, unreliable reports of outcomes and adverse events, and lack of rigorous statistical analysis were some of the major limitations of early human research with psychedelics. Although some of these methodological limitations have been addressed by modern psychedelic research, selection bias remains a limitation in recent studies, besides the continuous pursuit for active placebo, given difficulties blinding interventions due to unique pharmacological properties and psychotropic effects of psychedelic substances.

Ethics of Human Studies with Psychedelic Substances

In the U.S., protection of human participants is regulated by the Department of Health and Human Services [53]. Broadly, regulatory guidelines are based on the ethical principles of autonomy, respect to persons, beneficence, nonmaleficence, and justice and require voluntary participation in research through informed consent and equitable enrollment of participants in research studies [53–55].

The foundations for the ethical conduct of human research are based on multiple sources, including the Nuremberg Code, the Belmont Report, the Declaration of Helsinki, the International Ethical Guidelines for Biomedical Research Involving Human Subjects, and many other guidelines [56]. As human research has advanced throughout the years, ethical requirements for biomedical human research have also evolved. While early focus was primarily on informed consent as the cornerstone of ethical human research, more recently, broader frameworks of ethical requirements for studies in human subjects have emerged [56]. For instance, according to the National Institutes of Health (NIH) Clinical Center, Department of Bioethics, essential requirements for ethical human research are: (1) social value, (2) scientific validity, (3) favorable risk-benefit ratio, (4) fair selection of participants, (5) informed consent, (6) independent review, and (7) respect for participants (see Table 9.2) [56, 57].

Table 9.2 Framework of essential requirements for ethical human research

Essential requirements	Ethical principles	Examples of current challenges in human research with psychedelics
Social and scientific value	Limited resources, nonexploitation of subjects	Promising clinical benefits of psychedelic substances for psychiatric disorders including PTSD, treatment-resistant depression, anxiety and depressive symptoms in terminal cancer, and substance use disorders
Scientific validity	Limited resources, nonexploitation of subjects, feasibility, generalizability, and replicability of findings	Methodological limitations of early psychedelic research: Masking interventions, difficulty in finding suitable active controls, small sample sizes, open-label trials, highly structured study settings, and intensive behavioral and psychosocial interventions
Informed consent	Autonomy, decision-making capacity	Information about research protocols, study aims, risks, benefits, and alternatives
Independent review	Public accountability, minimizing potential conflicts of interest	Institutional review boards, funding agencies, data and safety monitoring boards (DSMB), review of proposals by independent and non-affiliated experts, ideally with some knowledge or background on psychedelic research, given uniqueness of these interventions
Risk-benefit ratio	Beneficence, nonmaleficence, nonexploitation	Potential risks of psychedelics: Behavioral and psychological distress, acute suicidality, potential for abuse and addiction liability, autonomic and cardiovascular side effects, substance interactions, unknown effects on brain development, unclear long-term effects
Respect to subjects	Autonomy	Acute behavioral effects of psychedelic substances may impair research participant's decision-making capacity and ability to withdraw their participation, particularly during the interventional session; privacy and confidentiality issues related to study participation
Fair subject selection	Justice	Overly restrictive exclusionary criteria, overly homogeneous study populations seem to limit equitable participation in psychedelic research and violate the principle of justice

The employment of broader ethical frameworks is especially relevant for human studies with psychedelics due to their unique challenges, ethical controversies, and methodological complexities. While most studies on psychedelics, to date, have focused their seminal questions on safety, efficacy, and mechanisms of action, which are absolutely critical for a better understanding of the potential risks and clinical benefits of these substances, the literature remains scarce regarding the ethics of human studies with psychedelic substances. In this section, we review each of the seven ethical principles proposed by NIH, discussing specific considerations for human research with psychedelics.

Social and Scientific Value

Social and scientific value are integral aspects of ethical research [56]. Because the resources are limited, human studies are expected to advance scientific knowledge and provide a significant impact to the society, even if research participants do not directly benefit from study interventions. A better understanding of the human biology, development of more effective therapeutics, improvements of health and well-being in individual or populational levels are some examples of acceptable goals of ethical human research on the grounds of social and scientific value.

Development of novel therapeutics for psychiatric disorders remains a daunting challenge, as many promising medications have failed in different stages of drug development, and current therapeutics have shown limited efficacy in some cases [58]. This resulted in skepticism and reduced enthusiasm of the pharmaceutical industry for investing in novel therapies for psychiatric conditions. Nonetheless, the search for more effective treatments for psychiatric disorders continues, and research with psychedelics is expanding [4, 59, 60]. For instance, the Food and Drug Administration (FDA) has recently granted a Breakthrough Therapy Designation for MDMA as an investigational drug for MDMA-assisted psychotherapy in patients with severe PTSD. By granting Breakthrough Therapy Designation, the FDA has accepted that MDMA-assisted psychotherapy may have clinical benefits and advantage over available pharmacological interventions for PTSD, and Phase 3 trials will be able to assess the efficacy and safety of this intervention in larger sample sizes. Recent studies suggest that some of the benefits of MDMA-assisted psychotherapy for PTSD are long-lasting and include decrease in avoidance and a new awareness of maladaptive patterns of emotional, behavioral, and cognitive responses to trauma and traumatic reminders, enabling a reappraisal of events and memory reconsolidation through new associations and more adaptive responses to traumatic events [23, 49]. It has been posited that the pro-social and empathogen effects of MDMA may also promote stronger therapeutic alliances leading to a greater sense of trust and safety during therapy, which ultimately may contribute to the observed clinical benefits as well [23, 49]. These and other promising findings on psychedelics warrant methodologically robust and well-designed studies to elucidate biological and neural mechanisms, therapeutic applications, and risks and potential benefits of these substances for psychiatric conditions. For instance, if the results Phase 3 trials indicate efficacy and acceptable safety, MDMA may emerge as a new evidence-based treatment for the treatment of PTSD.

Scientific Validity

Ethical human research is expected to follow rigorous scientific method, which consists of having clear research questions and applying accurate and reliable research methodology that is feasible, replicable, has an adequate sample size, and a sound data analysis plan in order to test the study hypotheses [56]. Without scientific validity, research cannot properly advance the scientific knowledge and answer the proposed research questions; thus, exposing research participants to unnecessary risks, besides wasting valuable and limited resources.

Concerning scientific validity, research on psychedelics has proven to be particularly challenging. Despite promising results, with a few exceptions, the majority of early studies on psychedelics displayed significant methodological problems, limiting scientific rigor and validity. For instance, the majority of early studies were descriptive reports or uncontrolled clinical trials, often encompassing small sample sizes and study populations that were not necessarily representative of the general population. Besides, early studies were not usually subject to Institutional Review Board (IRB) reviews. Moreover, heterogeneity of the treatment groups, inconsistent interventions, unreliable reports of outcomes and adverse events, lack of rigorous statistical analysis, and small study samples were some of the major limitations of early human research with psychedelics, limiting replicability and generalizability of findings.

Scientific methods employed in contemporary clinical research on psychedelics have dramatically improved, enhancing the scientific validity and reliability of findings. Nonetheless, open-label study designs, small and heterogeneous study samples, potential selection bias, and participants' expectancy favoring interventions continue to represent main methodological limitations of modern research on psychedelics. Furthermore, highly structured study settings and intensive psychosocial and behavioral interventions may favor psychedelic interventions. For example, study protocols usually consist of 1–3 day-long sessions of psychedelic substance in conjunction with psychotherapy over a 12-week treatment period, along with 8–12 preparatory sessions, which raises questions regarding the real pharmacological effects of these substances, especially in the absence of control groups for comparison, given the potential clinical benefits of intensive psychotherapy interventions.

The lack of studies comparing different psychedelic substances and inconsistency and heterogeneity of clinical scales and assessments represent additional limitations of human research with psychedelics. However, as a consequence of the unique psychoactive and physiological effects of psychedelics, masking the interventions remains one of the major methodological challenges for research with these substances. For the same reason, the use of placebos seems ineffective, and the pursuit for active controls continue. Medications with psychoactive effects, such as methylphenidate, diphenhydramine, and benzodiazepines, or subtherapeutic doses of the intervention psychedelic substance have been used in clinical protocols as active controls; however, the outcomes remain suboptimal, given that research participants and investigators can often predict treatment allocation accurately [3, 7, 26, 61]. As a result, masking the study interventions remains a challenge in human research with psychedelics, for which researchers are still in need to develop strategies to circumvent this limitation.

Favorable Risk-Benefit Ratio

Biomedical research involving human participants is justified if (1) the potential risks to the participants are minimized, (2) the potential benefits to the participants are maximized, and (3) the potential benefits to the individual or the society outweigh the risks [56]. These requirements are based on the ethical principles of beneficence, nonmaleficence, and nonexploitation of individuals.

Although psychedelic substances have often been described as safe and well-tolerated by the study participants, acute behavioral disturbances, psychological distress, impulsive behaviors, acute and prolonged psychosis, confusion, transient impaired reasoning and decision-making capacity, anxiety, depression, and acute suicidality have also been reported in the literature [42, 43]. In addition, autonomic and cardiovascular side effects, particularly due to serotonin and catecholamine toxicity (e.g., hypertension, tachycardia, tachyarrhythmias, hyperthermia, muscle rigidity, hyperreflexia, neurotoxicity) have been well described in the literature, warranting close monitoring in research protocols [42].

Significant co-interactions, particularly involving MDMA and psychotropic medications [27–29, 62], are also an area of concern and may limit more extensive use of these substances in clinical practice or even participation in research protocols. Some research protocols, however, have tapered and discontinued psychotropic medications prior to interventional sessions with psychedelic substances. Although these practices have been adopted to ensure safe administration of psychedelic substances, it also raises concerns about the risks of exacerbation of psychiatric symptoms and clinical decompensation, safety, and need for close monitoring and follow up by research teams, besides, more broadly, the ethical considerations of discontinuing FDA-approved treatments in favor of experimental substances. Furthermore, MDMA-induced neurotoxicity remains controversial. While several studies have suggested long-term neurotoxic effects from MDMA use, particularly on the nigrostriatal dopaminergic and serotonergic pathways [63–65], many authors have argued that the data suggesting MDMA-induced neurotoxicity came from studies with several methodological limitations, including retrospective study designs, study populations encompassing individuals with polysubstance use and heavy exposure to MDMA, as well as other potential confounders [66]. Although dose-exposure to psychedelic substances in clinical trials is significantly lower than recreational use, the long-term effects of a single- versus repeat-dose exposure to psychedelics, including MDMA, needs to be further clarified. The neuropsychopharmacological mechanisms of these substances and the impact on brain physiology and connectivity remain largely unknown, and concerns for safety involving healthy volunteers, particularly young adults, remain. Studies have shown that prenatal exposure to MDMA has been associated with neurodevelopmental delays [67, 68]; however, the consequences of the exposure to MDMA on brain development at later stages, such as in young adults between ages of 18 and 25, remain unclear. Moreover, the evidence on the effects of psychedelic substances in specific populations, such as individuals with a history of severe mental illness and/or substance use and addictive disorders, is also limited.

Evidence, to date, suggests that psychedelics are associated with low addiction liability [69, 70]. One exception is the class of phenylethylamines, including MDMA, given amphetamine-like properties and reinforcing effects, increasing the potential for abuse. Noteworthy, most data related to the abuse potential and addiction liability of psychedelics were obtained from studies employing a single-dose administration, and the risks of repeat-dose exposure are relatively unknown. Furthermore, the abuse potential of these substances and the risks of addiction, particularly among young adults, individuals with severe mental illness, or history of substance use disorders, remain unclear. Careful examination of pre-existing substance use and addictive disorders, active or in remission, prior to enrollment, and close monitoring of aberrant behaviors through clinical assessments and toxicology assays should be warranted, particularly in research studies involving MDMA. However, it also raises questions on whether persons with a history of psychedelic use disorder should be excluded from such studies, not only due to potential risks for the individual, since exposure to the substance involved in the disorder may trigger problematic use, but also considering study outcomes in the context of expectation bias (i.e., when an individual's expectations of an outcome influence one's perceptions or results from an intervention). These questions are particularly salient considering the challenges of blinding interventions in the absence of adequate active placebos.

Finally, the impact of federal regulations and schedule I classification of psychedelic substances by the CSA deserves some discussion. Current regulations have made research on these substances costly and challenging in many aspects. On the other hand, more flexible regulations—while allowing for more research on these substances—could also have unintended consequences in terms of broader societal impacts, for example, by changing the public's perception of harm from these substances, which may ultimately result in an increase in recreational use of psychedelics, similarly to the events observed in the 1960s. Conversely, if the benefits of only a single or a few doses of psychedelics may improve mental health outcomes, particularly for those in desperate need due to severe and refractory conditions, and the risks of psychedelic substances, when compared to other psychoactive agents, such as ketamine, benzodiazepines, opioids, or amphetamines, may be similar or even lower on the grounds of safety and abuse liability, then reviewing current regulations seems not only justifiable but pressing:

> "We have had an ever growing population of patients whose illness is seemingly refractory to standard therapies. (…) The interest in newer somatic treatments in part reflects a sense of desperation in treating this group of patients, many of whom may actually not be particularly responsive to somatic therapy."

Fair Selection of Participants

Fair subject selection and equitable research participation are also integral aspects of ethical research, based on the principle of justice [56]. For instance, several studies have excluded individuals with no previous experience with psychedelic substances from participation in research protocols with psychedelics to minimize risks of significant psychological distress in psychedelic-naïve persons. Although it might be reasonable from the safety standpoint, self-recruitment and prior psychedelic substance exposure could potentially lead to biased results, in part due to participants' beliefs and expectancies regarding the acute subjective effects of these substances. Therefore, rather than excluding psychedelic-naïve individuals to participate in studies with psychedelic substances, safeguards should be in place to minimize risks and potential adverse events, ensuring fair subject selection in studies with psychedelics [71–73]. Moreover, homogeneous and highly selected study populations limit generalizability of findings, whereas determining which patient populations might be more vulnerable to risks and potential harms of psychedelic exposure, such as individuals with psychotic disorders, should also be carefully considered. Furthermore, although early research suggested significant benefits of psychedelics for alcohol use disorders, and recent data have shown promising results of psychedelics for substance use disorders, such as psilocybin for alcohol and tobacco use disorders and ibogaine for the treatment of opioid withdrawal symptoms, long-term effects and safety of these substances in this population are still to be determined. In addition, many research protocols, to date, have excluded individuals with a history of addiction; therefore, given potential clinical benefits, preventing individuals with substance use disorders from participating in study protocols involving psychedelic substances may limit equitable participation in research, violating the ethical principle of justice.

Informed Consent

Informed consent has been one of the cornerstones of ethical clinical research. The essential elements of informed consent—competence, disclosure, understanding, and voluntariness—intend to ensure respect for the individual's autonomy in the decision-making process. During the consent process, potential participants should be provided clear and accurate information about the study objectives, proposed interventions, risks, benefits, and alternatives to study participation to ensure voluntary and uncoerced decision on whether or not to participate in the study [56]. It is important for individuals considering participation in clinical research to understand that, while clinical care addresses specific needs of an individual, primary goals of clinical research may vary and include developing new treatments, identifying determinants of health or causes of illnesses, with the ultimate goal to improve scientific knowledge, health, and well-being in a population's level, although not necessarily benefiting the research participant individually.

Although early research with psychedelics has not always fulfilled the above criteria for informed consent, modern research on psychedelics has been more diligent and committed to developing safeguards to respect individuals' autonomy. Guidelines for safety have been proposed to implement existing ethical frameworks, taking into consideration some of the unique aspects of research with psychedelics [74]. These guidelines suggest that, prior to interventions, research participants should receive adequate psychoeducation on the possible range of experiences and behavioral and physiologic reactions following psychedelic substance administration. This is to ensure that participants have enough information to determine whether the research protocol is consistent with their values and interests for a rational and autonomous decision. Besides, participants should also be fully aware of the study requirements, some of which may include discontinuation of current psychotropic medications, and risks, benefits, and consequences of such requirements as well. Lastly, as with any other human research studies or medical interventions, probing individuals by asking them questions to ensure that they have a clear understanding of the proposed research protocol, including risks and benefits of the study interventions, should be pursued to ensure a clear informed-decision making process while obtaining the consent forms [74].

An important caveat in studies on psychedelic substances is that decision-making capacity might be temporarily compromised during the active treatment session in the context of acute behavioral or psychological distress, including but not limited to acute psychotic symptoms (e.g., delusions and hallucinations), severe anxiety, or acute suicidality, resulting in temporary inability to withdraw participation from the study, for example, by leaving the premises of the research facility, due to safety concerns. Therefore, research participants should be made fully aware of this potential scenario prior to enrollment and understand the strategies that the research team may put in practice to ensure the safety of research participants and others.

Independent Review

Research protocols are independently reviewed by funding agencies, local or central Institutional Review Boards (IRB), and Data and Safety Monitoring Boards (DSMBs). In order to determine objectively whether the proposed research protocol is ethical, the risk-benefit ratio is favorable, and the study holds scientific validity, these independent panels should be composed of individuals who are not affiliated and have no conflicts of interest with the investigators or the study [53, 56]. Independent review is critical for social accountability, particularly given the historical examples of preventable or unnecessary harms and exploitation of individuals by research protocols. Furthermore, because of the unique challenges of conducting research with psychedelic substances, the review panel should preferably have some background knowledge or expertise on research with psychedelic substances.

Respect for Participants

Respect for the potential and already enrolled participants requires multiple actions: (1) privacy and confidentiality of the participants must be protected, (2) participants should be free to withdraw from the study at any point and without any penalties, (3) any new information gained during the study about the medication or the participant's clinical condition, should be shared with the participant, (4) the safety and welfare of the participant should be carefully monitored during study participation, and clinical care should be provided if needed, and (5) some mechanism to inform subjects of their contributions to clinical research and scientific knowledge should be in place [56].

Essentially, respect to participants is based on ethical principles of beneficence, nonmaleficence, autonomy, and welfare. To ensure the safety and welfare of participants enrolled in psychedelic research studies, recent guidelines emphasize the need for careful selection of participants, including screening for medical and mental health conditions and other relevant pre-existing factors that may increase the risks of adverse reactions to psychedelic substances [74]. Additionally, attention to the physical safety of the environment and the presence of trained staff, with whom the research participant would have already established rapport and trust through preparatory sessions, have also been recommended [74]. Participants should receive some form of follow-up after study participation to ensure that there are no physical or mental health adverse effects from study interventions, such as persistent hallucinations, delusions, paranoia, substance use, insomnia, or any other relevant adverse events not identified during the active treatment phase. Lastly, as with any other human research protocols, permitting withdrawal from the study, protecting subjects' privacy and confidentiality, as well as informing participants of newly discovered risks or benefits, should also be standard practices in human research with psychedelics.

Conclusion

Human psychedelic research is now at a crossroad with significant promise and many challenges. As we previously argued, applying a broad framework of ethical human research requirements will provide guidance in shaping future studies with psychedelic substances.

The social and scientific value of clinical research with psychedelics seems compelling as they emerge as potential treatments for a broad range of psychiatric disorders. Taken together, preliminary findings on psychedelics instill promise and hope to often pervasive, severe, and difficult-to-treat psychiatric conditions, usually associated with significant impairment and high rates of disability, in part due to suboptimal pharmacological and behavioral treatment response. Although evidence to date shows some promise, these studies have been limited by subject selection and small sample sizes, lack of control groups, and challenges blinding interventions. Therefore, additional research will be needed to replicate existing findings and further elucidate the neuropsychopharmacology, efficacy, and safety of psychedelic substances, especially in vulnerable populations. Conversely, clinical research with psychedelics faces many challenges, including limited funding, substantial regulatory constraints, and methodological difficulties, some of which might be intrinsic to psychedelic substances and their unique psychoactive effects. While overly strict regulations have somewhat hindered research on psychedelics, more flexible regulations and use of these substances in human research may change the public's perception of harm and increase the nonmedical use of psychedelics, reminiscent of the events in the U.S. 1950s and 60s. On the other hand, psychedelic substances may pose risks not higher than the risks associated with other psychoactive agents, such as ketamine, amphetamines, opioids, or benzodiazepines, which have been much less regulated by federal agencies and granted FDA-approval for different neuropsychiatric and medical use. Therefore, there is a need for ethically sound and methodologically rigorous studies on psychedelics for a better understanding of the risks, benefits, and potential long-term effects of psychedelics among genetically, racially, and socio-culturally diverse individuals and populations. Reassessing regulatory frameworks and legal constraints to lessen the barriers for future studies on psychedelic substances may also advance the field in that access to these compounds may be easier for researchers and institutions to conduct scientifically sound and methodologically reliable research. To date, financial and legal constraints have limited research on psychedelics, and concerns about stigma and other potential implications of conducting research on psychedelics have discouraged investigators and institutions. Thus, the level of existing evidence may also have been influenced by an investigator's expectancy. For example, Albert Hofmann reports, "During the the first years of its discovery, LSD brought me the same happiness and gratification that any pharmaceutical chemist would feel on learning that a substance she or he produced might possibly develop into a valuable medicine" [75]. Therefore, decreasing barriers to psychedelic research may result in a better understanding of the real benefits of these substances through independent research. Otherwise, many questions on psychedelic substances may well remain unanswered, data from anecdotal reports or methodologically limited studies will prevail, and substances with potential clinical applications may remain under-studied, under-utilized, and overly romanticized.

Key Points

1. As the need for more effective and novel therapies in psychiatric disorders remains, findings on psychedelic-assisted psychotherapy have been encouraging, particularly in post-traumatic stress disorder, treatment-resistant depression, substance use disorders, and anxiety and depressive symptoms in life-threatening conditions, such as terminal cancer.
2. As human research has evolved throughout the years, ethical requirements for human research have also expanded.
3. According to the National Institutes of Health (NIH) Clinical Center of Bioethics, the essential requirements for ethical research are: (1) social value; (2) scientific validity; (3) informed consent; (4) independent review; (5) favorable risk-benefit ratio; (6) respect to subjects; and (7) and fair subject selection.
4. Some of the acute behavioral effects of psychedelic substances may temporarily impair medical decision-making capacity, especially during the interventional sessions; overly restrictive exclusionary criteria may limit equitable participation in research with psychedelics and violate the principle of justice, and some

methodological limitations of human research with psychedelics, such as masking interventions, might be inherent to the unique psychoactive effects of psychedelics and ultimately difficult to circumvent.

5. Institutional Review Boards (IRBs), funding agencies, and Data and Safety Monitoring Boards (DSMBs) reviewing research protocols involving psychedelic substances should be composed of independent and non-affiliated experts, ideally with some knowledge or expertise on psychedelics, given the uniqueness of these interventions.
6. Overly strict regulations, limited funding, and difficult access to psychedelic substances may hinder human research with psychedelics.
7. More flexible federal regulations and widespread use of psychedelics in medical settings may change the public's perception of harm and result in an escalation of nonmedical use of psychedelic substances.

Questions to Consider

1. What kinds of safeguards might be necessary to facilitate the inclusion of psychedelic-naïve individuals (historically excluded from participation) in studies of psychedelic substances?
2. Prior psychedelic substance exposure and self-recruitment can potentially lead to biased results in studies involving psychedelic substances. What are some ways study investigators can address the balance between a research participant's familiarity and expectation with a study subject and a researcher's need for a non-biased sample?
3. How do psychedelic substances' interaction with decision-making capacity affect our understanding of informed consent?
4. What impact have federal regulations and schedule 1 classification of substances limited their scientific study? What changes might be useful in facilitating this research in the future?
5. How does a substance's reputation on the "street" affect our understanding of its utility in medical or psychiatric treatment? What about the reverse?

References

1. Nichols DE. Psychedelics. Pharmacol Rev. 2016;
2. Carhart-Harris RL. How do psychedelics work? Curr Opin Psychiatry. 2019;
3. Sellers EM, Romach MK, Leiderman DB. Studies with psychedelic drugs in human volunteers. Neuropharmacology. 2018;
4. Carhart-Harris RL, Goodwin GM. The therapeutic potential of psychedelic drugs: past, present, and future. Neuropsychopharmacology. 2017;
5. Ludwig A, Levine J, Stark L, Lazar R. A clinical study of LSD treatment in alcoholism. Am J Psychiatry. 1969;
6. Adinoff B, Griffiths R, Hendricks P, Garcia-Romeu A, Bogenschutz M, Ferguson SS, et al. A new era of treating substance use disorders with psychedelics. Am J Addict. 2019;
7. Nichols DE, Grob CS, Greer GR. Addressing the need for rigor in the rebirth of hallucinogen research. Drug Alcohol Depend. 2017.

8. Pahnke WN, Kurland AA, Goodman LE, Richards WA. LSD-assisted psychotherapy with terminal cancer patients. Curr Psychiatr Ther. 1969.
9. Fuentes JJ, Fonseca F, Elices M, Farré M, Torrens M. Therapeutic use of LSD in psychiatry: a systematic review of randomized-controlled clinical trials. Front Psychiatry. 2020.
10. Sewell RA, Halpern JH, Pope HG. Response of cluster headache to psilocybin and LSD. Neurology. 2006.
11. Carhart-Harris RL, Erritzoe D, Williams T, Stone JM, Reed LJ, Colasanti A, et al. Neural correlates of the psychedelic state as determined by fMRI studies with psilocybin. Proc Natl Acad Sci U S A. 2012.
12. Roseman L, Leech R, Feilding A, Nutt DJ, Carhart-Harris RL. The effects of psilocybin and MDMA on between-network resting state functional connectivity in healthy volunteers. Front Hum Neurosci. 2014.
13. Lebedev AV, Lövdén M, Rosenthal G, Feilding A, Nutt DJ, Carhart-Harris RL. Finding the self by losing the self: neural correlates of ego-dissolution under psilocybin. Hum Brain Mapp. 2015;
14. Carhart-Harris RL, Bolstridge M, Rucker J, Day CMJ, Erritzoe D, Kaelen M, et al. Psilocybin with psychological support for treatment-resistant depression: an open-label feasibility study. Lancet Psychiatry. 2016;
15. S. A, S. M, S. P, H. D, S. R, M. B. Psilocybin-assisted treatment for alcohol use disorder: A clinical perspective. Am J Addict. 2018.
16. Bogenschutz MP, Forcehimes AA, Pommy JA, Wilcox CE, Barbosa P, Strassman RJ. Psilocybin-assisted treatment for alcohol dependence: a proof-of-concept study. J Psychopharmacol. 2015;
17. Griffiths RR, Johnson MW, Carducci MA, Umbricht A, Richards WA, Richards BD, et al. Psilocybin produces substantial and sustained decreases in depression and anxiety in patients with life-threatening cancer: a randomized double-blind trial. J Psychopharmacol. 2016;
18. Moreno FA, Wiegand CB, Taitano EK, Delgado PL. Safety, tolerability, and efficacy of psilocybin in 9 patients with obsessive-compulsive disorder. J Clin Psychiatry. 2006;
19. Nielson EM, May DG, Forcehimes AA, Bogenschutz MP. The psychedelic debriefing in alcohol dependence treatment: illustrating key change phenomena through qualitative content analysis of clinical sessions. Front Pharmacol. 2018;
20. Garcia-Romeu A, Davis AK, Erowid E, Erowid F, Griffiths RR, Johnson MW. Persisting reductions in cannabis, opioid, and stimulant misuse after naturalistic psychedelic use: an online survey. Front Psych. 2020;
21. Vollenweider FX, Vollenweider-Scherpenhuyzen MFI, Bäbler A, Vogel H, Hell D. Psilocybin induces schizophrenia-like psychosis in humans via a serotonin-2 agonist action. Neuroreport. 1998.
22. Passie T, Benzenhöfer U. MDA, MDMA, and other "mescaline-like" substances in the US military's search for a truth drug (1940s to 1960s). Drug Test. Anal. 2018.
23. Mithoefer MC, Mithoefer AT, Feduccia AA, Jerome L, Wagner M, Wymer J, et al. 3,4-meth ylenedioxymethamphetamine (MDMA)-assisted psychotherapy for post-traumatic stress disorder in military veterans, firefighters, and police officers: a randomised, double-blind, dose-response, phase 2 clinical trial. Lancet Psychiatry. 2018;
24. Ot'alora GM, Grigsby J, Poulter B, Van Derveer JW, Giron SG, Jerome L, et al. 3,4-Methylenedioxymethamphetamine-assisted psychotherapy for treatment of chronic post-traumatic stress disorder: a randomized phase 2 controlled trial. J Psychopharmacol. 2018;
25. Danforth AL, Struble CM, Yazar-Klosinski B, Grob CS. MDMA-assisted therapy: a new treatment model for social anxiety in autistic adults. Prog Neuro-Psychopharmacol Biol Psychiatry. 2016;
26. Dyck E. "Hitting highs at rock bottom": LSD treatment for alcoholism, 1950-1970. Soc Hist Med. 2006;
27. Hysek CM, Simmler LD, Nicola VG, Vischer N, Donzelli M, Krähenbühl S, et al. Duloxetine inhibits effects of MDMA ("ecstasy") in vitro and in humans in a randomized placebo-controlled laboratory study. PLoS One. 2012;

28. Callaway JC, Grob CS. Ayahuasca preparations and serotonin reuptake inhibitors: a potential combination for severe adverse interactions. J Psychoactive Drugs. 1998;
29. Farré M, Abanades S, Roset PN, Peiró AM, Torrens M, O'Mathúna B, et al. Pharmacological interaction between 3,4-methylenedioxymethamphetamine (ecstasy) and paroxetine: pharmacological effects and pharmacokinetics. J Pharmacol Exp Ther. 2007;
30. Przekop P, Lee T. Persistent psychosis associated with Salvia divinorum use. Am J Psychiatry. 2009;
31. Vollenweider FX, Kometer M. The neurobiology of psychedelic drugs: implications for the treatment of mood disorders. Nat Rev Neurosci. 2010;
32. Osmond H. A review of the clinical effects of psychotomimetic agents. Ann N Y Acad Sci. 1957;
33. Brenan JPM, Schultes RE, Hofmann A. Plants of the gods: origins of hallucinogenic use. Kew Bull. 1980;
34. Schultes RE, Hofmann A, Rätsch C. Plants of the gods : their sacred, healing, and hallucinogenic powes. Phytochemistry. 2001;
35. Kauffman GB. LSD: my problem child; plants of the gods: origins of hallucinogenic use. J Chem Educ. 1982;
36. Kohn B, Hofmann A. LSD: my problem child. Antioch Rev. 1981;
37. Hofmann A. The discovery of LSD and subsequent investigations on naturally occurring hallucinogens. Discov Biol Psychiatry. 1970;
38. Grinspoon L, Bakalar JB. The psychedelic drug therapies. Curr Psychiatr Ther. 1981;
39. Frosch WA. Psychedelic drugs reconsidered. Am J Psychiatry. 1980;
40. Lee M, Shlain B. Acid dreams. Complet Soc Hist. LSD CIA, Sixties, Beyond. 2007.
41. Passie T. A complete international bibliography. Psycholytic Psychedelic Ther Res. 1931.
42. Strassman RJ. Adverse reactions to psychedelic drugs. A review of the literature. J Nerv Ment Dis. 1984.
43. Hendricks PS, Johnson MW, Griffiths RR. Psilocybin, psychological distress, and suicidality. J Psychopharmacol. 2015;
44. Nour MM, Evans L, Carhart-Harris RL. Psychedelics, personality and political perspectives. J Psychoactive Drugs. 2017.
45. U.S. Department of Justice: Drug Enforcement Administration. Title 21 United States Code (USC) Controlled Substances Act - Section 801. Title 21 United States Code Control. Subst. Act - Sect. 801. 1970.
46. Doblin RE, Christiansen M, Jerome L, Burge B. The past and future of psychedelic science: an introduction to this issue. J. Psychoactive Drugs. 2019.
47. Nutt DJ, King LA, Nichols DE. Effects of schedule I drug laws on neuroscience research and treatment innovation. Nat Rev Neurosci. 2013;
48. National Survey on Drug Use and Health. 2018 NSDUH annual national report | CBHSQ. Annu Rep 2018.
49. Mithoefer MC, Grob CS, Brewerton TD. Novel psychopharmacological therapies for psychiatric disorders: psilocybin and MDMA. Lancet Psychiatry. 2016;
50. The psychedelic renaissance: reassessing the role of psychedelic drugs in 21st century psychiatry and society. Choice Rev Online. 2013.
51. Tagliazucchi E, Carhart-Harris R, Leech R, Nutt D, Chialvo DR. Enhanced repertoire of brain dynamical states during the psychedelic experience. Hum Brain Mapp. 2014.
52. M.W. J, A. G-R, R.R. G. Long-term follow-up of psilocybin-facilitated smoking cessation. Am J Drug Alcohol Abuse. 2017.
53. Dhhs. Human Subjects Research (45 CFR 46). Code Fed Regul. 2009.
54. Vollmann J, Winau R. Informed consent in human experimentation before the Nuremberg code. Br Med J. 1996;
55. Rothman DJ. Ethics and human experimentation. N Engl J Med. 1987;
56. Emanuel EJ, Wendler D, Grady C. What makes clinical research ethical? J Am Med Assoc. 2000.
57. Emanuel E, Abdoler E, Stunkel L. Research ethics: How to treat people who participate in research. Methodol Issues Strateg Clin Res (4th ed). 2015.

58. Belouin SJ, Henningfield JE. Psychedelics: where we are now, why we got here, what we must do. Neuropharmacology. 2018.
59. Scott G, Carhart-Harris RL. Psychedelics as a treatment for disorders of consciousness. Neurosci Conscious. 2019.
60. Johnson MW, Griffiths RR. Potential therapeutic effects of psilocybin. Neurotherapeutics. 2017.
61. Sessa B. Shaping the renaissance of psychedelic research. Lancet. 2012;
62. Schmid Y, Rickli A, Schaffner A, Duthaler U, Grouzmann E, Hysek CM, et al. Interactions between bupropion and 3,4-methylenedioxymethamphetamine in healthy subjects. J Pharmacol Exp Ther. 2015;
63. Tai YF, Hoshi R, Brignell CM, Cohen L, Brooks DJ, Curran HV, et al. Persistent nigrostriatal dopaminergic abnormalities in ex-users of MDMA (ecstasy): an 18F-dopa PET study. Neuropsychopharmacology. 2011;
64. Curran HV. Is MDMA ('Ecstasy') neurotoxic in humans? An overview of evidence and of methodological problems in research. Neuropsychobiology. 2000.
65. Mueller F, Lenz C, Steiner M, Dolder PC, Walter M, Lang UE, et al. Neuroimaging in moderate MDMA use: a systematic review. Neurosci Biobehav Rev. 2016;
66. Krebs TS, Johansen PØR. Methodological weaknesses in non-randomized studies of ecstasy (MDMA) use: a cautionary note to readers and reviewers. Neuropsychopharmacology. 2012;
67. Singer LT, Moore DG, Min MO, Goodwin J, Turner JJD, Fulton S, et al. Motor delays in MDMA (ecstasy) exposed infants persist to 2 years. Neurotoxicol Teratol. 2016;
68. Singer LT, Moore DG, Fulton S, Goodwin J, Turner JJD, Min MO, et al. Neurobehavioral outcomes of infants exposed to MDMA (ecstasy) and other recreational drugs during pregnancy. Neurotoxicol Teratol. 2012;
69. Heal DJ, Gosden J, Smith SL. Evaluating the abuse potential of psychedelic drugs as part of the safety pharmacology assessment for medical use in humans. Neuropharmacology. 2018.
70. Hopfer C, Mendelson B, Van Leeuwen JM, Kelly S, Hooks S. Club drug use among youths in treatment for substance abuse. Am J Addict [internet]. 2006;15:94–9. http://www.ncbi.nlm.nih.gov/pubmed/16449098.
71. Nour MM, Carhart-Harris RL. Psychedelics and the science of self-experience. Br J Psychiatry. 2017;
72. Turton S, Nutt DJ, Carhart-Harris RL. A qualitative report on the subjective experience of intravenous psilocybin administered in an fMRI environment. Curr Drug Abuse Rev. 2014;
73. Lerner M, Lyvers M. Values and beliefs of psychedelic drug users: a cross-cultural study. J Psychoactive Drugs. 2006;
74. Johnson MW, Richards WA, Griffiths RR. Human hallucinogen research: guidelines for safety. J Psychopharmacol. 2008;
75. Hofmann A. LSD My Problem Child: Reflections on Sacred Drugs, Mysticism, and Science. Santa Cruz, CA: Multidisciplinary Association for Psychedelic Studies; 2009.

Further Reading

Carhart-Harris RL, Goodwin GM. The therapeutic potential of psychedelic drugs: past, present, and future. Neuropsychopharmacology. 2017;
Emanuel EJ, Wendler D, Grady C. What makes clinical research ethical? J Am Med Assoc. 2000;
Johnson MW, Richards WA, Griffiths RR. Human hallucinogen research: Guidelines for safety. J Psychopharmacol. 2008;
Nutt DJ, King LA, Nichols DE. Effects of Schedule I drug laws on neuroscience research and treatment innovation. Nat Rev Neurosci. 2013;
Schatzberg AF. Some comments on psychedelic research. Am. J. Psychiatry. 2020.

Chapter 10
In the Courts: Ethical and Legal Implications of Emerging Neuroscience Technologies Used for Forensic Purposes

W. Connor Darby, Michael MacIntyre, Richard G. Cockerill, Dustin B. Stephens, Robert Weinstock, and R. Ryan Darby

Tremendous growth in neuroscience research over recent years has led to the development of exciting neuroscience technologies that improve physicians' abilities to diagnose and treat various neurological and psychiatric disorders in the treatment setting as well as enhancing physicians' abilities to stratify risk for important health outcomes to be used in both patient care and the approach to forensic evaluees. While functioning in the treatment role, physicians are able to weigh and balance the utility of these new technologies with maximizing patient welfare as their guiding duty. When the neurologist or psychiatrist enters a forensic role, however, the primary duty is no longer to the individual being evaluated (i.e., the defendant, victim, plaintiff, witness, etc.).

Thus, forensic practitioners face a much different ethics calculus when making decisions regarding the use of emerging neuroscience technologies on individuals who are not their patients. Because the use of such technologies may lead to serious

W. C. Darby (✉) · M. MacIntyre · D. B. Stephens · R. Weinstock
Department of Psychiatry and Biobehavioral Sciences, David Geffen School of Medicine, Los Angeles, CA, USA
e-mail: wdarby@mednet.ucla.edu

R. G. Cockerill
Department of Psychiatry and Behavioral Neuroscience, University of Chicago, Chicago, IL, USA
e-mail: richard.cockerill@bsd.uchicago.edu

R. R. Darby
Department of Neurology, Vanderbilt University Medical Center, Nashville, TN, USA
e-mail: Richard.r.darby@vumc.org

L. W. Roberts (ed.), *Ethics and Clinical Neuroinnovation*,
https://doi.org/10.1007/978-3-031-14339-7_10

harm or consequences for evaluees in forensic settings as opposed to being used to benefit patients in treatment settings, forensic practitioners must be more sensitive to the unique risks for using such technologies in forensic settings. Chiefly, forensic practitioners need to be aware of the potential for coercing evaluees to undergo such testing when they would otherwise refuse in medical treatment settings as well as the possibility that the application of such technologies will distort the truth of a forensic opinion to be misleading to the trier of fact.

Additionally, new artificial intelligence (AI)-powered algorithms have the potential to drastically change and improve how psychiatrists stratify individual's risk for different types of violence including aggression toward others and self-injurious or suicidal behavior. Similar to advances in neuroimaging, the potential abuses and moral calculus of utilizing this technology are dependent on the specific role of the practitioner in each situation (i.e., treatment versus forensic role). We will explore both of these emergent technologies and the relevant ethical considerations in the two following sections.

Psychiatrists and neurologists encounter a multitude of new potential ethics dilemmas when they operate outside the traditional treatment role and assume a forensic role. Forensic psychiatry is a subspecialty of psychiatry in which scientific and clinical expertise is applied in legal contexts involving civil, criminal, correctional, regulatory or legislative matters, and in specialized clinical consultations in areas such as risk assessment or employment [1].

Per the American Academy of Psychiatry and the Law Ethics Guidelines for the Practice of Forensic Psychiatry, "When psychiatrists function as experts within the legal process, they should adhere to the principle of honesty and should strive for objectivity" [2]. This entails more than being subjectively honest in that a forensic practitioner believes what they are saying is true. Moreover, it requires that forensic psychiatrists are objectively truthful in that they are competent in their stated area of expertise, strive to combat their subjective biases with objective truths, and make considerable efforts to base their opinions on as much relevant data as possible.

Forensic psychiatrists gather data from reviewing relevant medical and psychiatric records, obtaining relevant collateral information, performing psychiatric evaluations, and performing or ordering relevant testing (e.g., psychological testing, labs, neuroimaging, etc.) [1]. Psychiatrists practicing in a forensic role enhance the honesty and objectivity of their work by basing their forensic opinions on all available data, qualifying any limitations of their data, and not distorting or misrepresenting the data [2]. It is also important to have knowledge of what is generally accepted in the field and to be as up to date as possible on scientific literature and emergent technologies that aid in the profession's understanding of underlying pathophysiological processes, diagnosis, monitoring, and treatment of psychiatric disorders. Thus, understanding and communicating the limits of new technologies that hold increasing promise to aid in forensic assessments are paramount in the pursuit of being as objective and ethical as possible.

Forensic evaluees are distinct from psychiatric patients. Treating providers are ethically bound to minimize potential harms to their patients in adhering to the principle of non-maleficence. Forensic practitioners, however, must be prepared for the very real possibility that their honest and objective reports will lead to harm and consequences for the person being evaluated (e.g., in criminal trials if an expert opines that a defendant does not meet the legal criteria to be incompetent to stand trial, not guilty by reason of insanity, or incompetent to be executed). Moreover, the societal value of forensic expert witness work in assisting the adjudication of civil disputes or criminal matters requires that the findings are not influenced by considerations of whether or not it will harm or benefit the evaluee or other parties with vested interests (e.g., the defendant or defendant's family, alleged victim or victim's family, plaintiff, etc.). Psychiatrists, however, must also be guided by a respect for persons principle that underscores the importance of not coercing, misleading, or using means of deception with forensic evaluees, even if this would yield relevant and probative data to maximize their truth-telling purposes [3]. Thus, forensic psychiatrists need to balance both the pursuit of truth and the autonomy of the evaluee.

Generally, treating psychiatrists should avoid stepping into the forensic role with their patients given the possibility of conflict of interests and compromising their ability to reach the most objective opinion possible [2]. For example, if a psychiatrist has been treating a patient for schizophrenia and that patient is later arrested for a crime committed while actively psychotic, it would create ethical conflicts for the treating psychiatrist to offer a forensic opinion that her patient was legally insane at the time of the crime. This is because the treating psychiatrist would have strong biases to help her patient that would be difficult to overcome. The traditional medical duties of advancing the patient's welfare would conflict with the primary duty principle in the forensic role of being objective and fostering truth. These conflicting duties would be challenging to balance and thus better to be avoided. The potential for unconscious or even conscious bias to jeopardize objectivity increases when a psychiatrist in a forensic role aligns herself too closely to being in a treatment role guided by the traditional physician ethics principles [4]. Appelbaum's solution [3] to this problem, which in certain cases may reflect a practitioner's unconscious bias to favor evaluees, was to delineate principles distinct to forensic psychiatrists: truth-telling and respect for persons and to assert that these principles should govern a forensic psychiatrist's ethical behavior in advancing justice rather than beneficence and non-maleficence that govern a treatment psychiatrist's ethical behavior to advance the patient's health or welfare.

Nonetheless, problems exist when forensic psychiatrists divorce themselves completely from traditional medical ethics principles and do not consider various ramifications of their forensic work for their evaluees. Weinstock and Darby have developed dialectical principlism as a method to analyze difficult ethics dilemmas by weighing and balancing competing ethics considerations based on the

Table 10.1 Duties of a physician working in different roles as described by Dialectical Principlism

	Forensic role	Treatment role	Research role	Managed care role
Primary duties	Advancing justice via: 1. Truth-telling 2. Respect for persons	Advancing patient welfare via: 1. Respecting autonomy 2. Beneficence 3. Non-maleficence	Advancing scientific knowledge	Appropriate allocation of resources
Secondary duties	1. Consideration of the evaluee's welfare 2. Consideration of the retaining attorney's case 3. Consideration of societal expectations for physicians 4. Consideration of personal values	Consideration of societal welfare via: 1. Protecting vulnerable third parties 2. Distributive justice	Safety and health of the research subjects	Welfare of the patient receiving care

practitioner's role, emphasizing that the calculus changes in different roles (e.g., treatment, forensic, research, managed care, etc.) [5–7]. Ethics duties are prioritized as primary versus secondary according to the role of the psychiatrist (See Table 10.1). A psychiatrist in a treatment role will have a primary duty centered on their patients' welfare with secondary duties to public welfare, society, hospitals, allocation of resources, among others. In dialectical principlism, competing obligations are weighed and balanced in order to help each practitioner determine the most ethical action. Primary duties have special weight in the balancing process leading them to outweigh all secondary duties most of the time. But unusually strong secondary duties in relatively rare contexts can outweigh primary ones becoming determinative of our most ethical action. For example, when a patient divulges in therapy that she is abusing her child, it is ethically advisable, and generally legally required, for the psychiatrist to breach confidentiality to notify child protective services, among other protective actions. This is an example of a secondary duty (i.e., safety considerations for vulnerable populations) trumping primary duties to the patient (i.e., autonomy and non-maleficence).

In contrast to the treatment role, the forensic psychiatrist's primary duty principles are derived from Appelbaum's model: truth-telling and respect for persons. Secondary duty principles, including Beauchamp and Childress's [8] four bioethical principles that are primary in the treatment role, are considered to guide how to maximize respect for the persons being evaluated as well as in rare contexts determining whether or not to accept cases that may be antithetical to the traditional goals of medicine and societal expectations of physicians (e.g., in the extreme testifying to aggravating circumstances in a capital case to assist the prosecution in obtaining a death sentence as opposed to life in prison without the possibility of parole).

Neuroimaging

Use of Neuroimaging in Forensic Settings

Progress in neuroimaging provides new tools for understanding normal human behavior and for diagnosing neuropsychiatric disorders that impair human behavior. In addition to scientific and medical applications, neuroimaging has increasingly been used in legal settings [9]. Structural brain scans using magnetic resonance imaging (MRI) and computed tomography (CT) are regularly accepted as evidence in courts across the United States [10]. Other advanced imaging modalities, including positron emission tomography (PET), single-photo emission computed tomography (SPECT), diffusion tensor imaging (DTI), and quantitative electroencephalography (qEEG), have all been admitted to courts as well [11].

Neuroimaging has three potential uses in legal settings. First, it can be used to support the clinical diagnosis of a defendant accused of a crime. For many neuropsychiatric disorders, including stroke, brain tumor, dementia, and multiple sclerosis, neuroimaging findings are a major component of the diagnostic criteria. In other disorders, including schizophrenia and concussion, neuroimaging differences may be present, but are not considered part of the diagnostic criteria. Therefore, neuroimaging evidence may support the diagnosis in some, but not all, neuropsychiatric diseases. In no instances is neuroimaging evidence sufficient to support a clinical diagnosis in the absence of corresponding clinical symptoms and/or neuropsychiatric examination findings. It must be further demonstrated that the neuropsychiatric disorder resulted in relevant behavioral impairment that diminish responsibility for a crime. Therefore, neuroimaging in this context may support the clinical diagnosis but is not sufficient to make a forensic determination.

Second, neuroimaging can provide mechanistic support for claims that a defendant has impaired behavioral capacities that diminish responsibility for a criminal act. This requires scientific evidence supporting the neuroanatomical localization of specific behavioral capacities to specific brain regions. It also requires establishing a temporal link between the estimated onset of the neurological injury and the onset of relevant behavioral changes in the defendant. This temporal link is particularly challenging in progressive disorders like dementia and multiple sclerosis, or in fluctuating disorders like psychosis or epilepsy. Because neuroimaging is often obtained far after the actual crime is committed, interpreting neuroimaging data in the context of temporal causality is a major limitation. A critical distinction must be made between neuroimaging findings at the time of testing associated with impaired behavioral capacities and the mental state specifically at the time of a crime. Evidence demonstrating impaired behavioral capacities can be used to infer the mental state of an individual at the time of a criminal act, but this inference is indirect. Finally, functional neuroimaging has the added complication of accounting for state-dependent effects. Sleep deprivation, caffeine use, effort, and psychiatric disorders related to the crime such as PTSD could result in functional neuroimaging differences distinct from changes that might be expected at the time of the crime.

The third use of neuroimaging is to infer the mental state of an individual at the time the imaging is actually performed. In this context, it has been proposed that neuroimaging might be used for lie detection, to determine the validity of eyewitness testimony, or measure implicit biases in witnesses, judges, or jurors.

In forensic settings, neuroimaging is typically obtained after significant time has passed from the incident being questioned. This limits the ability to draw strong conclusions between one's current brain scan and prior behavior. Because of this, some legal scholars have argued that brain imaging has limited application to determining criminal intent [12] and cannot answer legal questions of causation, criminal responsibility, or predicting future behavior [13]. Without the ability to make direct causal inferences, neuroimaging becomes less useful to the court, as the Eighth Circuit Court of Appeals succinctly explained in *Forrest v. Steele* (764 F.3d 848, 2014): "Generally speaking, a PET scan can reveal diminished energy usage in particular areas of the brain, thereby signifying damage. However, it cannot show the cause of damage, nor can it demonstrate the existence of diminished capacity, predict future behavior, or establish a person's state of mind." Although such evidence cannot alone determine the state of mind at the time of the crime or criminal responsibility, it sometimes can provide supportive evidence of an altered state of mind that may well be relevant for criminal responsibility.

Additionally, a close temporal relationship between a documented behavioral syndrome and neuroimaging changes, in the context of a clinically diagnosed neuropsychiatric disorder, strengthens the causal argument that a brain disease affecting behavior contributed to a criminal act. While no individual piece of evidence can definitively determine a defendant's mental state at the time of a crime, neuroimaging data can improve this causal inference by providing convergent evidence [14–16]. These indirect inferences represent the limited practical means of assisting the court to make such determinations. Assessment of mental state at the time of a crime is the essence of what is required in any psychiatric defense. Therefore, it is necessary to qualify the limitations of neuroimaging and not overstate its probative value in the assessment of mental state.

Validity and State of the Science

To prevent distorting the truth, anyone using neuroimaging in a forensic setting must be aware of the limitations of current science. Use of neuroimaging in court has to satisfy either the *Frye* or *Daubert* standards for admissibility of evidence (see Table 10.2). *Frye v. United States (293 F. 1013 (D.C. Cir. 1923))* requires the scientific evidence be "generally accepted" by the relevant scientific community, while *Daubert v. Merrel Dow Pharmaceuticals, Inc. (43 F.3d 1311 (ninth Cir. 1995))* provides five illustrative factors to guide a judge's decision to admit scientific evidence:

Table 10.2 A comparison of *Frye* and *Daubert* standards for admissibility of expert testimony

	Frye standard	*Daubert* standard
Case	*Frye v. United States* (293 F. 1013 (D.C. Cir. 1932))	*Daubert v. Merrel Dow pharmaceuticals* (43 F.3d 1311 (ninth Cir. 1995))
Questioned evidence in original case	Proposed systolic blood pressure deception test	"In vitro" and "in vivo" animal studies showing a drug may cause birth defects
Who decides admissibility?	Trial judge	Trial judge
Criteria to consider when admitting evidence	Evidence must be "sufficiently established to have gained general acceptance in the particular field in which it belongs"	1. Theory or technique is generally acceptable in scientific community 2. Evidence is peer-reviewed 3. Evidence is testable 4. Known or accepted error rates are acceptable 5. Research is independent from the specific legal case in which it is being used
States adopting the standard[a]	CA, IL, MN, NJ, NY, PA, WA	AL, AK, AZ, AR, CO, CT, DC, DE, FL, GA, HI, ID, IN, IA, KS, KY, LA ME, MD, MA, MI, MO, MS, MT, NC, NE, NH, NM, OH, OK, OR, RI, SC, SD, TN, TX, UT, VT, WV, WI,WY

[a]Warren R. Trazenfeld & Robert M. Jarvis, Daubert/Kumho Tire and the Legal Malpractice Expert Witness, 12 ST. MARY'S J. ON LEGAL MALPRACTICE & ETHICS 372 (2022). Available at: https://commons.stmarytx.edu/lmej/vol12/iss2/5

(1) whether the theory or technique is generally acceptable in the scientific community, (2) whether it is peer-reviewed, (3) whether it is testable, (4) whether the known or expected error rates are acceptable, and (5) whether it the research is independent from the specific litigation at hand. Furthermore, neuroimaging has been described as having a "methodological crisis" due to limited reproducibility across studies [17]. Limitations may result from small, insufficiently representative sample or non-specific findings [18]. Additionally, differences in computer software processing and statistical analysis can lead to unreliable results, even when using similar data [19, 20]. Finally, given the brain's complex organization into connected networks, certain clinical diseases or symptoms may localize better to a network than a specific brain region, leading to further inconsistency across studies [21, 22]. It is therefore important to use results that have been replicated, or to understand the reasons for a lack of replication, when using neuroimaging in forensic contexts.

Neuroimaging studies typically average differences in behavior and brain activity over multiple subjects and trials. When using such evidence, courts attempt to take this group data and apply it to individual cases, an issue termed "Group to Individual (G2i) inferences" [23]. Group data may provide a likelihood that a person's behavior is related to a brain injury but cannot be directly applied to any individual case with reasonable certainty. Moreover, some imaging studies look at specific populations, restricting generalizability of the results.

Neuroimaging in single subjects must address the two questions: (1) What is the validity of a neuroimaging abnormality detected in that subject; and (2) What is the likelihood that the neuroanatomical location of this abnormality relates to a specific behavioral change? Certain brain abnormalities have a very high likelihood of being a true abnormality, such as strokes or tumors. In such cases, the validity of a neuroimaging abnormality is not in question. In other instances, however, the validity of single-subject neuroimaging abnormalities is less clear. For example, voxel-based morphometry (VBM) and cortical thickness can be used to measure brain atrophy in single subjects by comparing patient MRIs to normal subjects without neurological or psychiatric diseases [24–26]. However, these approaches may have unexpectedly high false positive rates (i.e., suggesting brain damage in normal persons) depending on data analysis methods. Other authors have noted the limitations of a single-subject functional MRI to uncover evidence of behavioral aberration [27]. Despite these limitations, however, quantitative approaches to detect single-subject neuroimaging abnormalities are advantageous over unaided clinician interpretation of images in forensic settings, which is subject to observer bias. It has been shown that radiologists are more likely to detect a lesion if they have knowledge of a clinical abnormality; this would be expected to be highly prevalent in a courtroom, where expert testimony on imaging is required only after inappropriate behavior has occurred [28].

If evidence of a true neuroimaging abnormality is accepted, the next question is the likelihood that a neuroimaging abnormality is related to a specific behavioral change. A common overstatement of research occurs when one suggests the presence of an altered mental state based solely on the presence of abnormal brain imaging, a logic error known as reverse-inferencing [14]. To point, a large study found an atypical incidental finding in over 10% of asymptomatic patients receiving an MRI, suggesting that many neuroimaging abnormalities do not lead to significant behavioral change [29].

One study systematically studied the relationship between focal brain lesions and antisocial behavioral changes, including criminal behavior [30]. In 17 cases where a clear temporal association between lesion onset and behavioral change could be established, lesions occurred in several different locations, and no single brain region was affected in all cases. Because clinical symptoms can arise from other locations connected to a brain lesion and not only from the lesion itself, the authors used a new method called lesion network mapping to identify regions functionally connected to each specific lesion [30–34]. Using this approach, the authors found that all lesions were connected to the same common brain network [30]. Moreover, connectivity to this network was highly specific, as lesions that did not cause criminal behavior were not connected to this network [30]. This finding was replicated in a second group of 23 patients where lesions were suspected to have resulted in antisocial behavior including criminal behavior, but the temporal relationship between lesion and behavioral change was less clear [30]. Finally, the identified network associated with lesion-induced criminal behavior was shown to be involved in moral and value-based decision-making, cognitive processes associated with antisocial behavior [30].

Comparison to lesions not causing criminal behavior demonstrates that lesions outside of this network are less likely to result in an acquired antisocial behavior disorder. A similar approach was used to show that incidental lesions found in delusional patients with known psychiatric disorders causing psychosis were unlikely to be causal because they occurred outside of an identified delusions network [35]. However, the study did not include a group of patients with lesions occurring within the identified antisocial behavior network who did not go on to develop antisocial behaviors [30]. Thus, the likelihood that a lesion within this region will cause an acquired behavioral disorder is unknown.

An additional limitation of the above study is that it focused on focal brain lesions, in which determining an abnormality present is straightforward, but secondary effects on interconnected neural networks is less clear. Methods to quantitatively estimate the effect of brain atrophy on connected networks have also been developed, with important caveats regarding validity. Using atrophy network mapping, an approach similar to lesion network mapping, single-subject atrophy maps in Alzheimer's disease patients were connected to the same symptom-specific networks for delusions and memory as in patients with focal brain lesions [26]. This finding suggests that network mapping is a promising approach to determine brain–behavior relationships across different neuropsychiatric diseases with the same clinical symptoms. This approach has not yet been used to test whether locations of brain atrophy in patients with acquired antisocial behavior disorder, such as those with frontotemporal dementia, occur in regions connected to the same network identified in lesion-induced antisocial behavior [30, 36]. While the study by Darby and colleagues provides an important step towards a scientific basis for determining the likelihood that a neuroimaging abnormality is related to acquired behavioral abnormalities, people should be aware of the limitations before use in a forensic setting.

Data obtained from group imaging studies instead of single subjects can be useful to the court. In *United States v. Smith* (621 F. Supp. 2d 1207, 2009) the court said that educating the jury about research leads to a more accurate and fair legal proceeding, although "applying this research to the facts of the case is within the sole province of the jury" not the expert witness (see Box 10.1). Another practical use involves educating courts on group differences, potentially informing policy decisions and legal conclusions. Others have argued that group data should not play a major role since it is hard to draw specific conclusions, and findings are useful only insofar as they support other relevant data. Accordingly, although not a major part of most decisions, group neuroimaging data has been referenced in many important cases. For example, in *Roper v. Simmons* (543 US 551, 2005), *Graham v. Florida* (560 US 48, 2010), and *Miller v. Alabama* (567 US 460, 2012), the Supreme Court mentions group imaging data comparing the brains of adolescents and adults to support other arguments in making constitutional rules prohibiting capital punishment and life imprisonment without parole of minors.

Box 10.1 A Court Decides What an Expert May Say
United States v. Smith (621 F. Supp. 2d 1207, 2009)

Courts have long relied on eyewitness testimony to help uncover the facts of a particular case. Attorneys have attempted to use experts witnesses to discredit the credibility of eyewitnesses. Courts allow experts to educate the jury on issues that affect eyewitness testimony, such as the limitations of cross-racial identification or the effect of stress on the accuracy of a memory, but not on the actual credibility of the witness which is the ultimate issue. In *United States v. Smith,* Mr. Smith was arrested for bank robbery and eyewitness testimony was important evidence in the case. The defense hired Dr. Fulero, an expert on the science of eyewitness-identifications, to provide testimony. The court allowed Dr. Fulero to give his opinion about the science of eyewitness-identifications, but he was not allowed to testify about specific witnesses in the case. The court reasoned Dr. Fulero could educate the jury but that applying the research to the specific facts of the case was the "sole providence of the jury."

Ethical Issues in the Forensic Use of Neuroimaging

As previously stated, the ethical considerations of using neuroimaging shift significantly when used in a forensic setting rather than a clinical setting. In the legal system, imaging is not used to benefit individual patients, but rather to help the court answer questions about issues like culpability, liability, intentionality, truth, and punishment. A court may look to neuroimaging to help understand a number of questions: What was a defendant's mental state at the time of his or her act? Is a defendant lying? How accurate is a witness's memory? How biased is a witness [37]?

Weisenberg and colleagues note that neuroscience may have a "seductive allure" to provide explanations for behavior and personal responsibility not fully supported by current science [38]. The presence of neuroimaging without any additional information has been found to make scientific claims more convincing [39], though it has been argued that there is not enough empirical evidence to show neuroimages significantly bias perceptions of scientific validity [40]. Although neuroimaging has significant potential value in informing the diagnostic process, how that aids the legal system remains controversial. When testifying against the use of neuroimaging in court, prominent neurologist Helen Mayberg has claimed, "It is a dangerous distortion of science that sets dangerous precedents for the field" [41].

Use of Neuroimaging and Individual Autonomy

Advanced techniques, such as fMRI, DTI, perfusion imaging, PET, and SPECT, are currently utilized in limited settings when there is sufficient evidence of potential benefit to the patient [42]. However, in the court room, neuroimaging evidence has greater potential to harm, and the ethical considerations are very different. Courts must consider autonomy of the individual. It is unsettled if courts may compel neuroimaging or if a defendant's consent is required. There may be a temptation for the court or jury to judge a person based on his or her brain image and not the individual's behavior. The implications of finding structural brain defects also present ethical considerations. For instance, a person may not want to know if they have a structural brain abnormality; in addition, any neurological findings may have genetic implications for children or siblings that must be considered. Without safeguards in place, an imaging abnormality found in a defendant during a criminal trial could be later used to argue that person is not fit for their chosen career. These issues grow further complicated if neuroimaging is used for someone other than a defendant, such as imaging potential jury members to assess for bias or scanning a witness to detect lying.

Many concerns have been raised about how neuroimaging can infringe upon basic constitutional rights. If imaging progressed sufficiently to be able to share a subject's personal knowledge or beliefs, some argue this infringes on individual privacy [43]. Others argue that neuroimaging presented by the opposing side in a court case could be a violation of search and seizure protections [44]. As an example, research has been performed on the utility of functional MRI (fMRI) for lie detection. If used on a defendant to detect guilt, such practice would have significant implications with regard to an individual's Fifth Amendment right against self-incrimination [45]. The Supreme Court in *Schmerber v. California* (384 US 757, 1966) stated that lie detector tests may essentially be eliciting testimony and that "to compel a person to submit to testing in which an effort will be made to determine his guilt or innocence on the basis of physiological responses... is to evoke the spirit and history of the Fifth Amendment." The Supreme Court has not yet commented specifically on the use of fMRI.

fMRI and Lie Detection

Attempts to use fMRI studies to detect deception illustrate the limitations of neuroimaging and the importance of not overstating conclusions. In *United States v. Semrau* (693 F.3d 510, 2012), the court did not admit fMRI data as evidence of deception. The court found that "the error rate of real-life fMRI-based lie detection is unknown," and that no standards exist for how imaging should be obtained. Additionally, existing studies of fMRI and deception did not include subjects as old as the defendant in this case.

While certain regions of the brain have been associated with deception, these regions highly overlap with areas of the brain involved in executive control [46].

fMRI studies do not assess deception specifically, but rather the act of following an instruction to lie. This requires multiple tasks of executive control and may involve neural networks distinct from deception and lying. Furthermore, fMRI has not been able to distinguish the impact that incorrect memory may have on lying results [47]. For instance, one study found that fMRI brain activity is similar when a person recognizes a face and when a person simply believes she recognizes a face [48]. Should someone be punished for lying during an fMRI study if they simply remembered events incorrectly?

Impact on Judge and Jury

It is unclear how presenting neuroimaging will affect the judgment of an individual court. Evidence of a structural neurological cause of behavior may be interpreted by a judge as mitigating or as aggravating [49]. Due to this "double-edge" nature of neuroimaging, this could lead to a lower sentence because of reduced culpability or an increased sentence due to need for incapacitation and public safety if the condition does not have a treatment intervention.

Neuroimaging may inappropriately impact a jury, typically made up of people with minimal scientific background. In one study of not guilty by reason of insanity, neuroimaging evidence did not significantly influence mock jurors, but jurors not provided neuroimaging data believed it would have been the most helpful additional information [50]. Because brain images are visual evidence often presenting with strong, colorful impact, some have argued that they may be prejudicial or seem overly important to juries [51, 52]. In addition, the scientific implications of neuroimaging may be confusing. For example, color-coded DTI fiber-tracking maps may lead a jury to assume they are pictures of actual brain connections [53]. Furthermore, advanced images undergo computer processing and changing various parameters, and statistical thresholds can provide a different image that may be more compelling for one side's legal argument—a process cynically coined "dial-a-defect" [52]. Finally, if the science seems too complex, jurors will ignore potentially relevant information [50]. Ultimately, there is a balance between trying to explain science objectively while explaining it in terms a jury can understand.

Formal Guidelines for the Forensic Use of Neuroimaging Evidence

Given the nuance and complexity of neuroimaging and human behavior, Scarpazza et al. [14] proposed four rules for using neuroimaging in the court:

1. Neuroimaging results should be coupled with behavioral findings.
2. The criminal behavior cannot be considered a symptom.

Table 10.3 Use and abuse of neuroimaging in the courtroom

1.	Experts should present all relevant facts available in their testimony, ensure truthfulness and balance, and consider opposing points of view.
2.	Experts should specify known deviations from standard practice.
3.	Experts should have substantive knowledge and experience in the area in which they are testifying.
4.	Experts should use standard terminology and describe standardization methods and the cohort characteristic from which claims are determined, when applicable.
5.	Nonvalidated findings that are used to inform clinical pathology should be approached with great caution.
6.	Recognized appropriateness guidelines should be used to assess whether the imaging technique used is appropriate for the particular question.
7.	Experts should avoid drawing conclusions about specific behaviors based on the imaging data alone.
8.	Experts should be willing to submit their testimony for peer review.
9.	Experts should be prepared to provide a description of the nature of the neuroimages (e.g., representational/statistical maps when derived from computational postprocessing of several images) and how they were acquired.
10.	Raw images and raw data should be made available for replication if requested.
11.	Experts should be able to explain the reasoning behind their conclusions.
12.	False positive rates should be known and considered if the expert's testimony includes quantitative imaging.
13.	Experts should be prepared to discuss limitations of the technology and provide both confirming research and disconfirming studies.

Proposed Standards for Neuroradiology Imaging Testimony (From Meltzer et al., 2014, pg. 635)

3. Not every brain abnormality leads to behavioral symptoms.
4. Do not reason backwards.

Practical guidelines include always providing a descriptive diagnosis of any evaluee, clearly assessing causal links between symptoms and a crime, clearly describing how neuroimaging highlights a significant result, and using brain imaging only to assess anatomical-clinical correlations [14]. A multi-disciplinary expert conference, The Use and Abuse of Neuroimaging in the Courtroom, created guidelines for the appropriate use of neuroimaging in expert testimony (See Table 10.3) [42]. These should be reviewed by anyone planning to use neuroimaging in a forensic setting.

Artificial Intelligence in Medicine and Forensic Psychiatry

At the forefront of today's emerging technologies in medicine is artificial intelligence, commonly known as AI. Rapid advancement in the theoretical field, coupled with a massive increase in the amount of computing power available to researchers, has allowed AI-powered algorithms to tackle problems previously thought far too complex for machines. An illustrative example of such a problem is the interpretation of screening mammography. In January 2020, a large multi-institutional research team published a major paper describing an AI algorithm that consistently outperformed six experienced radiologists in the US and the UK in evaluating screening mammograms, resulting in significant improvements in test specificity and sensitivity over the human radiologists [54]. Importantly, this algorithm was based on a concept known as "deep learning."

Briefly, deep learning is a type of machine learning that utilizes artificial neural networks. A full discussion of this is beyond the scope of this chapter, suffice to say that deep learning, as compared to other methods of training AI algorithms, typically involves little human supervision. For example, an algorithm may be given a task, such as "identify rabbits," and then given reams of images, some of which contain rabbits. The algorithm receives feedback only based on whether its output is correct or false. Thus, it constructs, for itself, what features make a rabbit. Or, in the previously mentioned case, what features make cancer look like cancer on a mammogram. Critically, *how* the algorithm got from point A to point B often cannot be completely understood even by a trained computer scientist. This fact has major implications for physicians who work or will work with such algorithms (Fig. 10.1).

The strength of deep learning is solving complex problems with myriad contributing variables. Forensic psychiatry is rife with these kinds of problems. Two of the most important of these are assessments of suicide and violence risk, both of which actions are notoriously difficult to predict. Both actions are relatively rare events with numerous potentially relevant risk factors, which can be difficult to quantify. A 2016 meta-analysis found that traditional suicide risk assessments were only slightly better than chance at predicting future suicides among psychiatric patients [55]. Traditional risk assessment tools in the assessment of violence risk, like the Violence Risk Appraisal Guide (VRAG), have at best demonstrated similarly modest predictive value [56]. The modest predictive value of these current tools does not reflect that the risk assessment is inaccurate since whether or not there is a suicidal or violent act may depend on the presence or absence of intervening events that may or may not happen. But it is clear that there is significant room and need for improvement in forensic psychiatry to develop better tools to assess violence and suicide risk.

Fig. 10.1 Types of AI

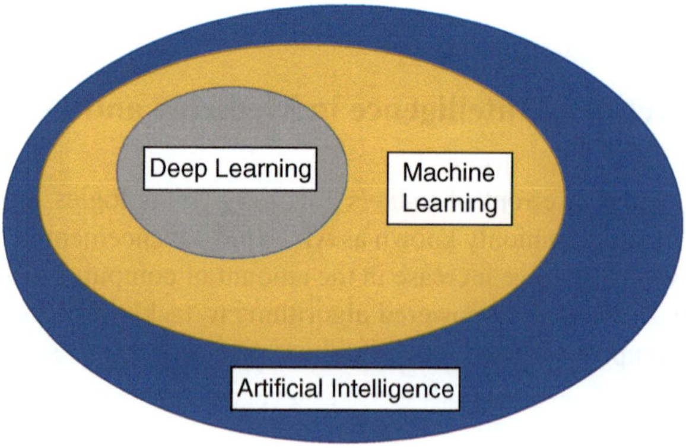

In this regard, AI may have a key role to play. In a prospective 2020 study, a deep learning-based algorithm was trained to assess suicide risk on a large population of patients in a major U.S. health system. The algorithm risk stratified patients into four risk groups, from "low" to "very high" risk. Those in the "very high" risk group had a relative risk of suicide of 59.02 when compared to the lowest risk group [57]. Controlled studies are certainly needed to properly compare different tools, but, if replicated, this would certainly represent a dramatic improvement over traditional assessments, which were found to have a pooled odds ratio of 4.84 in the aforementioned meta-analysis [56].

AI and deep learning are also being applied in a similar manner to violence risk assessment. A deep learning algorithm was recently developed to use retrospective clinical data, including nursing and physician notes, to predict future violent behavior in psychiatric inpatients. The area under the curve (AUC) for the performance of this algorithm in two different hospital settings was 0.80 and 0.76 [57]. In this context, AUC is a commonly used measure of the accuracy of a diagnostic test which plots the rate of false positive tests against the rate of true positives, then measures the area under the curve. Values over 0.7 are generally considered acceptable, values over 0.8 are considered good, and values over 0.9 are considered excellent. In comparison, the widely used Violence Risk Appraisal Guide (VRAG) was found to have an AUC of 0.72 in pooled data [56]. Because AI algorithms are capable of continuous self-improvement, it is not difficult to imagine future algorithms that are consistently superior to existing tools.

Ethical Issues

The rapidly increasing power of deep learning-based AI algorithms presents both enormous opportunity and carries significant risks for the field of forensic psychiatry. The ethics concerns raised by this technology fall under three main categories, which are explored below and approximate the bioethical principles developed by Beauchamp and Childress [8].

Protecting Autonomy

Informed consent: The ability to provide informed consent is critical to protecting any forensic evaluee's autonomy. For this to be possible, the subject in question must have an adequate understanding of the assessment they are agreeing to and be free of undue influence or coercion. With new technologies like AI, the knowledge gap between practitioner and subject may be even larger than in more typical clinical or forensic situations. Criminal defendants may also be under legal orders to undergo psychiatric evaluations or may believe declining to participate in any form of testing may negatively affect their legal outcomes. Thus, it is critical that clear and concise educational tools be developed for evaluees who might undergo

AI-assisted assessments. Whenever possible, evaluees should also be offered alternatives to such assessments if they are unable to provide adequately informed consent.

Loss of liberty: Laws governing civil detention of persons vary considerably across the United States. However, the mental health practitioner, most often a psychiatrist, is always central to such detainments. This practitioner is responsible for the decision to temporarily deny an individual his civil liberties and in most cases can be held responsible if the decision was made improperly. In a world where algorithms are center stage in such decisions, who, or what, can be held responsible if the decision is made incorrectly? Certainly, as AI algorithms begin to gain a foothold in forensic psychiatry, the outcomes will be subject to final review by a psychiatrist. However, it seems likely that as AI algorithms continue to improve and more efficiently manage higher volumes of forensic evaluations, there may come a time when such review is impracticable. In such a world, mechanisms for appealing assessments made by AI algorithms must be made understandable, accessible, and transparent.

Data privacy: Major concerns involving data privacy are raised by the use of AI algorithms in forensic psychiatry. Algorithms improve when they are exposed to higher quality and quantities of data. Location, biometric, search, and messaging data have all been proposed as inputs for AI algorithms. Some of these have already been used in Facebook's suicide prevention algorithm [58] The company has declined to publish details about the algorithm or the data generated from it. The "user agreements" millions signed when they joined Facebook, Twitter, or other social media do not constitute adequate informed consent. If such data are to be used to inform algorithms, those persons providing the data should be educated on how and why it is being used, including describing possible harms which may result. Consumers must be provided with accessible, convenient ways to "opt-out" of such programs.

Beneficence and Non-Maleficence

Balancing preventing tragedy with limiting false positives: Civil or criminal detention of individuals based on *future* violence or suicide risk is done to protect society from rare but catastrophic events, such as homicide. This practice is not new. Clinical psychiatrists routinely hold patients involuntarily in hospitals based on violence risk assessments. Judges give harsher sentences to those defendants thought to be at highest risk for future violence. No doubt, many of these individuals may not have committed a violent act had they been released earlier. But society has decided that it is worth some "false positives" in the form of needlessly prolonged detention to protect us from catastrophic harms.

How might AI change this calculus? As noted in the previous section, psychiatrists perform only modestly better than chance in assessing violence risk. It is conceivable that, in time, AI algorithms will significantly outperform psychiatrists in this arena. If false positives and false negatives decline, should violence risk assessments be more widely used? Should more people be detained based on the results of these assessments? It will be critically important for policymakers to carefully consider these questions as the technology continues to advance. Each detention, even those involving individuals correctly deemed to be a highest risk for violent behavior, carries consequences for the individual detained as well as his family, his friends, and for the broader community, and for this reason such technology should be used with utmost caution.

Overruling algorithms: AI algorithms continually improve and refine themselves based on new data. In fact, it is not inconceivable that they will someday outperform human practitioners in suicide or violence risk assessments. If this is clearly the case, the role of human reviewers of such algorithms would require reexamination. Consider again the previously mentioned breast cancer screening algorithm. Imagine a scenario in which the algorithm, which is already performing better than many practicing radiologists, identifies an atypical sign on X-ray that it determines to be suspicious for neoplasm. The human radiologist reviews the image and decides this was simply "machine error," recommending against biopsy. Six months later the same patient returns with advanced breast cancer. It is easy to imagine a similar scenario arising for a psychiatrist who overrules a suicide or violence risk assessment algorithm. If the patient later commits a violent act, where does the blame lie? If algorithms consistently and clearly outperform practitioners in the future, it may be incumbent upon those practitioners to reassess and redefine their roles in the context of rapidly evolving AI technology.

Fostering Justice

Algorithms may propagate racial inequity in the legal system: Algorithms are increasingly being used in criminal justice, from predictive mapping software, which helps police allocate resources to high-risk neighborhoods or individuals, to recidivism risk assessment tools used to aid judges in sentencing. A major criticism of these tools lies in the fact that they are only as good as their input data. Thus, if police reports, probation officer documentation, policing practices, and an individual's conviction history all are subject to pre-existing biases, and the underlying algorithms being utilized rely on these data to generate assessments, then those assessments will further propagate such systemic biases [59]. The sheen of objectivity offered by algorithms may disguise these latent biases from casual observers.

Using algorithms to correct bias: Conversely, algorithms may play a constructive role in addressing and correcting for systemic racial biases in the criminal justice system. A well-developed AI algorithm could conceivably detect subtle biases in criminal justice data and adjust for them. This may have the effect of reducing biases in the system from policing to sentencing [60]. For example, an AI tool for detecting racial or gender-based bias is being developed by a research group at Columbia and Penn State Universities. This program can generate hypotheses, like the predicted salary for an individual at a given institution, using a multitude of predictive factors. It then references predicted salary against actual salary. The difference may be explained by racial or gender bias [61]. The power of AI algorithms lies in their ability to analyze enormous amounts of data to make future predictions, but they are only as good as the inputted data used to drive such algorithms such as deep learning. For this reason, input data must be carefully curated and selected to avoid untoward future impact of rapidly evolving technologies such as artificial intelligence in the forensic setting.

Overall, AI and other technologies have the potential to have a transformative impact on the field of forensic psychiatry. However, the power of such technology also presents significant risks. As the field rapidly evolves, it will be critically important to identify and analyze the ethical implications of the use of this new technology.

Key Points

1. In a legal setting, neuroimaging may support a clinical diagnosis, provide a neural mechanism for claims of impaired behavioral capacities, or potentially elucidate the mental state of an individual at the time imaging is performed (not necessarily elucidating the mental state of an individual at the time of a crime).
2. Neuroimaging results must be presented and interpreted together with relevant behavior and neuropsychiatric symptoms.
3. Neuroimaging studies are based on group data and alone do not provide definitive conclusions about an individual's mental status.
4. Expert witnesses utilizing neuroimaging studies must remain up to date on the current state of the science, validity of different modalities, and limitations; they should not overstate the significance of their observations and make efforts to qualify their opinions understanding the potential for this type of testimony to be given more credence than warranted.
5. Artificial intelligence and deep learning are rapidly transforming the practice of forensic psychiatry.
6. The use of these novel tools raises significant ethics concerns.
7. Expert witnesses asked to interpret assessments by AI algorithms should have a basic understanding of both the technology itself and associated ethics concerns.

Questions to Consider

1. What would be necessary in a neuroimaging study to inform a court of an individual's thoughts, intents, morality, or free-will?
2. How would you present an explanation of a neuroimaging study, including its limitations, while ensuring the explanation would be understood by non-scientists?
3. For which legal cases is neuroimaging most helpful? Least helpful?
4. What is deep learning? How is it related to artificial intelligence?
5. Are there cases in which the products of AI algorithms should be ignored by an expert witness? What ethical issues are raised by refusing to use this technology?
6. AI algorithms are also being applied to suicide risk assessment (see Linthicum et al. on the Additional Reading list). What ethical issues apply to both uses of this technology? Which issues are unique?

References

1. Glancy GD, Ash P, Bath EP, Buchanan A, Fedoroff P, Frierson RL, et al. AAPL practice guideline for the forensic assessment. J Am Acad Psychiatry Law. 2015;43(2 Suppl):S3–53.
2. AAPL. Ethics Guidelines for the Practice of Forensic Psychiatry; 2005. https://aapl.org/ethics.htm.
3. Appelbaum PS. A theory of ethics for forensic psychiatry. J Am Acad Psychiatry Law. 1997;25(3):233–47.
4. Darby WC, Weinstock R. Facing the elephant in the evaluation room: confronting bias and striving for objectivity (part I). Am Acad Psychiatry Law Newslett. 2020;45(1):21–6.
5. Darby WC, Weinstock R. Facing the elephant in the evaluation room: confronting bias and striving for objectivity (part II). Am Acad Psychiatry Law Newslett. 2020;45(2). Pages 16, 20, and 22.
6. Darby WC, Weinstock R. Resolving ethics dilemmas in forensic practice. In: Griffith E, editor. Ethics dilemmas in forensic psychiatry and psychology practice. New York, NY: Columbia University Press; 2018. p. 7–22.
7. Weinstock R. Dialectical principlism: an approach to finding the most ethical action. J Am Acad Psychiatry Law. 2015;43(1):10–20.
8. Beauchamp TL, Childress JF. Principles of biomedical ethics. 7th ed. New York, NY: Oxford University Press; 2013.
9. Gaudet LM, Marchant G. Under the radar: neuroimaging evidence in the criminal courtroom. Drake Law Rev; 2016.
10. Moriarty JC. Flickering admissibility: neuroimaging evidence in the U.S. courts. Behav Sci Law. 2008;26(1):29–49.
11. Farahany NA. Neuroscience and behavioral genetics in US criminal law: an empirical analysis. J Law Biosci. 2015;2(3):485–509.
12. Morse SJ. The future of neuroscientific evidence. In: Henderson C, Epstein J, editors. The future of evidence: how science and technology will change the practice of law. Chicago, IL: American Bar Association; 2011. p. 137–63.
13. Morse SJ. Brain imaging in the courtroom: the quest for legal relevance. AJOB Neurosci. 2016;5(2):24–7.
14. Scarpazza C, Ferracuti S, Miolla A, Sartori G. The charm of structural neuroimaging in insanity evaluations: guidelines to avoid misinterpretation of the findings. Transl Psychiatry. 2018;8(1):227.
15. Darby RR. Neuroimaging abnormalities in neurological patients with criminal behavior. Curr Neurol Neurosci Rep. 2018;18(8):47.
16. Darby RR, Edersheim J, Price BH. What patients with behavioral-variant frontotemporal dementia can teach us about moral responsibility. AJOB Neurosci. 2016;7(4):193–201.
17. Kellmeyer P. Ethical and legal implications of the methodological crisis in neuroimaging. Camb Q Healthc Ethics. 2017;26(4):530–54.
18. Moriarty JC, Langleben DD, Provenzale JM. Brain trauma, PET scans and forensic complexity. Behav Sci Law. 2013;31(6):702–20.
19. Poldrack RA, Baker CI, Durnez J, Gorgolewski KJ, Matthews PM, Munafo MR, et al. Scanning the horizon: towards transparent and reproducible neuroimaging research. Nat Rev Neurosci. 2017;18(2):115–26.
20. Eklund A, Nichols TE, Knutsson H. Cluster failure: why fMRI inferences for spatial extent have inflated false-positive rates. Proc Natl Acad Sci U S A. 2016;113(28):7900–5.
21. Darby RR, Joutsa J, Fox MD. Network localization of heterogeneous neuroimaging findings. Brain. 2019;142(1):70–9.
22. Darby RR, Fox MD. Reply: heterogeneous neuroimaging findings, damage propagation and connectivity: an integrative view. Brain. 2019;142(5):e18.
23. Faigman DL, Monahan J, Slobogin C. Group to individual (G2i) inference in scientific expert testimony. Univ Chicago Law Rev. 2014;81(2):417–80.

24. Scarpazza C, Sartori G, De Simone MS, Mechelli A. When the single matters more than the group: very high false positive rates in single case voxel based morphometry. NeuroImage. 2013;70:175–88.
25. Scarpazza C, Nichols TE, Seramondi D, Maumet C, Sartori G, Mechelli A. When the single matters more than the group (II): addressing the problem of high false positive rates in single case voxel based morphometry using non-parametric statistics. Front Neurosci. 2016;10:6.
26. Tetreault AM, Phan T, Orlando D, Lyu I, Kang H, Landman B, et al. Network localization of clinical, cognitive, and neuropsychiatric symptoms in Alzheimer's disease. Brain. 2020;143(4):1249–60.
27. Ganis G, Rosenfeld JP, Meixner J, Kievit RA, Schendan HE. Lying in the scanner: covert countermeasures disrupt deception detection by functional magnetic resonance imaging. NeuroImage. 2011;55(1):312–9.
28. Hugh TB, Dekker SW. Hindsight bias and outcome bias in the social construction of medical negligence: a review. J Law Med. 2009;16(5):846–57.
29. Vernooij MW, Ikram MA, Tanghe HL, Vincent AJ, Hofman A, Krestin GP, et al. Incidental findings on brain MRI in the general population. N Engl J Med. 2007;357(18):1821–8.
30. Darby RR, Horn A, Cushman F, Fox MD. Lesion network localization of criminal behavior. Proc Natl Acad Sci U S A. 2018;115(3):601–6.
31. Darby RR, Joutsa J, Burke MJ, Fox MD. Lesion network localization of free will. Proc Natl Acad Sci U S A. 2018;115(42):10792–7.
32. Fox MD. Mapping symptoms to brain networks with the human connectome. N Engl J Med. 2018;379(23):2237–45.
33. Darby RR, Laganiere S, Pascual-Leone A, Prasad S, Fox MD. Finding the imposter: brain connectivity of lesions causing delusional misidentifications. Brain. 2017;140(2):497–507.
34. Boes AD, Prasad S, Liu H, Liu Q, Pascual-Leone A, Caviness VS Jr, et al. Network localization of neurological symptoms from focal brain lesions. Brain. 2015;138(Pt 10):3061–75.
35. Ferguson MR, Yu CK, Poliakov AV, Friedman SD, McClellan JM. Capgras syndrome: neuroanatomical assessment of brain MRI findings in an adolescent patient. Brain. 2017;140(7):e43.
36. Darby RR, Dickerson BC. Dementia, decision making, and capacity. Harv Rev Psychiatry. 2017;25(6):270–8.
37. Rosen J. The brain on the stand. The New York Times Magazine. 2007.
38. Weisberg DS, Keil FC, Goodstein J, Rawson E, Gray JR. The seductive allure of neuroscience explanations. J Cogn Neurosci. 2008;20(3):470–7.
39. McCabe DP, Castel AD. Seeing is believing: the effect of brain images on judgments of scientific reasoning. Cognition. 2008;107(1):343–52.
40. Farah MJ, Hook CJ. The seductive allure of "seductive allure". Perspect Psychol Sci. 2013;8(1):88–90.
41. Hughes V. Science in court: head case. Nature. 2010;464(7287):340–2.
42. Meltzer CC, Sze G, Rommelfanger KS, Kinlaw K, Banja JD, Wolpe PR. Guidelines for the ethical use of neuroimages in medical testimony: report of a multidisciplinary consensus conference. AJNR Am J Neuroradiol. 2014;35(4):632–7.
43. Boundy M. The government can read your mind: can the constitution stop it? Hastings Law J. 2012;63(6):1627–44.
44. Shen FX. Neuroscience, mental privacy, and the law. Harv J L & Pub Pol'y. 2013;36:653.
45. Kraft CJ, Giordano J. Integrating brain science and law: neuroscientific evidence and legal perspectives on protecting individual liberties. Front Neurosci. 2017;11:621.
46. Farah MJ, Hutchinson JB, Phelps EA, Wagner AD. Functional MRI-based lie detection: scientific and societal challenges. Nat Rev Neurosci. 2014;15(2):123–31.
47. Gamer M, Klimecki O, Bauermann T, Stoeter P, Vossel G. fMRI-activation patterns in the detection of concealed information rely on memory-related effects. Soc Cogn Affect Neurosci. 2012;7(5):506–15.
48. Rissman J, Greely HT, Wagner AD. Detecting individual memories through the neural decoding of memory states and past experience. Proc Natl Acad Sci U S A. 2010;107(21):9849–54.

49. Aspinwall LG, Brown TR, Tabery J. The double-edged sword: does biomechanism increase or decrease judges' sentencing of psychopaths? Science. 2012;337(6096):846–9.
50. Schweitzer NJ, Saks MJ. Neuroimage evidence and the insanity defense. Behav Sci Law. 2011;29(4):592–607.
51. Kulich R, Maciewicz R, Scrivani SJ. Functional magnetic resonance imaging (FMRI) and expert testimony. Pain Med. 2009;10(2):373–80.
52. Brown T, Murphy E. Through a scanner darkly: functional neuroimaging as evidence of a criminal defendant's past mental states. Stanford Law Rev. 2010;62(4):1119–208.
53. Feigl GC, Hiergeist W, Fellner C, Schebesch KM, Doenitz C, Finkenzeller T, et al. Magnetic resonance imaging diffusion tensor tractography: evaluation of anatomic accuracy of different fiber tracking software packages. World Neurosurg. 2014;81(1):144–50.
54. McKinney SM, Sieniek M, Godbole V, Godwin J, Antropova N, Ashrafian H, et al. International evaluation of an AI system for breast cancer screening. Nature. 2020;577(7788):89–94.
55. Large M, Kaneson M, Myles N, Myles H, Gunaratne P, Ryan C. Meta-analysis of longitudinal cohort studies of suicide risk assessment among psychiatric patients: heterogeneity in results and lack of improvement over time. PLoS One. 2016;11(6):e0156322.
56. Douglas T, Pugh J, Singh I, Savulescu J, Fazel S. Risk assessment tools in criminal justice and forensic psychiatry: the need for better data. Eur Psychiatry. 2017;42:134–7.
57. Zheng L, Wang O, Hao S, Ye C, Liu M, Xia M, et al. Development of an early-warning system for high-risk patients for suicide attempt using deep learning and electronic health records. Transl Psychiatry. 2020;10(1):72.
58. Singer N. In screening for suicide risk, Facebook takes on tricky public health role. The New York Times; 2018.
59. van Eijk G. Socioeconomic marginality in sentencing: the built-in bias in risk assessment tools and the reproduction of social inequality. Punish Soc. 2017;19(4):463–81.
60. Simmons R. Quantifying criminal procedure: how to unlock the potential of big data in our criminal justice system. Mich St L Rev. 2016;947:947–1017.
61. Penn State. Using artificial intelligence to detect discrimination. ScienceDaily. ScienceDaily; 2019. www.sciencedaily.com/releases/2019/07/190710121649.htm.

Further Reading

Berlin L. Neuroimaging, expert witnesses, and ethics: convergence and conflict in the courtroom. AJOB Neurosci. 2014;5(2):3–8. https://doi.org/10.1080/21507740.2014.880089.

Kang HW, Kang HB. Prediction of crime occurrence from multi-modal data using deep learning. PLoS One. 2017;12(4):e0176244. https://doi.org/10.1371/journal.pone.0176244. Published 2017 Apr 24.

Linthicum KP, Schafer KM, Ribeiro JD. Machine learning in suicide science: applications and ethics. Behav Sci Law. 2019;37(3):214–22. https://doi.org/10.1002/bsl.2392.

Meltzer CC, Sze G, Rommelfanger KS, Kinlaw K, Banja JD, Wolpe PR. Guidelines for the ethical use of neuroimages in medical testimony: report of a multidisciplinary consensus conference. AJNR Am J Neuroradiol. 2014;35(4):632–7. https://doi.org/10.3174/ajnr.A3711.

Morse SJ. Brain imaging in the courtroom: the quest for legal relevance. AJOB Neurosci. 2016;5(2):24–7. https://doi.org/10.1080/21507740.2014.880090.

Tortora L, Meynen G, Bijlsma J, Tronci E, Ferracuti S. Neuroprediction and A.I. in forensic psychiatry and criminal justice: a neurolaw perspective. Front Psychol. 2020;11:220. https://doi.org/10.3389/fpsyg.2020.00220. Published 2020 Mar 17.

Chapter 11
Into the Wild: Reflecting on Neuroethics as Innovation Moves from the Laboratory to Society

Diana Saville, Albert Kim, Juan Enriquez, Karen Rommelfanger, Michael McCullough, Calvin Nguyen, and Abraham Dada

Neuroscience startups today are beginning to rewrite the human experience. With emerging technologies like embedded brain–computer interfaces (BCI) and sophisticated imaging, sensing, and stimulating tools, we are entering an age in which we will gain unprecedented control over the brain, seen by many as the last frontier in scientific discovery. The rate at which we can now scale these technologies will easily outpace our ability to thoughtfully protect society's interests, as it already has with other technologies. As neuroscience innovations move out of the lab and into society, the importance of real-world ethical discussions scales exponentially.

Brain researchers, ethicists, makers, and funders need to come together and iterate the framework for thinking about mapping and altering thoughts, memories,

D. Saville (✉) · A. Kim · C. Nguyen
BrainMind, Menlo Park, CA, USA
e-mail: dsaville@brainmind.org; aldk@stanford.edu; cnguyen@brainmind.org

J. Enriquez
VC Fund, Newton, MA, USA
e-mail: jenriquez@excelvm.com

K. Rommelfanger
Department of Neurology and Psychiatry and Behavioral Sciences, Emory University School of Medicine, Atlanta, GA, USA
e-mail: krommel@emory.edu; director@instituteofneuroethics.org

M. McCullough
Department of Emergency Medicine, University of California, Menlo Park, CA, USA
e-mail: mmccullough@brainmind.org

A. Dada
University of California, San Francisco School of Medicine, San Francisco, CA, USA
e-mail: abraham.dada@dreem.com

L. W. Roberts (ed.), *Ethics and Clinical Neuroinnovation*,
https://doi.org/10.1007/978-3-031-14339-7_11

emotions, and behaviors. Within the next decade, we will have the first map of a full human brain [1], and soon thereafter, thousands and millions of brains. Then, as occurred in genomics, we will develop methods to alter what has been mapped. These kinds of discoveries have implications for altering our very humanity. This innovation space is where ethical theory meets praxis.

These unprecedented, rapid changes will not be primarily driven by academia. They will occur within startups and large tech and biotech companies [2]. Google, Facebook, and IBM all have large and growing neuroscience research departments, and they are funneling academic talent from the nation's top universities. Leading biotech companies have dedicated venture arms to invest in emerging neurotech startups [3]. Neuralink and Kernel are the first of a rising trend of emerging neuro startups funded at $100 M or more by billionaire brain science enthusiasts, and BrainMind has catalogued over 1700 neuro startups of varying sizes internationally. Ideas can incubate in the lab for decades; but investors typically expect returns in 5 years or less [4].

The stakes with neuroscience R&D are high and the decision-makers are diffused across the public and private sectors. We can learn lessons from the ethical failures of social media and internet companies by focusing on ethics now and create a dialogue-generated consensus on issues surrounding data privacy, user autonomy, and unintended uses of technologies that could encroach on human rights and human dignity.

Why Now?

Historically, neuroscience startups had come to represent a black hole for brilliant ideas and significant amounts of investment capital. For evidence, look no further than the Alzheimer's therapeutics development track record, which has seen only a single new disease-altering drug approval since 2003 [5], with recent controversial U.S. Food and Drug Administration (FDA) approval of Aduhelm [6]. Humorously, commentators have coined Eroom's Law (Moore's Law backwards) to describe the exponentially increasing cost of bringing new medicines to market. The cost of bringing a single new chemical entity to market has increased from $1.1 billion in 2003 to $2.8 billion in 2013 (in 2018 US dollars) [7]. And even within this rarefied environment, CNS treatments stand out as a gross outlier. The cost and failure rate of neuroscience treatments has been estimated at $800 billion in the United States alone [8, 9].

Those brave pioneers who invested in this space were beset by failure after failure. Worse, many of these losses occur not in Phase 1 clinical trials where the cost and investment are manageable but during widespread and very expensive Phase 3 trials where the investments are into the hundreds of millions of dollars; this destroys expectations and companies and may destroy the future of research for many conditions [10].

As a result, many investors see this space as dangerously premature for investment, evidenced by the highly publicized R&D investment exodus of pharmaceutical giants like Amgen, AstraZeneca, Bristol-Myers Squibb, Pfizer, GlaxoSmithKline, and Novartis from neurotherapeutics starting in 2010 [11–13]. Yet, the neuromodulation market is beginning to grow rapidly at an annual rate of 11%, and investment into the space has doubled in the past 2 years [14]. This is, in part, because advances in machine learning and a greatly increased capacity to store and analyze data have enabled a boom in the use and advancement of older technologies like electroencephalography (EEG), deep brain stimulation (DBS) [15], and trans-cranial magnetic stimulation (TMS) [16].

Meanwhile, philanthropists in neuroscience are beginning to act more like investors. Institutions like the FB Heron Foundation, Cohen Veterans Bio, and the Rainwater Charitable Trust, the Dementia Discovery Fund, the Alzheimer's Drug Discovery Foundation, and the Cure Alzheimer's Fund support translational R&D and targeted startups and may even generate a return to the foundations. With these new energies headed toward translation of brain science out of the lab, it is now more important for philanthropists to understand the thought process of neuro startups.

The countdown has begun. Forecasted by a few early movers, the neuroinnovation investment land grab awaits. Philanthropists are beginning to fill some early-stage funding gaps, but trends with large philanthropic organizations move relatively slowly. With respect to neuroethics, this means we have time, albeit limited, to discuss, understand, debate, and importantly, act.

The Potential

What if the cost curves of understanding-intervening in the brain began to follow the cost curves of genomics instead of the traditional medical cost curves? (Fig. 11.1).

Tools like wearable EEG, eye trackers (ETs), electrocardiography (ECGs), galvanic skin response (GSR), and facial electromyography (EMG) [17] do seem to be following this logic, getting cheaper, smaller, and responding to more demand. As we get a better understanding of the brain, as brain diseases of the elderly are now beginning to affect one of the greatest concentrations of wealth on the planet, and as longevity becomes a topic of increasing concern in a world bracing to care for its largest disproportion of an aging population [18, 19], one sees a serious uptick in aggressive investment into brain startups. One already sees more examples of large, private sector, non-traditional pharma startups in neurotech and neurotherapeutics, even as the pharma divests in R&D [12]:

- Neuralink—BCI startup developing microelectrode implants for directly reading brain signals, funded by Elon Musk to the tune of $100 million [20].
- Kernel—Startup offering "neuroscience as a service" in addition to developing magnetoencephalography (MEG) technology in a portable form factor, led and funded with an initial $100 million investment by Bryan Johnson, and now taking on additional VC dollars [21].
- CTRL-Labs—An engineering startup which reads neural signals to predict movement intentions, was acquired by Facebook for between $500 million and $1 billion [22].

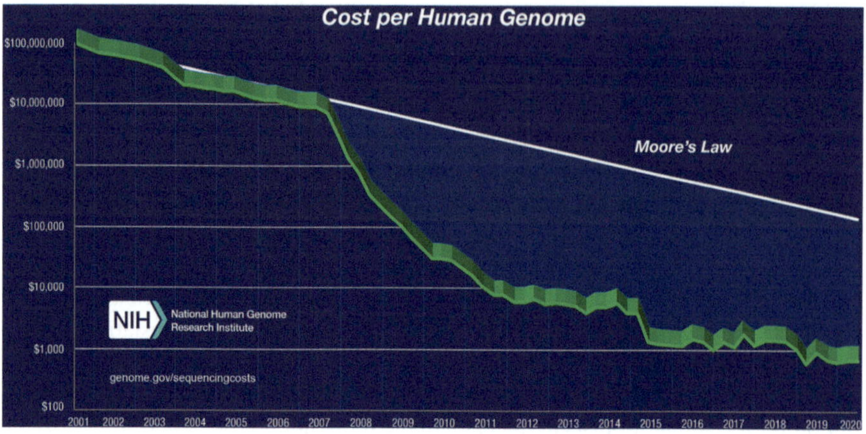

Fig. 11.1 Cost curves of human genomics

- Facebook recently announced that the CTRL-Labs technology was being integrated into their exclusive controller for augmented Reality environments.
- Facebook also recently published some of their internal research findings on noninvasive indirect measurement of neural activity [23].
- Pear Therapeutics, the first ever FDA-approved digital therapeutic offering app-based cognitive behavioral and insomnia therapy, which was co-founded with investor and philanthropist Stephen Kennedy Smith, a member of the Kennedy family [24].
- Project Amber, launched by the X unit of Google's parent company Alphabet, billed as a mental health project, spent 3 years developing a company headset to measure brain waves [25].
- Jazz Ventures is one of the first venture funds to exclusively focus on companies at the intersection of neuroscience and digital technology. Most venture capitalists (VCs) don't employ Chief Scientists, but Jazz brought in noted neuroscientist and entrepreneur Adam Gazzaley to support the vetting process. Jazz is a relatively small fund, but other larger players in the tech and biotech world, including Khosla Ventures, Arch Ventures, and increasing numbers of pharma-backed funds are beginning to invest $1 billion annually in neurotechnology startups as of 2017 [26].

The Potential Cost

One thing to consider is what occurs as the Silicon Valley ethos of "move fast and break things" enters neurotech. There are both opportunity and financial costs of moving too slowly, but there are too many cautionary tales of entrepreneurs whose ventures were decimated by lapses in ethical judgment or ignorance of the ethical implications of their hasty and unconsidered decision-making.

We also want to avoid either-or thinking: the cost of preventing unethical behavior should not be equated to withholding and preventing societally beneficial research from coming to market. There is an undeniable societal need to address pervasive brain disorders that cause substantial global suffering.

- 1 in 5 adults experience mental illness every year in the US alone [27].
- 20% of patients diagnosed with depression are treatment-resistant [28]
- 70% of patients with depression are still symptomatic after treatment [29]
- Across the field of mental health, treatment efficacy has not improved in 25 years [30].
- In America, the costs of all mental disorders and neurological illness is over $760 billion per year [31].
- Direct medical costs of Alzheimer's in America will increase by $1 trillion in the next 30 years [32], portending the insolvency of Medicare.

Worldwide, nearly 50 million people have Alzheimer's or related dementia [33]. With a drug development failure rate of nearly 99.6% coupled with ballooning regulatory expenses that discourage investment, these individuals and their families must simply wait for the field to move forward [34].

Neuroethicists also agree that neuroscience is an ethical imperative [35] and we have a real opportunity to accelerate discovery through neuroethics as a partner with neuroscience. But neuroethics is most powerful as a thought partner throughout the discovery process, not just after things go awry.

A great example of this opportunity is found in clinical trial design. A few years ago, Dr. Helen Mayberg, a neuroscientist and neurologist at the Icahn School of Medicine at Mount Sinai, was developing an experimental DBS intervention for treatment-resistant major depressive disorder. The industry sponsored BROADEN trial, which was based on her promising research, was halted by the company. Patients enrolled in the trial were offered rechargeable batteries so they would have the option to continue treatment, but they needed to have a clinician who would agree to ongoing clinical care. In other similar situations involving implanted neurostimulators, the patients fared worse. NeuroVista had run a trial for a seizure-predicting device, but the company had to cease operations when it ran out of funding in 2013. Patients benefiting from the therapy had to be explanted as there was no company to support their device. Similarly, when Northstar Neuroscience's stroke rehabilitation trial was declared unsuccessful in 2008, the company went bankrupt. All patients in Northstar trials had to be explanted, including those in a small parallel study in depression, even if they were doing well on the treatment. This is a heartbreaking outcome for patients who, through no fault of their own, lose access to a treatment that is giving them benefit. Clinicians participating in these trials are confronted with the horrible experience of having to abruptly discontinue a therapy that had been working for many of their patients who are in need. These are just a few of the myriad issues that need to be considered when planning, running, and finishing a trial [36].

Neuroethics can also help to avoid the massive cost of making uninformed decisions that can destroy or delay not just an individual's research but also that of an entire field. A prime example of the ethical cautionary tale comes from psychedelics. The world is currently witnessing a renaissance in psychedelics research, with psilocybin designated as a "Breakthrough Therapy" by the FDA for severe depression [37] and 3,4-methylenedioxy-methamphetamine (MDMA)-assisted psychotherapy winning the same status for post-traumatic stress disorder (PTSD) [38]. However, this therapeutic potential is being realized almost 60 years late. Early and particularly notorious ethical missteps in psychedelics research resulted in academic research in this field being effectively cancelled for decades.

We can't ignore the fracturing cost of waiting to bring neuroethics to the fore on a global scale. With an increasing share of research, innovation, and deployment of leading neurotechnologies taking place internationally, leading neuroethicists have already advocated for cultural humility and have alerted the neuroscience community that awareness of underlying differences in cultural values will be critical in advancing impactful neuroscience discovery.

To get a sense of how complex these ethical entrepreneurial conversations could get, imagine you are a VC or an endowment manager asked to fund the following business plans (put the likelihood of success aside for a moment, assume they could be successful):

- A technology that can erase unwanted memories at will, insert memories at will, or allow you to share someone else's feelings/state of mind. After considering safety issues, one might consider what would happen if this technology came into the wrong hands. Like facial-recognition technology, is it inevitable that technology like this would be used and abused at the nation-state level over time?
- A new deep brain stimulation technique for the treatment of amyotrophic lateral sclerosis (ALS) which completely alters the patient's personality, causing destructive or antisocial traits. Should the treatment be pursued regardless? Is this ethical for the caregiver?
- A gene editing company that can raise the IQ of a customer by 10–20 points. Is it ethical to fund a human speciation event? What of those who cannot afford the treatment? Will they be left behind? Will directed evolution become a human right? (Note that companies like Neuralink can't even progress with gene therapy for neurons because there are not enough ethical guidelines for engineers to work within set ethical boundaries: engineers are banned from working on anything related to this as of now.
- A crime-prevention neurotechnology product that can scan for and eliminate forbidden desire in a convicted pedophile, or perhaps in persons who haven't yet committed any crimes. When is it ok to reprogram a brain?
- A consumer neurotechnology that can alter mood, cravings, and self-confidence. Should we limit the range of human experience? When?

Now let's move from hypothetical to practical: DBS is a highly effective treatment for Parkinson's disorder, and it has been proven to alleviate impulse control disorders, one of the many adverse effects of traditional medication [39]. In some patients, however, it has been shown to regress and even cause new cases of impulse disorders [40–42]. Loss of sense of identity is an emerging consideration for BCI-mediated therapies, but concerns related to identity changes have already surfaced with pharmaceuticals like selective serotonin reuptake inhibitors (SSRIs) where people who are getting treatment for mental health disorders report not feeling “like themselves” when the treatment is working [43]. There is also a growing conversation around potential treatments for autism spectrum disorder (ASD). This is seen, by some, as a human rights issue of societal acceptance of non-neurotypical persons.

We can alter our mood, our cognitive function, and potentially our memories with electrical stimulation; at what point does this desire to correct our state of mind impinge on our sense of identity, and how do we adequately inform the user about this possibility? Neuroethics can help us collectively think through unintended consequences of groundbreaking technologies.

Throughout history, there has been a consistent drive to “enhance” humans and the drive and allure to enhance the brain has drawn significant attention from investors and entrepreneurs to an eager public. Neuroethicists continue to help guide and navigate the increasingly muddy value conflicts of enhancement, access, and justice. Technologies that can modify foundational human capacities—often adapted from their initial clinical applications (i.e., attention-deficit/hyperactivity disorder (ADHD) medications and transcranial direct-current stimulation (tDCS))—can introduce new kinds of equity divides not just between individuals, but could deepen the economic divide between countries [44].

Jumping in

Neuroethics is not just another word for regulation or compliance. Neuroethics offers a deeper conversation and framework for foresight and proactive, creative problem-solving that happens during, and not just after the development of a technology. It offers the potential to obviate the need for much of the regulation that has been seen as slowing innovation over the years. Perhaps, it could empower us to have a hand in developing the world we want to have, as opposed to being passive bystanders to market forces.

Amongst some researchers, neuroethics may be seen as just another word for regulation, and the topic whose value is met with skepticism or disdain and avoided altogether. An important counterpoint to this thinking is that neuroscience itself is seen as an ethical imperative (as stated in the NIH BRAIN Neuroethics Principles). We have the opportunity to use neuroethics to advance and accelerate high-impact neuroscience.

Entrepreneurs are now leaping ahead of many academics, intentionally working to enhance their engagement with neuroethics. Founders and executives may be approaching neuroethics from a defensive and/or opportunistic position, but many are beginning to see the value and practicality, the business case for neuroethics engagement.

Instead of having discussions on what is an appropriate use for an emerging technology after it is broadly deployed (think Congress interrogating Facebook after the Cambridge Analytica/data privacy debacle), a growing list of neuroscience entrepreneurs across clinical and commercial applications are embracing neuroethics early in their company lifecycle and in their professional lives:

- Ariel Garten, the founder of InteraXon, which developed EEG-sensing headbands for commercial use, founded the Center for Responsible Brainwave Technologies, a professional association which created a set of ethical principles for brain data management.
- Tim Mullen, the Founder and Chief Executive Officer (CEO) of Intheon, a middleware platform for brain data analysis, is active in Institute of Electrical and Electronics Engineers (IEEE) development of neuroethics frameworks, and is a founding advisor to BrainMind's Neuroethics Initiative.
- Marc Chevillet, former Research Director for Facebook's Building 8 BCI program, has spoken publicly about his pro-neuroethics and pro-transparency stance on his research program, and Stephanie Thacker, formerly of Defense Advanced Research Projects Agency (DARPA), has launched an internal neuroethics initiative including offering responsible neuroscience funding at Facebook.
- Ramses Alcaide, the founder and CEO of Neurable, a full-stack commercial neurotechnology tools company, launched a blog publishing neuroethics content on his company website in 2020.
- Dario Gil, Director of IBM research is actively advocating for careful consideration of neuroethics issues [45].
- Ana Maiques, the Founder and CEO of Neuroelectrics, a clinical-grade neuroscience device company, has participated at the International BRAIN Initiative Global Neuroethics Summit, has participated in thought leadership events with the International Neuroethics Society, and advises the BrainMind Neuroethics Initiative.
- Thomas Reardon, the founder and CEO of CTRL-labs, a neural interface company that predicts movement intent, sat on a NeurotechX Neuroethics panel in 2020, and is participating in BrainMind's neuroethics programming.
- Dan Rizzuto, founder and CEO of NIA Therapeutics, was a panelist at the International Neuroethics Society Annual Meeting and has participated in BrainMind Neuroethics programming.
- Many other examples abound, with entrepreneurs like Philip Sabes, cofounding scientist of Neuralink, Matt Angle, CEO of Paradromics, and Tan Le, CEO of Emotiv, speaking and writing publicly on ethics related to their work.

Plenty of entrepreneurs have altruistic motivations for incorporating neuroethics into their work. Some find the topic intellectually stimulating and have a long history of engaging on the subject.

Others are likely coming from a defensive position or thinking of policy readiness, responding to previous scrutiny, and taking a careful look at how to move forward developing technologies that could be seen as intrusive by consumers. Simply put, it's insurance against bad public relations. When Geoff Ling led DARPA's Biotechnologies Office, internal neuroethics principles were put in place because leadership saw the need to address questions of privacy, agency, and identity early, to avoid criticism later on. Still other entrepreneurs are looking to harness neuroethics for competitive advantage. Many large companies engage not so much in ethics, but in regulation as a means of increasing barriers to entry by creating huge capital requirements to clear regulatory hurdles. This effectively squeezes out or blocks competition. Similarly, many companies are simply seeing the writing on the wall and are looking for insurance against future regulation. They know that regulators will step in eventually, so they might as well write the rules of the game, so they can play it better than the rest.

We argue here that neuroethics is not to be conflated with regulation, and that if introduced early, neuroethics can be an accelerator, not a hurdle. For example, companies that are seen as ethical enjoy a tremendous boost in brand value. Social justice, in particular, is now an established pillar of brand value. Another reason entrepreneurs are looking to engage with neuroethics for better corporate governance. Building this type of thinking into the company culture allows for more efficient management of the organization, as employees often have to make decisions in the absence of direct guidance.

Preliminary studies also show that neuro-entrepreneurs have been grappling with significant neuroethical concerns not addressed satisfactorily with current regulation. They are concerned about issues like misuse of neural data and responsible ownership particularly in non-clinical domains and unintended uses or access to data and technologies that could lead to stigma and discrimination (personal communication, KSR) [47].

Companies that do care about ethics often fail in their efforts to address those concerns because they only have a legal department, for example, so they don't have the tools to address ethical issues/gray zone problems. And they want guidance, not necessarily regulation.

Encouragingly, a handful of prescient international non-governmental organizations (NGOs) have already begun thinking carefully about neuroethics frameworks for makers, technologists, and private sector stakeholders. Two of these groups have recently published neuroethics frameworks meant to address private sector considerations:

- In 2019, the Organisation for Economic Co-operation and Development (OECD) published their *Recommendation on Responsible Innovation in Neurotechnology* is the first international ethics standard in this field, which engaged many private sector stakeholders in carefully addressing ethical issues while taking care to assure the promotion of innovation.
- In 2020, the IEEE Brain Neuroethics Subcommittee published a plan to establish a *Neuroethics Framework* to define, and surface ethical, legal, and social issues for emerging neurotechnologies. IEEE is the largest professional network of engineers globally.

In 2018, the Global Neuroethics Summit Delegates published *Neuroethics Questions to Guide Ethical Research in the International Brain Initiatives*, offering a toolkit for scientists, and potentially a valuable tool for private sector groups.

A major challenge for entrepreneurs is that neuroethics doesn't currently fit into the established workflow of company. The community still lacks a clear incentive structure for engaging neuroethics and are still seeking creative ways to integrate neuroethics into their current operations.

BrainMind's Multi-sectoral Neuroethics Initiative

BrainMind is a nonprofit platform and private community of scientists, entrepreneurs, investors, philanthropists, and policymakers collaborating to accelerate neuroscience research and entrepreneurship that will most benefit humanity (brainmind.org). BrainMind is especially focused on high-impact ideas that might otherwise languish in the investment valley of death or on the publication shelves of academia. Because the organization is developing new philanthropic and investing approaches to promote the positive impact of brain science, and has assumed a curatorial role in the community, ethical frameworks need to be strongly integrated into organizational decision-making and ecosystem norms.

With a vision to form vital connective tissue between the lab and society, BrainMind's Neuroethics initiative engages its members in considering how to maximize the potential benefits and minimize the unintended hazards that accompany rapid innovations in brain science.

In the near future, BrainMind plans to host an international, multi-sectoral neuroethics summit focused on the development, distribution, and use of existing and near-term neurotechnology innovations at the Asilomar Conference Grounds in California. This gathering will kick off a decadal review and ongoing engagement of researchers and private sector stakeholders developing major innovations in the field. A defining feature of the initiative is its capacity to engage multilateral players in the private sector, including industry stakeholders, investors, and entrepreneurs.

BrainMind's desired outcomes for the Neuroethics Asilomar Program are as follows:

1. Conserve and direct energy to the technologies that have the greatest potential for meaningful impact for people as guided by ethical principles vs. profit motive.
2. Minimize the risk of unintended negative consequences of powerful technologies.
3. Establish a venue for collaborative navigation of complex or ambiguous ethical situations.

BrainMind has begun the Asilomar planning process by organizing advisory committee discussions with small multi-stakeholder groups. With support from the Dana and Kavli Foundations, in February 2020, BrainMind convened its first multi-sectoral neuroethics advisory committee meeting at Duke University. This summit brought together a group of the world's leading neuroscientists, neuro- and bioethicists, entrepreneurs, policymakers, and investors in order to explore engaging ethical frameworks among diverse stakeholders in the advancement of neurotechnologies and affirmed the opportunity for BrainMind to form a "practical layer" between conceptual neuroethics frameworks and its functional application in research and company projects emerging in neuroscience.

Since then, BrainMind has been conducting follow-on advisory sessions across stakeholders groups in a virtual setting in order to build momentum for the initiative and create buy-in among influential private sector stakeholders. The desire among entrepreneur participants for neuroethics strategy advising has emerged as a powerful trend in these interactions.

Asking the Right Questions

In its curator role, BrainMind continues to carefully build community with entrepreneurs and entrepreneurial endeavors to provide a platform for raising capital and other forms of support. Ethical curation, and an ethical investing framework, will be the backbone of well-integrated neuroethics programming for BrainMind's activities, and some form of advisory services provided to BrainMind's members, with the right organizational partnerships, may be an emerging opportunity to explore.

So what kinds of questions should BrainMind's ecosystem members, the scientists, entrepreneurs, investors, and philanthropists committed to bringing impactful science out of the lab and into society, be asking if they want to find themselves on the right side of history?

Perhaps a place to start would be to modify some of the questions for ethical neuroscience in the International Brain Initiatives [44] for the context of company creation:

1. What are the possible effects of this product or service on a person's sense of identity?
2. How can human brain data (e.g., images, neural recordings, etc.), and the privacy of users be protected by the company?
3. Is there moral significance of neural systems that are under development in the company?
4. How could this company's product or service impact or reduce autonomy, and who will have responsibility for these effects?
5. In which markets and contexts might the product or service be deployed, and which applications should be considered misuse?

Most broadly, these questions all come to the heart of the central question: When is it ok to look inside and/or modify the human brain? They go to the core of what it is to be human and what it means to have a "healthy" brain. Of course, we have no universally agreed-upon definitions of either, which is why this is so interesting and important.

Finally, there are some practical questions to answer in the next few years:

- Guidelines have some utility, but how do you build a forcing function to implement them across all these stakeholders groups, and what incentives would be effective in this effort for each group?
- What are the practical tools from neuroethics as a problem-solving framework that can be used by each group?
- Ethics differs in different countries and even within them: we need to understand what individual versus community emphasis, rights, and values will make sense to implement across cultures.

As an initial exploration of these questions, BrainMind is having those early conversations with private sector leaders, both nationally and internationally, while surfacing alternative mechanisms of idea formation beyond the profit-driven VC model. To help clear away misconceptions about neuroethics and make the topic more accessible in general, BrainMind is also developing multimedia education on neuroethics, including filmed interviews with innovation-friendly voices in the field, and visual and animated web content. BrainMind's first measure of success will be to make all ecosystem members conversant and comfortable on neuroethics topics in their work life. Longer-term, neuroethics will become a standard part of the decision-making toolkit for all leaders translating brain science out of the lab and bringing it to humanity.

Concluding Thoughts

While not without significant challenges, we have a great opportunity to align private sector stakeholders on common values of reducing potential harm of neuroscience innovations entering the commercial sphere. Industry is moving faster than academia, and neuroethics needs to equally be fast-tracked and engaged in this domain. Many private sector leaders agree on the need to establish guardrails before the stakes are too high, and it will be easier to start in the design stage than to walk it back on a product ready to ship. This is why being proactive and strategic with capacity building can make ethics second nature and nimble enough to keep up with the innovation process.

References

1. Geddes L. Human brain mapped in unprecedented detail [Internet]. Nature News. Nature Publishing Group; 2016 [cited 2021 Jan27]. https://www.nature.com/news/human-brain-mapped-in-unprecedented-detail-1.20285.
2. Fernandez A. Good news: the digital revolution meets the human brain (as evidenced by the explosion of neurotech patent filings) [Internet] SharpBrains; 2015 [cited 2021Jan27]. https://sharpbrains.com/blog/2015/06/09/good-news-the-digital-revolution-meets-the-human-brain-as-evidence-by-the-explosion-of-neurotech-patent-filings/.
3. http://analytics.dkv.global/data/pdf/NeuroTech/NeuroTech-Landscape-Overview-Teaser.pdf.
4. Zider B. How venture capital works [internet]. Harv Bus Rev; 2014 [cited 2021 Jan27]. https://hbr.org/1998/11/how-venture-capital-works.
5. Cummings JL, Morstorf T, Zhong K. Alzheimer's disease drug-development pipeline: few candidates, frequent failures. Alz Res Therapy. 2014;6:37. https://doi.org/10.1186/alzrt269.
6. Food and Drug Administration. FDA grants accelerated approval for Alzheimer's drug. June 7, 2021. https://www.fda.gov/news-events/press-announcements/fda-grants-accelerated-approval-alzheimers-drug.
7. Wouters OJ, McKee M, Luyten J. Estimated research and development investment needed to bring a new medicine to market, 2009–2018. JAMA. 2020;323(9):844–53. https://doi.org/10.1001/jama.2020.1166.
8. Wiley. Neurological diseases cost the US nearly $800 billion per year. ScienceDaily; 2017. www.sciencedaily.com/releases/2017/03/170328105855.htm.
9. Gribkoff VK, Kaczmarek LK. The need for new approaches in CNS drug discovery: Why drugs have failed, and what can be done to improve outcomes. Neuropharmacology. 2017;120:11–9. https://doi.org/10.1016/j.neuropharm.2016.03.021. Epub 2016 Mar 12. PMID: 26979921; PMCID: PMC5820030.
10. Fogel DB. Factors associated with clinical trials that fail and opportunities for improving the likelihood of success: a review [internet]. Contemp Clin Trials Commun; 2018 [cited 2021 Jan30]. https://www.ncbi.nlm.nih.gov/pmc/articles/PMC6092479/

11. Bell, J. Big pharma backed away from brain drugs. Is a return in sight? 2020. https://www.biopharmadive.com/news/pharma-neuroscience-retreat-return-brain-drugs/570250/.
12. Carroll J. Amgen is joining the biopharma exodus out of neurosciences R&D, cutting 180 R&D jobs in reorganization [internet]. Endpoints News; 2019 [cited 2020 Dec 8]. https://endpts.com/amgen-is-joining-the-biopharma-exodus-out-of-neurosciences-rd/.
13. Abbot, A. Novartis to SHUT brain research facility; 2011. https://www.nature.com/news/novartis-to-shut-brain-research-facility-1.9547.
14. Marketsandmarkets.com. Neuromodulation market by technology (deep brain stimulation, spinal cord stimulation, Transcranial magnetic stimulation) and by application (depression, Parkinson's, tinnitus, Alzheimer's, Epilepsy, Ischemia, Obesity) – Trends and Global Forecast to 2020; 2015.
15. Habets JG, Heijmans M, Kuijf ML, Janssen ML, Temel Y, Kubben PL. An update on adaptive deep brain stimulation in Parkinson's disease. Mov Disord. 2018;33:1834–43. https://doi.org/10.1002/mds.115.
16. S. M. Sasubilli, A. Kumar and V. Dutt, "Machine learning implementation on medical domain to identify disease insights using TMS. In: 2020 International conference on advances in computing and communication engineering (ICACCE), Las Vegas, NV, USA; 2020, pp. 1–4. https://doi.org/10.1109/ICACCE49060.2020.9154960.
17. Alvino L, Pavone L, Abhishta A, Robben H. Picking your brains: where and how neuroscience tools can enhance marketing research. Front Neurosci. 2020;14:577666. https://doi.org/10.3389/fnins.2020.577666.
18. United Nations, Department of Economic and Social Affairs, Population Division (2017). World Population Ageing 2017 - Highlights (ST/ESA/SER.A/397).
19. Nasser HE. The graying of America: more older adults than kids by 2035 [internet]. The United States Census Bureau; 2021 [cited 2021 May12]. https://www.census.gov/library/stories/2018/03/graying-america.html.
20. Markoff J. Elon Musk's Neuralink wants 'Sewing machine-Like' robots to wire brains to the internet [internet]. The New York times; 2019 [cited 2020 Dec 8]. https://www.nytimes.com/2019/07/16/technology/neuralink-elon-musk.html.
21. Statt N. Kernel is trying to hack the human brain - but neuroscience has a long way to go [internet]. The Verge; 2017 [cited 2020 Dec 8]. https://www.theverge.com/2017/2/22/14631122/kernel-neuroscience-bryan-johnson-human-intelligence-ai-startup.
22. Wagner K. Facebook to buy startup for controlling computers with your mind [internet]. Bloomberg.com; 2019 [cited 2020 Dec 8]. https://www.bloomberg.com/news/articles/2019-09-23/facebook-to-buy-startup-for-controlling-computers-with-your-mind.
23. Sie EJ, Tiecke T, Chen H, Saung E-F, Catoen R, Chevillet MA, et al. High-sensitivity multispeckle diffuse correlation spectroscopy [internet]. Facebook Research; 2020 [cited 2020 Dec 8]. https://research.fb.com/publications/high-sensitivity-multispeckle-diffuse-correlation-spectroscopy/.
24. Pear obtains FDA clearance of the first prescription digital therapeutic to treat disease [internet]. Pear Therapeutics; 2017 [cited 2020 Dec 8]. https://peartherapeutics.com/fda-obtains-fda-clearance-first-prescription-digital-therapeutic-treat-disease/.
25. Farr C. Google sister company X is working on monitoring depression using the brain's electrical signals [internet]. CNBC; 2020 [cited 2020 Dec 11]. https://www.cnbc.com/2020/11/02/alphabet-x-project-amber-uses-eeg-to-monitor-depression.html.
26. Sideris B., Moniz C., Longo K. Trends in neuro device investing: advances in neurology drive new activity. Santa Clara (CA): Silicon Valley Bank; 2018. 17p.
27. National Institute of Mental Health. Mental Illness [Internet]. National Institute of Mental Health. U.S. Department of Health and Human Services; 2020 [cited 2020 Dec 11]. https://www.nimh.nih.gov/health/statistics/mental-illness.shtml.
28. Crown WH, Finkelstein S, Berndt ER, et al. The impact of treatment-resistant depression on health care utilization and costs. J Clin Psychiatry. 2002;63(11):963–71. https://doi.org/10.4088/jcp.v63n1102.
29. Al-Harbi KS. Treatment-resistant depression: therapeutic trends, challenges, and future directions [Internet]. Patient Preference and Adherence. Dove Medical Press; 2012 [cited 2020 Dec 11]. https://www.ncbi.nlm.nih.gov/pmc/articles/PMC3363299/

30. Insel TR, Landis SC. Twenty-five years of progress: the view from NIMH and NINDS [internet]. Neuron Cell Press; 2013 [cited 2020 Dec 11]. https://www.sciencedirect.com/science/article/pii/S0896627313008982.
31. BrainFacts. Global burden of neurological and mental disorders [internet]. BrainFacts.org; 2014 [cited 2020 Dec 11]. https://www.brainfacts.org/diseases-and-disorders/mental-health/2014/global-burden-of-neurological-and-mental-disorders.
32. https://www.ajmc.com/view/economic-burden-of-alzheimer-disease-and-managed-care-considerations.
33. BrightFocus Foundation. Alzheimer's disease: Facts & Figures [internet]. BrightFocus Foundation; 2019 [cited 2020 Dec 11]. https://www.brightfocus.org/alzheimers/article/alzheimers-disease-facts-figures.
34. Cummings JL, Morstorf T, Zhong K. Alzheimer's disease drug-development pipeline: few candidates, frequent failures [internet]. Alz Res Therapy; 2014 [cited 2021 Jan 30]. https://alzres.biomedcentral.com/articles/10.1186/alzrt269#B13.
35. Greely H. T., et. al. (2018). Neuroethics guiding principles for the NIH BRAIN initiative. J Neurosci38(50). https://doi.org/10.1523/JNEUROSCI.2077-18.2018, 10586
36. Hendriks S, Grady C, Ramos KM, et al. Ethical challenges of risk, informed consent, and Posttrial responsibilities in human research with neural devices: a review. JAMA Neurol. 2019; https://doi.org/10.1001/jamaneurol.2019.3523.
37. Saplakoglu Y. FDA calls psychedelic psilocybin a 'Breakthrough Therapy' for severe depression [internet]. LiveScience Purch; 2019 [cited 2020 Dec 14]. https://www.livescience.com/psilocybin-depression-breakthrough-therapy.html.
38. Valentino T. MDMA-assisted psychotherapy for PTSD trials progressing through phase 3 [internet]. Psych Behav Health Learn Net; 2020 [cited 2020 Dec 14]. https://www.psychcongress.com/article/mdma-assisted-psychotherapy-ptsd-trials-progressing-through-phase-3.
39. Demetriades P, Rickards H, Cavanna AE. Impulse control disorders following deep brain stimulation of the subthalamic nucleus in Parkinson's disease: clinical aspects [internet]. Parkinson's Disease Hindawi; 2011 [cited 2020 Dec 14]. https://www.hindawi.com/journals/pd/2011/658415/.
40. Smeding HMM, Goudriaan AE, Foncke EMJ, Schuurman PR, Speelman JD, Schmand B. Pathological gambling after bilateral subthalamic nucleus stimulation in Parkinson disease [internet]. J Neurol Neurosurg Psychiatry BMJ Publishing Group Ltd; 2007 [cited 2020 Dec 14]. https://jnnp.bmj.com/content/78/5/517.short.
41. Knobel D, Aybek S, Pollo C, Vingerhoets FJ, Berney A. Rapid resolution of dopamine dysregulation syndrome (DDS) after subthalamic DBS for Parkinson disease (PD): a case report. [Internet] LWW; 2008 [cited 2020 Dec 14]. https://journals.lww.com/cogbehavneurol/Abstract/2008/09000/Rapid_Resolution_of_Dopamine_Dysregulation.10.aspx.
42. Lim S-Y, O'Sullivan SS, Kotschet K, Gallagher DA, Lacey C, Lawrence AD, et al. Dopamine dysregulation syndrome, impulse control disorders and punding after deep brain stimulation surgery for Parkinson's disease [internet]. J Clin Neurosci Churchill Livingstone; 2009 [cited 2020 Dec 14]. https://www.sciencedirect.com/science/article/abs/pii/S0967586809000666
43. Sansone RA, Sansone LA. SSRI-induced indifference [internet]. Psychiatry (Edgmont (Pa.: Township)). Matrix Medical Communications; 2010 [cited 2020 Dec 14]. https://www.ncbi.nlm.nih.gov/pmc/articles/PMC2989833/
44. Global Neuroethics Summit Delegates, Rommelfanger KS, Jeong SJ, Ema A, Fukushi T, Kasai K, Ramos KM, Salles A, Singh I. Neuroethics questions to guide ethical research in the international brain initiatives. Neuron. 2018;100(1):19–36. https://doi.org/10.1016/j.neuron.2018.09.021.
45. Gil D. The ethical challenges of connecting our brains to computers. Scientific American; 2020. https://www.scientificamerican.com/article/the-ethical-challenges-of-connecting-our-brains-to-computers/.

Part III
Neuroethics and Innovation: Inquiry Informed by the Roberts Valence Model

Chapter 12
Introduction to Our Project: Understanding Ethically Salient Perspectives of Diverse Societal Stakeholders in Innovative Neuroscience Research on Mental Disorders

Laura Weiss Roberts, Katie Ryan, Jane Paik Kim, and Laura B. Dunn

Advancements in neuroscience hold great promise for reducing the burden of many of the most disabling conditions that threaten human health on a global scale, including mental illnesses and addictions. Increasingly, exceptionally innovative science inspires hope that these devastating brain-based disorders may be prevented, treated, and even cured. With its inception in 2014, the National Institutes of Health's (NIH's) Brain Research through Advancing Innovative Neurotechnologies (BRAIN) Initiative® began its 10-year project aimed at revolutionizing our understanding of the human brain through the accelerated development and applications of new techniques and technologies [1]. Through its funding of over hundreds of different scientists and engineers over the past 7 years, the BRAIN Initiative has encouraged researchers to better understanding how the brain works and how disease occurs, inciting hope for patients who live with brain-based diseases and disorders in the process.

Alongside the pursuit of innovative neuroscience come a suite of novel ethical considerations and challenges. Concerns surrounding the deepest questions about what defines humanity and personhood, what forms of novel inquiry may exceed ethically acceptable limits in society, and how to perform ethically sound studies with volunteers who may be vulnerable to exploitation in the research situation represent just a few of the ethical dimensions and implications of neuroinnovation. When the NIH's BRAIN Initiative announced in 2016 that it would be funding neuroethics research projects with the goal of identifying and analyzing the ethical

L. W. Roberts · K. Ryan (✉) · J. P. Kim
Department of Psychiatry and Behavioral Sciences, Stanford University School of Medicine, Stanford, CA, USA
e-mail: robertsl@stanford.edu; kryan2@stanford.edu; janepkim@stanford.edu

L. B. Dunn
Department of Psychiatry, University of Arkansas for Medical Sciences, Little Rock, AR, USA
e-mail: ldunn@uams.edu

L. W. Roberts (ed.), *Ethics and Clinical Neuroinnovation*,
https://doi.org/10.1007/978-3-031-14339-7_12

issues implicit in innovative neuroscience research [2, 3], our Stanford University-based team developed a proposal for a project intended to accelerate neuroscience toward lessening the burden of mental illness and addiction through hypothesis-driven empirical ethics inquiry. This project, entitled "Enabling ethical participation in innovative neuroscience on mental illness and addiction: towards a new screening tool enhancing informed consent for transformative research on the human brain," was among the first four neuroethics projects funded by NIH as part of the BRAIN Initiative.

Central to our approach is the engagement of diverse stakeholders to gain greater understanding of the ethically salient dimensions of innovation in society—in this case, innovative neuroscience that focuses on important aspects of public health tied to mental disorders and addiction. The project has been led by one of us (LWR), following on years of similar hypothesis-driven investigative work exploring differences and similarities of stakeholders regarding ethical aspects of research and innovation that engage individuals who belong to vulnerable populations. Co-investigators for the project include Laura B. Dunn, M.D., Jane Paik Kim, Ph.D., Mildred Cho, Ph.D, and Casey Halpern, M.D. The combined expertise of the investigators on the team is quite diverse, with representation from two psychiatrists, a biostatistician, a bioethicist, and a neurosurgeon, and is further supported by an interdisciplinary research team with members with backgrounds in psychology, sociology, neuroimaging, public health, literature, and art history. The team additionally has enlisted an advisory board, which consists of bioethicists, physicians, and neuroscience researchers from universities across the United States. These individuals provide expertise and guidance regarding the development of interview and survey instruments and the interpretation of findings over the course of the project.

The chapters in this section detail the development, design, and deployment of our team's hypothesis-driven empirical inquiry into the ethics of clinical neuroinnovation and present initial findings from select portions of this BRAIN Initiative project. In order to provide the appropriate context for these findings, an overview of the rationale and methods for each aim of this project are discussed below.

Project Rationale

Innovative neuroscience research is imperative to address the suffering associated with mental disorders, including addiction. Studying these conditions poses great ethical challenges, however addressing these challenges after the fact or as a postscript could lead to potential harm to participants and a lack of public trust in research, thereby slowing advancement and innovation in the field. It is our team's belief that by preemptively identifying ethical issues in this emerging field, and giving the most vulnerable stakeholders a voice, innovative neuroscience inquiry can be accelerated and the burden of mental illness and addiction can be alleviated. The rationale for this project is therefore grounded in our team's prior empirical work with stakeholders in psychiatry research, as well as a novel theoretical model of ethical research participation.

Value of a Stakeholder Approach

Previous work completed by our team has demonstrated the value of approaches that are predicated on collecting the views of various stakeholders through surveys and semi-structured interviews (See Table 12.1). In a series of studies over the past two decades, Drs. Laura Weiss Roberts, Laura B. Dunn, and Jane Paik Kim, in collaboration with other investigators in psychiatry and bioethics, have used various

Table 12.1 Examples of stakeholder-based empirical ethics projects undertaken by our team

Topic	Awarding agency, type	Representative papers	Method used
Stakeholder perspectives on ethical challenges in algorithmic medicine [in progress]	National Center for Advancing Translational Sciences, R01	Kim [4]	Semi-structured interviews ($n = 40$) & online survey ($n = 420$)
Interactions between law enforcement and unhoused individuals with mental illness [in progress]	Dollard foundation	Lane-McKinley et al. [5]	[in progress]
Willingness of mothers to enroll children in research	Stanford University Department of Psychiatry and Behavioral Sciences	Kim et al. [6]	Online survey via MTurk ($n = 126$)
Research decision-making by caregivers of people with Alzheimer's	National Institute on Aging, R01	Dunn et al. [7, 8]; Overton et al. [9]	Surveys and in-depth interviews ($n = 142$)
Ethical issues in deep brain stimulation (DBS) research	Greenwall Foundation	Bell et al. [10]; Christopher et al. [11]; Dunn et al. [12]; Fisher et al. [13]; Leykin et al. [14]	Semi-structured interviews ($n = 26$)
Psychiatric genetics research ethics	National Institute of Mental Health, R01	Roberts and Kim, [15]; Roberts et al. [16–18]; Rostami et al. [19]	Structured interview ($n = 182$) & online survey ($n = 386$)
Psychiatric genetic research consent process intervention	Institutional funding	Kim et al. [20]	Simulated informed consent process and follow-up written survey ($n = 79$)
Use of genetic information in the workplace	Department of Energy, Small & Large Grants	Hoop et al. [21]; Roberts et al. [22–24]	Written survey and structured interview ($n = 63$) & online survey ($n = 570$)
Psychiatric research ethics: science & safeguards study	National Institute of Mental Health, K02	Roberts et al. [25–29]	Structured interview ($n = 60$) & written survey ($n = 69$)

(continued)

Table 12.1 (continued)

Topic	Awarding agency, type	Representative papers	Method used
Informed consent & surrogate decision-making in clinical care	National Alliance for Research on Schizophrenia and Depression	Roberts and Kim [30, 31]	Written survey ($n = 52$)
Vulnerability and informed consent in clinical research	National Institute of Mental Health, R01	Kim and Roberts [32]; Roberts and Kim [33–35]	Written survey and structured interview ($n = 181$)
Vulnerability and informed consent in clinical research: educational intervention study	National Institute of Mental Health/ National Institute on Drug Abuse, R01	Roberts et al. [36]; Roberts et al. [37–39]	Randomized educational intervention and follow-up written survey ($n = 83$)
Research participation and participant safeguards	National Alliance for Research on Schizophrenia and Depression	Kaminsky et al. [40]; Roberts et al. [41–44]; Warner et al. [45]	Structured interview ($n = 63$) & written survey ($n = 73$)

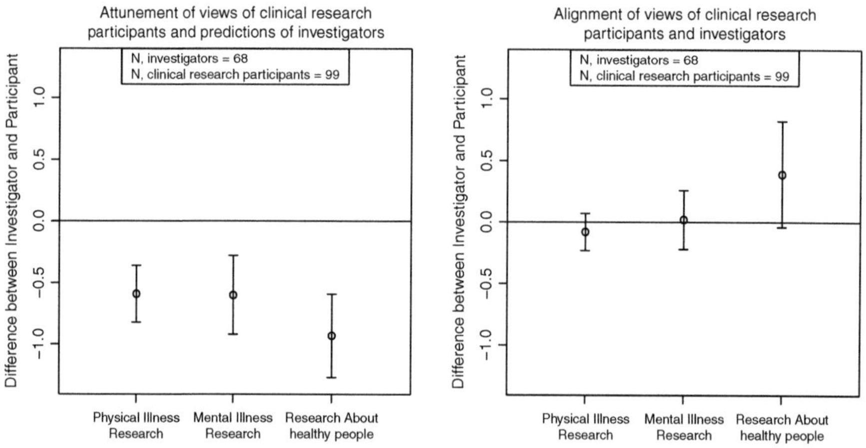

Fig. 12.1 Estimates of attunement and alignment, adjusted by covariates for the domain of "views regarding the importance of medical research." Whiskers indicate 95% confidence intervals. Reprinted from Journal of Psychiatric Research, 52, Roberts, LW, Kim JP, Do investigators understand ethically important perspectives of clinical research participants? A "piggy-back" study of attunement and alignment in serious illness research, Pp 36–43, Copyright 2014 with permission from Elsevier

stakeholder approaches to together provide substantial empirical data evaluating the abilities of people with mental illnesses to provide informed consent to research and correlate and predictors of these abilities [46–51], the impact of educational interventions on capacity to consent among people with mental illness [52, 53], the impact of differing levels of risk and compensation on potential participants' willingness to participate in hypothetical research protocols [54], and tools for assessing abilities of people with mental illness to consent to research [55] (see Figs. 12.1, 12.2, 12.3, and 12.4). These hypothesis-driven studies demonstrated the feasibility

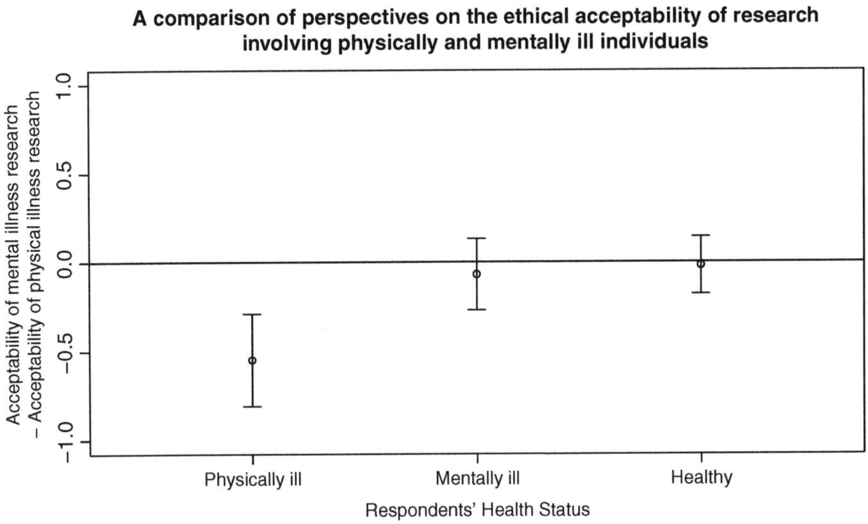

Fig. 12.2 A comparison of perspectives on the ethical acceptability of mental illness research and the ethical acceptability of physical illness research. Adapted from Roberts, L. W., & Kim, J. P. Giving voice to study volunteers: comparing views of mentally ill, physically ill, and healthy protocol participants on ethical aspects of clinical research. Journal of Psychiatric Research 2014;56:90–97

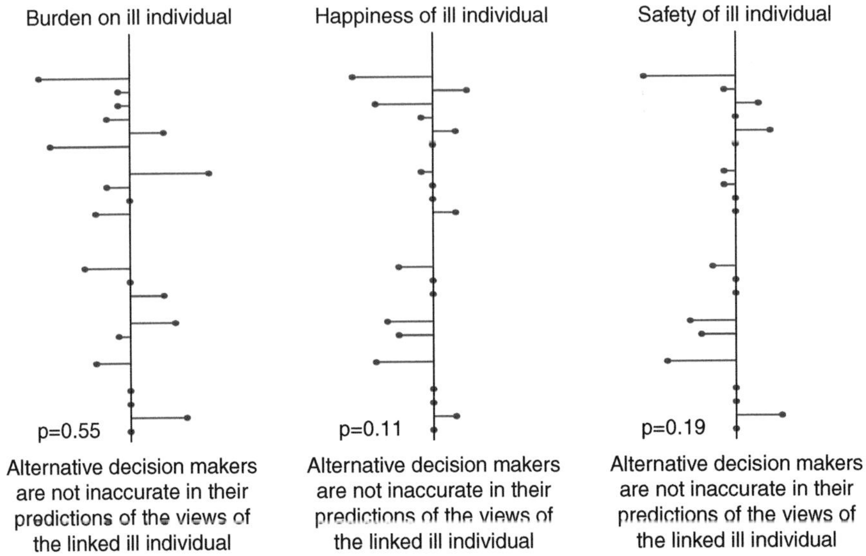

Fig. 12.3 Alternative decision-makers' predictions are attuned to the perspectives of ill individuals. Differences between the perspectives of ill individuals and the predictions of preferred alternative decision-makers (i.e., "attunement") were tested with two-sided paired t-tests and are graphically depicted here. *P* values were not adjusted for multiple comparisons as conceptual areas of interest (i.e., burden, happiness, safety) were identified a priori. Adapted from Roberts, L. W., & Kim, J. P. Attunement and alignment of people with schizophrenia and their preferred alternative decision-makers: an exploratory pilot study comparing treatment and research decisions. Journal of Psychiatric Research, 2015;71:70–77

	Healthy		Ill		*P* value
	N = 47		N = 100		
	strength of endorsement	sd	strength of endorsement	sd	
The researcher(s) tried to make sure:					
I felt comfortable	4.52	0.72	4.74	0.62	0.08
I really wanted to be in the study	3.91	1.11	4.35	0.99	**0.03**
I did NOT feel pressured	4.28	0.96	4.68	0.66	**0.01**
I felt I have choice about whether to drop out	4.76	0.64	4.75	0.62	0.08

Anchors for the survey items were as follows: 1 = Strongly Disagree; 3 = Equally agree/disagree; 5 = Strongly Agree.

Fig. 12.4 Comparison of informed consent questionnaire items directly assessing experiences of voluntarism in consenting to clinical research for both healthy and ill individuals. Reprinted from Journal of Psychiatric Research, 103, Roberts LW, Kim JP, Does informed consent given by healthy individuals when enrolling in clinical research feel less voluntary than for ill individuals? Pp 33–37, Copyright 2018 with permission from Elsevier

of performing empirical ethics research using a stakeholder-based approach, as well as the testability of ethics hypotheses regarding perspectives, attitudes, motivations, behavioral intentions, and decision-making. With this background and understanding, it was decided that a foundational aspect of our BRAIN Initiative project would involve an empirical line of inquiry directly with the stakeholders in neuroscience research—i.e., neuroscience researchers, Institutional Review Board (IRB) members, ethicists, patients with mental illness or addiction, and family members of patients with mental illness or addiction—in order to gain deeper insight into the ethical issues and processes that influence research decision-making.

Novel Theoretical Model of Ethical Research Participation: The Roberts Valence Model for Ethical Engagement in Research

The rationale for our project was additionally grounded in the understanding that ethical engagement of potentially vulnerable volunteers in human studies is predicated on rigorous, authentic informed consent processes that enable positive influences on participation decisions and appropriately safeguard against negative influences [56]. Prior work has identified factors that influence research participation decision-making [57, 58]. Some affect a potential participant's decision-making favorably and appropriately; we call these "positive valence" factors (e.g., altruism; salience of the condition under study; accurate understanding of study procedures, risks, and benefits). Other influences are more ethically problematic; they may sway an individual toward participation by factors that have "negative valence" (e.g., desperation; lack of resources; threats to voluntarism) [59, 60]. Past research has examined positive and negative valence factors in isolation from one another or in relatively small combinations of factors [61–65]. The Roberts Valence Model for Ethical Engagement in Research, represented in Figs. 12.5, 12.6, 12.7, and 12.8, takes into consideration a fuller array of positive and negative valence factors in combination amongst research participation decisions.

It is expected that a decision to participate in research will be influenced by a number of factors, some "positive" and some "negative." The presence of negative

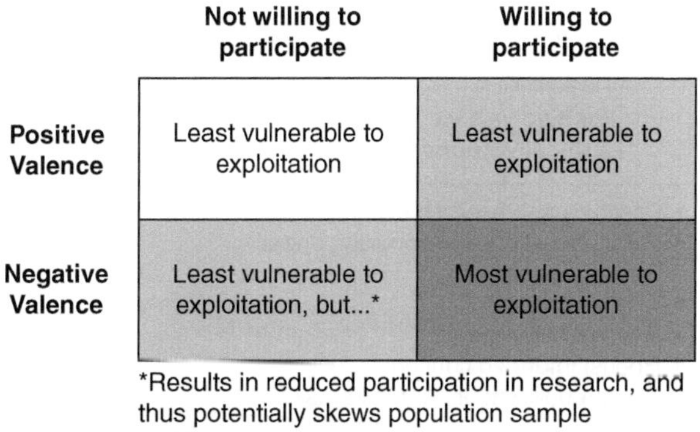

Fig. 12.5 Application of Roberts Valence Model of ethical research participation—a 2x2 construct

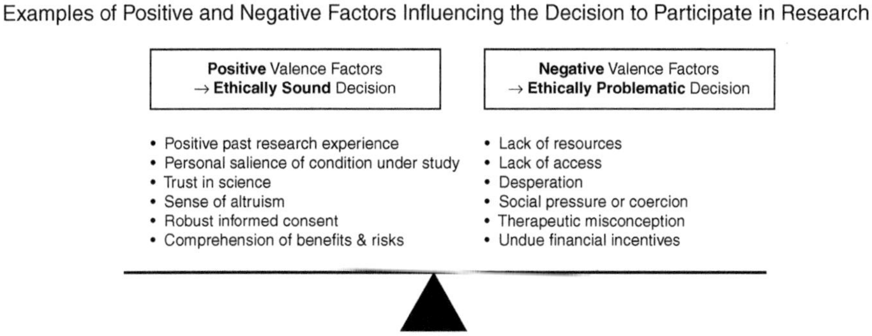

Fig. 12.6 Roberts Valence Model with examples of positive and negative factors influencing the decision to participate in research. Adapted from Roberts LW, Kasun M, Termuehlen G. Ethics in the mental health professions. IN: Roberts LW, Termuehlen G, eds. Professionalism and Ethics Q&A Self Study Guide for Mental Health Professionals, second Edition. Washington, DC: American Psychiatric Association Publishing: 2021, pg 112. Used with permission

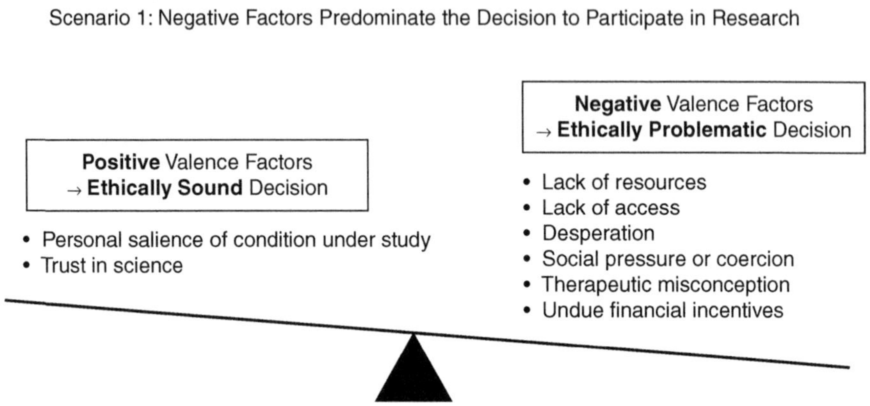

Fig. 12.7 Roberts Valence Model applied to a scenario in which negative factors predominate the decision to participate in research. **Negative** valence factors are predominant in the decision to participate, rendering the overall choice to enroll in research **ethically problematic**

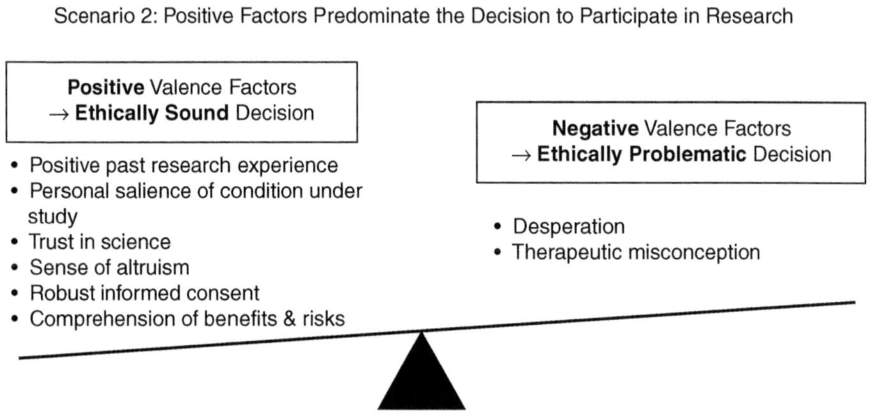

Fig. 12.8 Roberts Valence Model applied to a scenario in which positive factors predominate the decision to participate in research. **Positive** valence factors are predominant in the decision to participate, rendering the overall choice to enroll in research **more likely to be ethically sound**

valence factors is not in and of itself ethically problematic, but overly weighted negative valence factors are problematic. In ethically sound decision-making, negative valence factors will be at least balanced by positive valence factors. Ideally, positive valence factors will shape the decision to participate in research. Researchers can "tip the scale" through robust study-specific safeguards that ensure that positive factors outweigh negative factors.

By applying the Roberts Valence Model to our BRAIN Initiative project, we aimed to examine the influence of positive and negative valence factors on participation decisions of people with mental illness and addiction across a range of innovative neuroscience research. Understanding these valence factors is even more crucially important in cutting-edge research with as yet poorly understood risks and benefits, and in research that involves vulnerable populations. Such efforts lessen the likelihood that volunteers' potential sources of vulnerability (e.g., desperation, misplaced hope, poor understanding, intractable pain, and coercion) are exploited in human research. More positively, attention to engagement with potential volunteers through optimal informed consent interactions and processes can ensure ethical participation of volunteers and enhance trust in science.

Notably, the Roberts Valence Model points to interventions and safeguards of value in ensuring ethical research participation. Because all risks cannot be eliminated or protected against, the safeguards themselves must be particularly well founded, especially when involving potentially vulnerable populations. At the same time, safeguards should not be so prohibitive that they hinder research due to biases about people with mental illness and addiction [66, 67].

Project Methods

The overarching goal of this BRAIN Initiative project was to encourage ethical engagement and innovation in neuroscience research in two main parts: first, by mapping a topography of salient ethical issues in highly innovative neuroscience research related to mental illness and addiction; and second, empirically examining influences on research decision-making in innovative neuroscience research in order to test a new evidence-informed conceptual model—the Roberts Valence Model of Ethical Research Participation. These goals are reflected in the methods for each project aim, demonstrated in Fig. 12.9 and discussed in detail below.

Prospective approval from the Stanford Institutional Review Board (IRB) for this project was received in October 2017 and was continuously maintained throughout the project period. All human subject participants engaged in an informed consent process before the start of any research procedures, and all collected data was de-identified prior to analysis and publication.

Aim 1

The first aim of this BRAIN Initiative project focused on *identifying the distinct ethical issues encountered in innovative neuroscience research related to mental illness and addiction* through the completion of semi-structured interviews with stakeholders. This stakeholder approach was based on our team's understanding that those best-situated to provide detail and insight regarding current, emerging, and possible future ethical issues are those whose careers and professional experiences encourage the development of first-hand views and opinions regarding neuroscience research ethics—namely, neuroscience researchers, ethicists, and IRB members. The rationale of interviewing these professional stakeholders was that it would yield novel data to map this new ethical terrain of innovative neuroscience research.

Fig. 12.9 Sequential design and methods for project aims 1 and 2

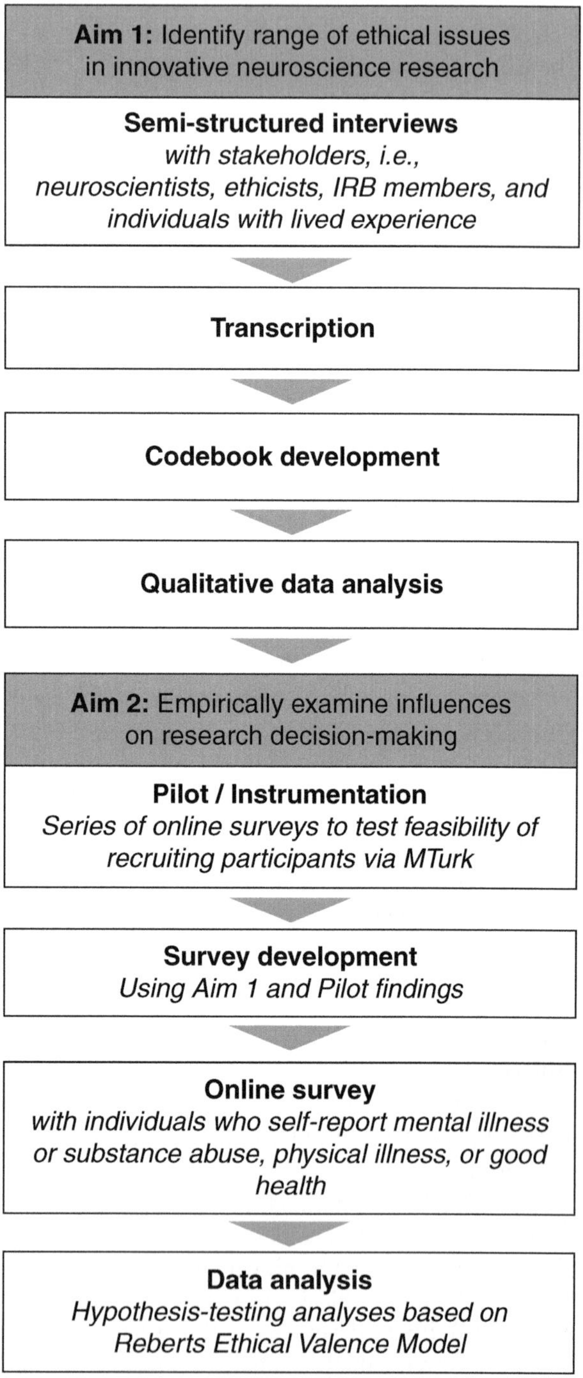

Furthermore, as reflected in the entries in Table 12.1, our prior empirical work has been centrally motivated by the need to give greater emphasis to underrepresented voices in research, including individuals who may be living with or at-risk for mental illness and individuals with multiple sources of vulnerability in the research situation. In order to elevate the voices of these populations, individuals who were living with a mental illness or addiction and immediate family members of individuals who were living with a mental illness or addiction were also included in this project aim. These individuals with lived experience belong to the groups who are most directly impacted by the processes of innovative neuroscience research, as it is these individuals who volunteer to take on the burden and risk of innovative research, and who stand to gain the most from advancements in treatment and care.

During this exploratory Aim 1, over 60 semi-structured interviews with stakeholders were conducted. The "professional" population consisted of over 40 neuroscience researchers, ethicists, and IRB members from Stanford and other universities across the United States. The "lived experience" population included over 20 individuals living with a mental illness or addiction and immediate family members of individuals living with a mental illness or addiction. Stakeholder interviews were designed to be semi-structured in order to facilitate exploration of unanticipated issues and in-depth understanding of the core topics being examined. One-on-one interviews, which typically lasted between 50 and 90 minutes, allowed for the eliciting of diverse, in-depth, and independent information from participants.

Interviews with "professional" participants (researchers, ethicists, and IRB members) were completed by a project co-investigator in-person at Stanford University or via a video call when necessary. Participants were queried regarding three organizing themes: (1) Experiences relevant to research ethics (e.g., specific examples of participant-related issues; experiences with institutional safeguards); (2) Perspectives on policy and implementation issues in neuroscience research; and (3) Differences between neuroscience research and other types of health research.

Interviews with "lived experience" populations (individuals living with mental illness or addiction and family members of individuals living with mental illness or addiction) were completed in-person at Stanford University by a trained team member. These interviews, while still administered in a semi-structured format, provided additional structured context regarding the field of neuroscience research for participants to reference throughout the interview. After a brief introduction to the field of neuroscience research, participants were queried regarding the following topics: (1) Interest in and knowledge of neuroscience; (2) Hopes and fears for neuroscience research; (3) Attitudes toward participation in medical research; (4) Opinions regarding hypothetical scenarios that included various real world neuroscience research projects.

As discussed in more detail in Chap. 13, the audio and video recordings of all interviews were transcribed and then qualitatively coded and analyzed. This exploratory analysis identified key issues, claims, and concerns about the field of neuroscience research, portions of which are reported in the following chapters (See Chaps. 14, 15, and 16).

Aim 2

The second aim of this BRAIN Initiative project was to *empirically examine influences on research decision-making in innovative neuroscience research in order to test a new evidence-informed conceptual model—the Roberts Valence Model of Ethical Research Participation.* To fulfill the project's second aim, we developed a 463-item online survey to examine factors both negative and positive theorized to influence research decision-making by people with mental illness and addiction in the context of innovative neuroscience research, as compared with individuals with physical illness and in good health.

The qualitative findings from the first aim of the project provided insight into stakeholder perspectives, which informed the content of the structured survey in this quantitative second aim. The survey included over 20 validated personality, attitudes, and health instruments that evaluated relevant aspects of participants' experiences and perspectives on research and measured an array of both positive and negative valence factors theorized to influence willingness to participate. The main outcome measure of the survey included a series of research vignettes, which presented details regarding various innovative neuroscience research projects and served as the stimuli to which participants were asked to respond with respect to perceptions of risk and willingness to participate. These research vignettes were carefully developed by the team by drawing on our past work and applying findings from Aim 1. The use of these hypothetical research vignettes allowed our team to manipulate specific dimensions of the research in order to examine protocol-specific influences on decision-making in research. The Aim 2 survey was distributed online via Amazon MTurk (see Chap. 17) to nearly 1000 participants from across the United States living with mental illness or addiction, physical illness, or in good health.

Supplement

In September 2018, this project was awarded a one-year administrative supplement by the National Institute of Aging. The supplement allowed for the expansion of the Aim 1 and Aim 2 populations to include stakeholders in the field of innovative neuroscience research on Alzheimer's disease and related dementias (AD/ADRD). Over 30 AD/ADRD researchers, patients living with mild AD/ADRD and family members of individuals living with AD/ADRD were interviewed to supplement the Aim 1 populations from the parent award. An additional 240 individuals (60 individuals who were at-risk of AD/ADRD, 60 caregivers of individuals with AD/ADRD, and 120 controls) were recruited via Amazon MTurk and completed a modified Aim 2 survey regarding decision-making in innovative AD/ADRD research [68]. The results of this supplemental project are not represented within the scope of this book, and instead will be submitted to peer-reviewed publications for review and consideration.

Preliminary Findings

The following chapters in this section present qualitative findings from the Aim 1 semi-structured interviews, as well as quantitative findings obtained while piloting our survey methods for the Aim 2 online survey. Chap. 13 discusses detailed methods for the completion of our Aim 1 semi-structured interviews and delves into our process for developing and refining the codebook that was used to perform qualitative coding and analysis. Chapters 14–16 present the findings from our Aim 1 qualitative analysis, divided into thematic chapters that we feel best represent the voices and intentions of those we interviewed. Chap. 17 provides background regarding Amazon MTurk, a scalable online workforce that our team employed to recruit the populations necessary for the Aim 2 quantitative surveys, while the Appendix goes on to present findings that emerged while piloting our survey methods for the Aim 2 survey.

Looking Forward

With the completion of this foundational project, we look forward to applying our findings to the future development of a novel screening tool, which will be adaptable to a wide range of clinical research protocols and will aim to help investigators efficiently identify and address both positive and negative valence factors affecting participants' consideration of, or consent to, specific research protocols. In turn, this information will facilitate greater effort by investigators to provide and tailor additional participant safeguards on empirical and individualized bases (e.g., further teaching regarding study risks; clarifying the investigative or innovative nature of the research; helping participants better distinguish between research and treatment aspects; and helping participants identify other resources for treatment). We plan to expand research efforts in four lines of work:

1. Further testing of an evidence-informed conceptual model of ethical participation in research (the Roberts Valence Model) in additional populations, e.g., broader range of illnesses; greater diversity of age and ethnicity, and other research contexts, e.g., types of studies; different settings;
2. Implementation testing of the Roberts Ethical Valence screening tool in a range of studies;
3. Development and testing of interventions aimed at target areas identified by the tool;
4. Creation and dissemination of best practice recommendations from new knowledge, insights, and wisdom of neuroscientists, neuroethicists, and IRB members entrusted with safeguarding human subjects, generated from qualitative insights from interviews.

Innovative research fundamentally provides the possibility of transformative change, but as the parameters of research expand, the need for nuanced efforts to observe, anticipate, and minimize potential ethical issues becomes only more paramount. Our project, detailed here and in the following chapters, uses a stakeholder approach to engage and explore these concerns empirically. Informed by what we learn from this project, we support the development of innovative research tools to support future innovative research endeavors.

References

1. https://braininitiative.nih.gov/
2. https://braininitiative.nih.gov/brain-programs/neuroethics
3. Greely HT, Ramos KM, Grady C. Neuroethics in the age of brain projects. Neuron. 2016;92:637–41.
4. Kim JP. Letter to the editor: machine learning and artificial intelligence in psychiatry: balancing promise and reality. J Psychiatr Res. 2021;136:244–5.
5. Lane-McKinley K, Tsungmey T, Roberts LW. The Deborah Danner story: officer-involved deaths of people living with mental illness. Acad Psychiatry. 2018;42(4):443–50.
6. Kim JP, Rostami M, Roberts LW. Attitudes of mothers regarding willingness to enroll their children in research. J Empir Res Hum Res Ethics. 2020;15(5):452–64.
7. Dunn L, Fisher S, Hantke M, Appelbaum P, Dohan D, Young J, Roberts L. "Thinking about it for somebody else": Alzheimer's disease research and proxy decision makers' translation of ethical principles into practice. Am J Geriatr Psychiatry. 2013;21(4):337–45.
8. Dunn L, Hoop J, Misra S, Fisher S, Roberts L. "A feeling that you're helping": proxy decision making for Alzheimer's research. Narrat Inq Bioeth. 2011;1(2):107–22.
9. Overton E, Appelbaum P, Fisher S, Dohan D, Roberts L, Dunn L. Alternative decision makers' perspectives on assent and dissent for dementia research. Am J Geriatr Psychiatry. 2013;21(4):346–54.
10. Bell E, Racine E, Chiasson P, Dufourcq-Brana M, Dunn L, Fins J, Ford P, Glannon W, Lipsman N, Macdonald M, McAndrews M, Mathews D. Beyond consent in research: revisiting vulnerability in deep brain stimulation for psychiatric disorders. Camb Q Healthc Ethics. 2014;23(3):361–8.
11. Christopher PP, Leykin Y, Appelbaum PS, Holtzheimer PE III, Mayberg HS, Dunn LB. Enrolling in deep brain stimulation research for depression: Influences on potential subjects' decision making. Depress Anxiety. 2012;29(2):139–46.
12. Dunn LB, Holtzheimer PE 3rd, Hoop JG, Mayberg H, Roberts LW, Appelbaum PS. Ethical issues in deep brain stimulation research for treatment-resistant depression: focus on risk and consent. AJOB Neurosci. 2011;2:29–36.
13. Fisher CE, Dunn LB, Christopher PP, Holtzheimer PE, Leykin Y, Mayberg HS, Lisanby SH, Appelbaum PS. The ethics of research on deep brain stimulation for depression: decisional capacity and therapeutic misconception. Ann N Y Acad Sci. 2012;1265:69–79.
14. Leykin Y, Christopher PP, Holtzheimer PE, Appelbaum PS, Mayberg HS, Lisanby SH, Dunn LB. Participants' perceptions of deep brain stimulation research for treatment-resistant depression: risks, benefits, and therapeutic misconception. AJOB Prim Res. 2011;2(4):33–41.
15. Roberts LW, Kim JP. Receptiveness to participation in genetic research: a pilot study comparing views of people with depression, diabetes, or no illness. J Psychiatr Res. 2017;94:156–62.
16. Roberts LW, Kim JP, Tsungmey T, Dunn LB. Do human subject safeguards matter to potential participants in psychiatric genetic research? J Psychiatr Res. 2019;116:95–103.
17. Roberts LW, Tsungmey T, Kim JP, Hantke M. Views of the importance of psychiatric genetic research by potential volunteers from stakeholder groups. J Psychiatr Res. 2018;106:69–73.
18. Roberts LW, Dunn LB, Kim JP, Rostami M. Perspectives of psychiatric investigators and IRB chairs regarding benefits of psychiatric genetics research. J Psychiatr Res. 2018;106:54–60.
19. Rostami M, Dunn LB, Kim JP, Roberts LW. Safeguarding participants in psychiatric genetic research: perspectives of key stakeholder groups. Ethics Hum Res. 2019;41(6):12–22.

20. Kim JP, Ryan K, Roberts LW. Attitudes regarding enrollment in a genetic research project: an informed consent simulation study comparing views of people with depression, diabetes, and neither condition. J Empir Res Hum Res Ethics. 2020;14(4):328–37.
21. Hoop JG, Roberts LW, Hammond KAG. Genetic testing of stored biological samples: views of 570 U.S. workers. Genet Test Mol Biomarkers. 2009;13(3):331–7.
22. Roberts LW, Warner TD, Erickson JA. A preliminary study of employees' views of genetic research: perceived harm, risk, and willingness to participate. AJOB Prim Res. Taylor & Francis Group. 2012;3(4):72–80.
23. Roberts L, Barry L, Warner T. Potential workplace discrimination based on genetic predisposition: views of workers. AJOB Prim Res. 2011;2(3):1–12.
24. Roberts LW, Geppert CM, Warner TD, Green Hammond KA, Rogers M, Smrcka J. Roberts BB perspectives on use and protection of genetic information in work settings: results of a preliminary study. Soc Sci Med. 2005;60(8):1855–8.
25. Roberts LW, Warner TD, Hammond KG, Dunn LB. Assessments by patients with schizophrenia and psychiatrists of relative risk of research procedures. Psychiatr Serv. 2006;57(11):1629–35.
26. Roberts L, Hammond K, Hoop J. An inverse relationship between perceived harm and participation willingness in schizophrenia research protocols. Am J Psychiatry. 2006;163(11):2002–4.
27. Roberts LW, Warner TD, Hammond KG, Hoop JG. Views of people with schizophrenia regarding aspects of research: study size and funding sources. Schizophr Bull. 2006;32(1):107–15.
28. Roberts LW, Dunn LB, Green Hammond KA, Warner TD. Do research procedures pose relatively greater risk for healthy persons than for persons with schizophrenia? Schizophr Bull. 2006;32(1):153–8.
29. Roberts L, Green Hammond K, Warner T, Lewis R. Influence of ethical safeguards on research participation: comparison of perspectives of people with schizophrenia and psychiatrists. Am J Psychiatry. 2004;161(12):2309–11.
30. Roberts LW, Kim JP. Are individuals living with mental illness and their preferred alternative decision-makers attuned and aligned in their attitudes regarding treatment decisions? J Psychiatr Res. 2016;78:42–7.
31. Roberts LW, Kim JP. Attunement and alignment of people with schizophrenia and their preferred alternative decision-makers: an exploratory pilot study comparing treatment and research decisions. J Psychiatr Res. 2015;71:70–7.
32. Kim JP, Roberts LW. Demonstrating patterns in the views of stakeholders regarding ethically-salient issues in clinical research: a novel use of graphical models in empirical ethics inquiry. AJOB Empir Bioeth. 2015;6:33–42.
33. Roberts LW, Kim JP. Does informed consent given by healthy individuals when enrolling in clinical research feel less voluntary than for ill individuals. J Psychiatr Res. 2018;103:33–7.
34. Roberts LW, Kim JP. Healthy individuals' perspectives on clinical research protocols and influences on enrollment decisions. AJOB Empir Bioeth. 2017;8(2):89–98.
35. Roberts LW, Kim JP. Giving voice to study volunteers: comparing views of mentally ill, physically ill, and healthy protocol participants on ethical aspects of clinical research. J Psychiatr Res. 2014;56:90–7.
36. Roberts LW, Warner TD, Dunn LB, Brody JL, Hammond KAG, Roberts BB. Shaping medical students' attitudes toward ethically important aspects of clinical research: results of a randomized, controlled educational intervention. Ethics Behav. 2007;17(1):19–50.
37. Roberts L, Warner T, Hammond K. Coexisting commitments to ethics and human research: a preliminary study of the perspectives of 83 medical students. Am J Bioeth. 2005;5(6):W1–7.
38. Roberts L, Geppert C, Warner T, Green Hammon K, LP L. Bioethics principles, informed consent, and ethical care for special populations: curricular needs expressed by men and women physicians-in-training. Psychosomatics. 2005;46(5):440–50.
39. Roberts L, Warner T, Green Hammond K, Brody J, Kaminsky A, Roberts B. Teaching medical students to discern ethical problems in human clinical research studies. Acad Med. 2005;80(10):925–30.
40. Kaminsky A, Roberts L, Brody J. Influences upon willingness to participate in schizophrenia research: an analysis of narrative data from 63 people with schizophrenia. Ethics Behav. 2003;13(3):279–302.

41. Roberts LW, Warner TD, Anderson CT, Smithpeter MV, Rogers MK. Schizophrenia research participants' responses to protocol safeguards: recruitment, consent, and debriefing. Schizophr Res. 2004;67:283–91.
42. Roberts L, Warner T, Brody J, Nguyen K, Roberts B. What is ethically important in clinical research? A preliminary study of attitudes of 72 psychiatric faculty and residents. Schizophr Bull. 2003;29:607–13.
43. Roberts LW, Warner TD, Nguyen K, Geppert C, Rogers M, Roberts BB. Schizophrenia patients' and psychiatrists' perspectives on ethical aspects of symptom reemergence during psychopharmacological research participation. J Clin Psychopharmacol. 2002;170(1):58–67.
44. Roberts LW, Warner TD, Brody JL. Perspectives of patients with schizophrenia and psychiatrists regarding ethically important aspects of research participation. Am J Psychiatry. 2000;157:67–74.
45. Warner T, Roberts L, Nguyen K. Do psychiatrists understand research-related experiences, attitudes, and motivations of schizophrenia study participants? Compr Psychiatry. 2003;44(3):227–33.
46. Dunn LB, Lindamer LA, Palmer BW, Schneiderman LJ, Jeste DV. Methods to enhance understanding of informed consent in older patients with psychotic disorders [abstract]. Am J Geriatr Psychiatry. 2002;10:36–7.
47. Palmer BW, Dunn LB, Heaton RK, Nayak G, Jeste DV. Decision making capacity in older patients with psychotic disorders [abstract]. Am J Geriatr Psychiatry. 2002;10:36–7.
48. Saks ER, Dunn LB, Lindamer LA, Palmer BW, Marshall PJ, Jeste DV. The 'appreciation' component of capacity to consent to psychiatric research [abstract]. Am J Geriatr Psychiatry. 2002;10:37.
49. Dunn LB, Jeste DV. Problem areas in the understanding of informed consent for research: study of middle-aged and older patients with psychotic disorders. Psychopharmacology. 2003;171:81–5.
50. Palmer BW, Dunn LB, Appelbaum PS, Jeste DV. Correlates of treatment-related decision-making capacity among middle-aged and older patients with schizophrenia. Arch Gen Psych. 2004;61:230–6.
51. Palmer BW, Dunn LB, Depp CA, Eyler LT, Jeste DV. Decisional capacity to consent to research among patients with bipolar disorder: comparison with schizophrenia patients and healthy subjects. J Clin Psychiatry. 2007;68:689–96.
52. Dunn LB, Lindamer LA, Palmer BW, Schneiderman LJ, Jeste DV. Enhancing comprehension of consent for research in older patients with psychosis: a randomized study of a novel consent procedure. Am J Psychiatry. 2001;158:1911–3.
53. Dunn LB, Lindamer LA, Palmer BW, Golshan S, Schneiderman LJ, Jeste DV. Improving understanding of research consent in middle-aged and elderly patients with psychotic disorders. Am J Geriatr Psychiatry. 2002;10:142–50.
54. Dunn LB, Kim DS, Fellows IE, Palmer BW. Worth the risk? Relationship of incentives to risk and benefit perceptions and willingness to participate in schizophrenia research. Schizophr Bull. 2009;35:730–7.
55. Hickman NJ, Prochaska JJ, Dunn LB. Screening for understanding of research in the inpatient psychiatry setting. J Empir Res Hum Res Ethics. 2011;6(3):65–72.
56. Emanuel EJ, Wendler D, Grady C. What makes clinical research ethical? JAMA. 2000;283:2701–11.
57. Tsungmey T, Kim JP, Dunn LB, Ryan K, Lane-McKinley K, Roberts LW. Negative association of perceived risk and willingness to participate in innovative research protocols. J Psychr Res. 2020;122:9–16. https://doi.org/10.1016/j.jpsychires.2019.12.010.
58. Kim JP, Mondal S, Tsungmey T, Ryan K, Dunn LB, Roberts LW. Influence of dispositional optimism on ethically-salient research perspectives: a pilot study. Ethics and Human Research. 2022;44(3):12–23.
59. Roberts LW. Informed consent and the capacity for voluntarism. Am J Psychiatry. 2002;159:705–12.

60. Roberts LW. Mental illness and informed consent: seeking an empirically derived understanding of voluntarism. Report: Lippincott Williams & Wilkins; 2003. Report No.: 545; 16. Disord. 2004;80:115–23.
61. Muroff JR, Hoerauf SL, Kim SY. Is psychiatric research stigmatized? An experimental survey of the public. Schizophr Bull. 2006;32:129–36.
62. Roberts LW, Warner TD, Brody JL, Roberts B, Lauriello J, Lyketsos C. Patient and psychiatrist ratings of hypothetical schizophrenia research protocols: assessment of harm potential and factors influencing participation decisions. Am J Psychiatry. 2002;159:573–84.
63. Christopher PP, Leykin Y, Appelbaum PS, Holtzheimer PE 3rd, Mayberg HS, Dunn LB. Enrolling in deep brain stimulation research for depression: influences on potential subjects' decision making. Depress Anxiety. 2012;29:139–46.
64. Warner TD, Roberts LW, Nguyen K. Do psychiatrists understand research-related experiences, attitudes, and motivations of schizophrenia study participants? Compr Psychiatry. 2003;44:227–33.
65. Roberts LW, Hammond KA, Warner TD, Lewis R. Influence of ethical safeguards on research participation: comparison of perspectives of people with schizophrenia and psychiatrists. Am J Psychiatry. 2004;161:2309–11.
66. Michels R. Are research ethics bad for our mental health? N Engl J Med. 1999;340:1427–30.
67. Michels R. Research on persons with impaired decision making and the public trust. Am J Psychiatry. 2004;161:777–9.
68. Dunn LB, Kim JP, Rostami M, Mondal S, Ryan K, Waraich A, Roberts LW, Palmer BW. Stakeholders' perspectives regarding participation in neuromodulation-based dementia intervention research. J Empir Res Hum Res Ethics. 2021;17(1–2):29–38.

Further Reading

Kim JP, Roberts LW. Demonstrating patterns in the views of stakeholders regarding ethically-salient issues in clinical research: a novel use of graphical models in empirical ethics inquiry. AJOB Empir Bioeth. 2015;6:33–42.

Roberts LW, Kim JP. Giving voice to study volunteers: comparing views of mentally ill, physically ill, and healthy protocol participants on ethical aspects of clinical research. J Psychiatr Res. 2014;56:90–7.

Roberts LW, Kim JP. Attunement and alignment of people with schizophrenia and their preferred alternative decision-makers: an exploratory pilot study comparing treatment and research decisions. J Psychiatr Res. 2015;71:70–7.

Roberts LW, Dunn LB, Kim JP, Rostami M. Perspectives of psychiatric investigators and IRB chairs regarding benefits of psychiatric genetics research. J Psychiatr Res. 2018;106:54–60.

Roberts LW, Kasun M, Termuehlen G. Ethics in the mental health professions. In: Roberts LW, Termuehlen G, editors. Professionalism and Ethics Q&A Self Study Guide for Mental Health Professionals. 2nd ed. Washington, DC: American Psychiatric Association Publishing; 2021.

Chapter 13
Qualitative Phase: Codebook Development

Laura Turner-Essel and Katie Ryan

Introduction

This chapter provides an overview of the methodology used within an ongoing research project funded by the National Institutes of Health's (NIH's) Brain Research through Advancing Innovative Neurotechnologies (BRAIN) Initiative. This project, entitled "Enabling ethical participation in innovative neuroscience on mental illness and addiction: Toward a new screening tool enhancing informed consent for transformative research on the human brain," was awarded in September of 2017 to a team of ethics, psychiatry, and neuroscience researchers at Stanford University. Laura Weiss Roberts, MD, MA, served as the Principal Investigator, while Laura B. Dunn, MD, Jane Paik Kim, PhD, Mildred Cho, PhD, and Casey Halpern, MD, provided support and expertise as co-investigators. The investigators were additionally supported by a team of 10 members with backgrounds in psychology, sociology, public health, and neuroimaging.

The stated overarching goal of this project, which is discussed in greater detail in Chap. 12, was to accelerate neuroscience toward lessening the burden of mental illness and addiction through stakeholder-based empirical ethics inquiry in two main parts (See Fig. 13.1):

- Part 1: Determine the distinct ethical issues and problems encountered in innovative neuroscience related to mental illness and addiction through semi-structured interviews with stakeholders.
- Part 2: Examine factors that influence decision-making in the context of innovative neuroscience research by surveying people with mental illness and addiction and comparing their decision-making with that of individuals with physical illness and healthy controls.

L. Turner-Essel · K. Ryan (✉)
Department of Psychiatry and Behavioral Sciences, Stanford University School of Medicine, Stanford, CA, USA
e-mail: lte@stanford.edu; kryan2@stanford.edu

L. W. Roberts (ed.), *Ethics and Clinical Neuroinnovation*,
https://doi.org/10.1007/978-3-031-14339-7_13

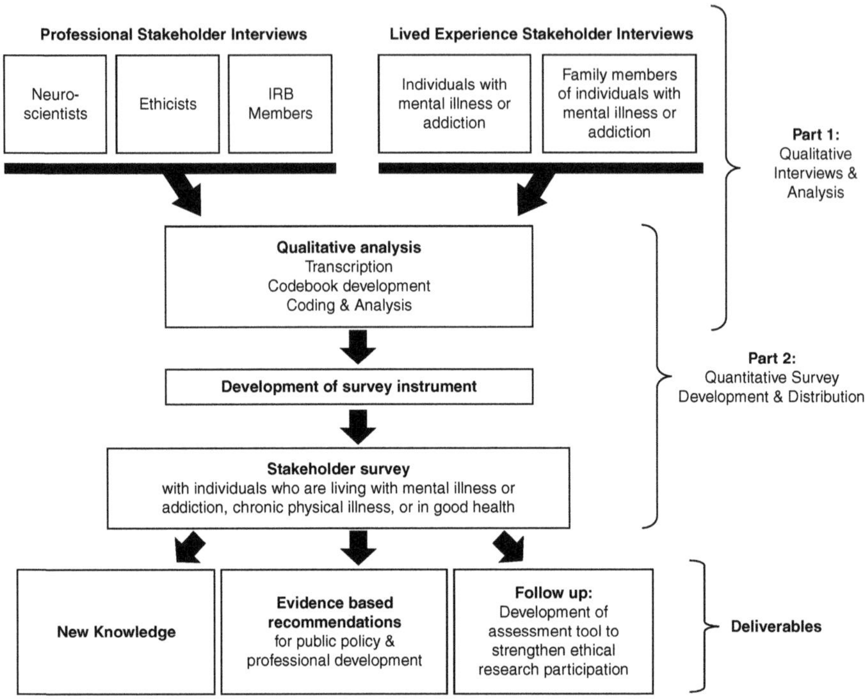

Fig. 13.1 Structural overview of project, demonstrating stakeholder-based empirical ethics approach

This chapter and the three chapters that follow (Chaps. 14, 15, and 16) represent a portion of the work that has been completed through the first part of the project, in which the research team performed semi-structured interviews with 66 stakeholders in order to learn about the ethical issues that are perceived and encountered in innovative neuroscience research. Within this chapter, we discuss the specifics of the qualitative methodologies used throughout the completion, transcription, and analysis of these stakeholder interviews, while Chaps. 14, 15, and 16 will go on to present and discuss initial findings obtained from the analysis of a select number of these interviews.

Research Approach

To address our research question of the ethical issues associated with innovative neuroscience research, we chose a qualitative approach that would allow us to discover various facets of this understudied topic. Qualitative methods are particularly useful when researchers are seeking to uncover aspects of a phenomenon that may be prevalent but underrepresented by traditional models. In this case, the "phenomena" in question are the ethical implications of an evolving science that may not yet be well captured by the existing discourse on medical ethics. Qualitative inquiry allows for the discovery of new ideas that emerge throughout the research process, without the imposition of pre-selected and limited variables upon participants [1].

Morrow [2] argues that a qualitative approach is ideal in the event that variables of interest are not easily identifiable or have yet to be identified, since it would be very difficult to devise a quantitative instrument in such a case. Certainly, this fits the topic of the current study, wherein very little research has explored the perspectives of multiple stakeholders regarding ethical aspects of cutting-edge neuroscience. Thus, an inductive qualitative method was determined to be the most appropriate and efficient way to address the research questions and deepen our understanding.

Beyond its exploratory value, a qualitative approach offers a dynamic and fluid approach that can be more responsive to participants' expressed ideas. The emergent nature of this study allowed the research team to be open to novel connections, encouraging participants to expand on their responses in order to create space for ideas to emerge that were not initially considered. It also offered the opportunity to examine the research topic more in depth, beyond the surface, and to highlight the voices and perspectives of those not often included in dialogue regarding ethics (i.e., patients and their family members).

Research Methodology

A specific qualitative tradition—phenomenology— was utilized to capture participants' experiences with and perceptions of ethical issues in neuroscience, and to understand how they make sense of these experiences and perceptions. Although in the field of psychiatry, the term phenomenology often refers to a descriptive listing of abnormal mental states or symptoms, most likely influenced by Jaspers' *General Psychopathology* [3], the term actually holds a quite different meaning in the world of philosophy. Here we are referring to the movement spearheaded by scholars like Heidegger and Husserl who sought to move away from all reductionist accounts of human experience [4] and to instead explore how human beings make sense of and assign meaning to elements of their day-to-day existence.

Phenomenology as rooted in an interpretivist paradigm seeks to explore the conscious lived experience of phenomena, particularly ways in which phenomena are perceived in everyday life [5]. In this case, we sought to understand participants' perceptions of what ethical issues are involved with innovative neuroscience and how this impacts the way they think about participation in research studies. According to the conception of early twentieth century philosopher Husserl, the examination of everyday experience requires a stepping back from one's typical, unreflective immersion in a taken-for-granted existence [6]. Husserl proposed a "phenomenological attitude" in which one redirects thoughts away from an unreflective immersion in the world to an examination of how phenomena are subjectively experienced. This reflection on one's own psychic life constituted Husserl's original method of phenomenological inquiry. Other philosophers brought diverse interpretations and applications of the phenomenological method [7]. Heidegger [8], for instance, considered phenomenological inquiry an interpretative, rather than purely descriptive, process. He argued that real meaning in one's experiences could only be found through deciphering those experiences, uncovering the meanings concealed by phenomena's surface-level appearances, and thus linked phenomenology to philosophical hermeneutics (the study of the theory and practice of interpretation). Moustakas's [5] transcendental or psychological phenomenology is focused less on the interpretations of the researcher and more on a description of the experiences of participants. In addition, Moustakas focuses on one of Husserl's concepts, "bracketing," in which investigators set aside their own experiences, as much as possible, to take a fresh perspective toward the phenomenon under examination. In this tradition, the research approach consists of identifying a phenomenon to study, bracketing out one's experiences, and collecting data from several persons who have experienced the phenomenon. The researcher then analyzes the data by reducing the information to significant statements or quotes and combines the statements into categories. Following that, the researcher develops a *textural description* of the experiences of the persons (what participants experienced), a *structural description* of their experiences (how they experienced it in terms of the conditions, situations, or context), and a combination of the textural and structural descriptions to convey an overall *essence* of the experience [9].

Participants

One major aim of this study was to gain a well-rounded perspective of the ethical issues pertaining to innovative neuroscience research. As new technologies and treatment methods are developed, numerous segments of society—researchers, research approvers, potential research participants and their loved ones, psychiatry patients, scholars that specialize in research ethics, the general public—are each likely to view or encounter these emerging technologies from a slightly different vantage point. It is useful to understand the various lenses (based on personal and professional experiences, social contexts, personal fears and hopes, etc.) that inform each of these groups' attitudes toward this area of research. Additionally, it is important to consider that each may offer valuable insight into the types of ethical concerns that we may anticipate as rapid advancements continue to occur in the field.

Given the goals of the study, it was crucial to include the perspectives of multiple stakeholders.

Participants in Part 1 of the study represented five groups with both personal and professional interest in innovative neuroscience research:

1. *Neuroscientists (n = 22).* Faculty, affiliated faculty, and executive committee members who were members of the Stanford Neurosciences Institute (SNI) as of October 2017 were eligible for preliminary inclusion in the study. Any Stanford University faculty member who had been awarded a BRAIN Initiative grant between 2014 and 2018 was also eligible for preliminary inclusion. Purposeful sampling allowed members of the research team to identify and recruit professionals based on their expertise in the field of innovative neuroscience research, while also ensuring a representation of diverse research experience.
2. *Ethicists (n = 12).* Members of the research team who have a background in ethics used a snowball sampling technique to identify neuro- and bioethicists from across the United States for preliminary inclusion.
3. *IRB Members (n = 10).* All faculty-level members of a medical Stanford Institutional Review Board (IRB) committee in the 2017–2018 year were eligible for preliminary inclusion.
4. *Lived Experience—Patients (n = 12).* Individuals who were over 18 years of age and self-identified as living with a mental illness or addiction were eligible for preliminary inclusion.
5. *Lived Experience—Family Members (n = 10).* Individuals who were over 18 years of age and self-identified as having a family member with a mental illness or addiction were eligible for preliminary inclusion.

Professionals (neuroscientists, IRB members, and ethicists) were recruited via email. Individuals with lived experience of mental illness and family members were recruited by flyers posted in the Stanford clinics and on the Psychiatry department webpage. Demographic data was collected for all participants.

Instruments

The main data collection method of this study was the semi-structured interview, a common method in much health care-related research. Semi-structured interviews elicit people's own views and descriptions sometimes not anticipated by the researcher. They are commonly used when the aim is to gain information on the perspectives, understandings, and meanings constructed by people regarding the events and experiences of their lives [10].

Interview guides were developed separately for each stakeholder group. These guides consisted of a series of open-ended questions for interviewers to probe with participants, in order to explore participant experiences and attitudes. The guides allowed those members of the research team that would be conducting interviews to pursue the same basic lines of inquiry with each participant and manage the interviews in a systematic and consistent way, while still maintaining enough flexibility to encourage elaboration if a participant expressed particularly strong interest in a given question or topic area. Such structure with room for fluidity assists the researcher in developing a rapport with the interviewee, and thus eliciting more authentic and engaged responses. Though questions in the interview guide were developed based on previous work in the realm of medical ethics, interviewers held these concepts lightly and recognized that they were subject to reformulation and/or rejection as the study progressed [10].

For interviewees in the professional groups, demographics surveys were administered in-person or via email, depending on how the interview was conducted. Surveys contained ten items which assessed participants' race, ethnicity, gender, education level, IRB experience, and federal grant review committee experience.

For patients and family members, demographic and health surveys were administered via an iPad when participants arrived for their scheduled interview. Surveys contained between 44 and 52 items (depending on stakeholder group) which assessed participants' age, race, ethnicity, gender, income, education level, marital status, experience with research participation, and closeness of the relationship with the family member experiencing mental illness (for family members stakeholder group). Participants also completed the PROMIS Scale v1.2—Global Health and 3 questions from SF-12 to assess their general health, and the 14-item Perceived Stress Scale and Life Orientation Test—Revised to assess mental health.

Interview Process

Professionals

Semi-structured interviews were conducted in-person (in the subject's office) or via videoconferencing. In-person interviews were recorded using a handheld audio recording device. Virtual interviews were recorded using the "Audio Only" recording function in the videoconferencing software. All professional interviews were conducted between February 2018 and May 2019. Participants were compensated with retail gift cards in appreciation for their time and effort.

Patients and Family Members

Semi-structured interviews were conducted in-person at one of two Stanford Psychiatry locations between June 2019 and October 2019. All interviews were audio and video recorded. Participants were compensated with retail gift cards and an additional flat rate gift card to cover the costs of their transportation/parking at the interview locations.

Analysis

Analysis might be described as interpretation, making sense of data, or transforming data. Our goal in analyzing the interview data for this project was to identify overarching themes among the ideas presented by our participants and to develop the "big picture" meaning of our data. Thematic analysis involves the search for and identification of common threads that extend across an entire interview or set of interviews [11]. Thus, a thematic analysis approach was chosen to provide a set of defined steps for assessing data, determining themes, and reporting findings in a clear and credible fashion.

Coding

In qualitative inquiry, a code is defined as a word or short phrase that symbolically assigns a summative, salient, essence-capturing, and/or evocative attribute to a portion of language-based or visual data [12]. The development of codes is the initial step in analyzing interview data. To ensure meaningful labels, codes are assigned to chunks of data, usually phrases, sentences, or paragraphs that are connected to a specific topic or context [13]. The purpose is to develop a coding system that will enable the conversion of the data into meaningful and specific units of information (codes or categories).

Codes can be developed a priori from existing theory or concepts (theory-driven); they can emerge from the raw data (data-driven); or they can grow from a specific project's research goals and questions (structural) [14]. Regardless, code development is an iterative process as it requires constant revisiting of theory (theory-driven codes) and repeated examination of the raw data (data-driven codes).

For this project, codes were both theory-derived and data-driven. The research team conducted an initial literature search on ethics in medicine, neuroethics, and ethics in innovative technology [15–22]. Team members also reviewed landmark ethical documents such as the Belmont Report [23] and the U.S. Common Rule, which have a direct impact upon biomedical researchers. This review of the literature helped to elucidate theoretical perspectives that could inform our initial set of codes.

A codebook is a set of codes, definitions, and examples used as a guide to help analyze interview data. Codebooks are essential to analyzing qualitative research because they provide a formalized operationalization of the codes [24]. Even so, like codes, codebooks are developed through an iterative process that may necessitate revising definitions as the researchers gain clearer insights about the interview data and core concepts emerge throughout the course of the project.

The codebook developed for this study was based on the research team's past work in medical and psychiatric ethics, as well as specific principles found throughout the aforementioned foundational literature pertaining to national and international ethical regulations. In order to compile a list of preliminary codes, the team highlighted themes repeatedly mentioned in this literature. Each team member derived 15–20 potential codes from the articles and wrote a short definition for each code. Potential codes and definitions were reviewed and discussed as a group, with the goal of compiling a list of meaningful, clear, and distinct summative labels related to ethics in neuroscience research. The result was the first draft of the codebook. The list of codes derived from this initial literature review are depicted in Table 13.1.

The actual process of coding is an integral part of the interview data analysis process which involves the assignment of codes (operationalized in a codebook) to raw data. This allows researchers to engage in data reduction and simplification. It also allows for data expansion (making new connections between concepts),

Table 13.1 List of codes derived from our initial literature review

Name	Description
Conflict of interest	Any reference to potential or perceived conflicts of interest within research
Culture of academia	Any reference to cultural aspects and norms about specific academic fields of study; any reference to career development; any reference to how researchers value their own work or the work of other academics; any reference to "hype" amongst researchers within a specific field
Current or past practices	Any reference to specific ways that things are done, or ways that things have historically been done, in a specific field. Can include criticism or praise
Funding agencies	Any reference to the processes that fund research
Important	Any text that the coder believes could be crucial to our research that is not currently contained within our designated codes
Industry	Any reference to the commercialization of technologies or procedures; any reference to an investigator's relationships (financial, professional, personal) with non-academic groups or organizations
Influences on decision-making	Any reference to reasons why an individual may or may not choose to participate in a research project (e.g., desperation; incentives)
Informed consent	Any specific reference to the process of informed consent
Innovation	Any reference to first-of-its-kind procedures, methods, tools, or technologies; any reference to specific risks and benefits associated with innovative procedures, methods, tools, or technologies; any discussion around the definition of "innovation" or "innovative" procedures
Institutional regulations	Any reference to institutional regulations, rules, policies, or culture that impact the process of research
Justice	Any reference to the process of selecting subjects; any reference to the distribution of burden and benefits of research
Nontherapeutic use	Any reference to augmentation, enhancement, or the use of neurotechnologies or procedures for any reasons that are not therapeutic; any reference to "dual-use," or the possibility of utilizing the same technology or procedure for both beneficial applications (e.g., clinical) and harmful misuse (e.g., bioterrorism)

Table 13.1 (continued)

Public perception	Any reference to opinions about brain research and neurotech that are experienced on a larger, more public scale; any reference to ideas such as "science fiction" or public "hype" in regard to brain research; any reference to public confidence in research; any reference about how public perceptions may impact research
Recent findings	Any reference to specific findings that have recently emerged from the investigator's work
Research vs. treatment	Any reference to the difference between research and treatment or the therapeutic misconception, either from the perspective of the researcher (e.g., the tension they feel between their duties as a clinician vs. as a researcher) or from the perspective of the participant (e.g., the lack of understanding the differences between participating in a research project on their illness and receiving approved treatment for their illness)
Resources	Any references to resources that are needed for the completion of research (e.g., personnel, equipment, lab space; funds)
Respect for persons	Any reference to the autonomy of patients or research participants; any reference to protections for those with diminished autonomy, including alternative decision makers; any reference to the capacity of an individual to make their own decisions regarding participation in research; any reference to the privacy or confidentiality of patients or research participants
Risks and benefits (beneficence)	Any reference to risks and/or benefits of participating in research. Risks/ benefits can include those that are psychological, physical, legal, social, or economic
Scientific validity	Any reference to the validity of specific research in terms of underlying assumptions, methodology, or analysis; any reference to the competence of the researcher to perform valid research; any reference to the official review of the design, conduct, or analysis of research by an individual or board that is not associated with the research team; any reference to the robustness of evidence, sample sizes, or amount of data that can be collected
Sharing of knowledge	Any reference to the sharing of research methods or results, either within the research community or with the public at large
Uniqueness of the brain	Any reference to how the brain is different than other organs, or why brain research is different than other types of medical research; any reference to scientific uncertainty about how the brain functions
Value of the research	Any reference to the potential of specific research to produce meaningful results; any reference to the prioritization of different types of research based on perceived value; any reference to "clinical equipoise" (e.g., there must be some evidence that the new treatment is better than standard therapy), or directing research to serve a clearly defined need

transformation (converting data into meaningful units), and reconceptualization (rethinking theoretical associations; [25]). Through coding, researchers make connections between ideas, see patterns across participants/sources, and link concepts found in the data with overarching narratives in the research literature.

When beginning to code interview data, the first step is to engage in the process of open coding or "breaking data apart and delineating concepts to stand for blocks of raw data" [26, p. 195]. At this point, several members of the team utilized the initial list of codes to joint code one transcribed interview from the neuroscientist stakeholder group. Coders independently coded the same text and then met to compare the consistency of their code application. Applying these theory-driven codes to the interview data, the team was able to clarify and revise some codes to better reflect the sentiments within the data and eliminate or clarify codes that proved to be ambiguous or redundant. One example was the use of the "Conflict of Interest" code, which some coders had initially interpreted rather broadly but which the team eventually determined would apply specifically to situations in which financial or other personal considerations compromise a researcher's professional judgment.

In discussing this transcript as a group, team members also highlighted ideas that emerged directly from the data but were not yet represented in the code list. After further discussion and consensus about appropriate labels, definitions, and examples of these emergent ideas, these data-derived codes were added to the codebook.

The more specificity in a codebook, the easier it is for coders to distinguish between codes and to determine examples from non-examples of individual codes. In addition, the more detailed the codebook, the more consistency there will be among coders when using it to code interviews. Thus, Macqueen et al. [24] suggest up to six components that can be included within the structure of codebooks to help provide clear guidelines for coders. To this end, the research team developed inclusion/exclusion criteria for each code (what it is/what it isn't) to help clarify and delineate any potentially overlapping categories, in addition to labels, definitions, and exemplar quotes for each code.

A larger group of researchers from the team used this revised codebook to collaboratively code the transcript a second time, again adding and refining codes based on group review and discussion. This inclusion of multiple perspectives and lenses in the coding process helped to establish inter-coder reliability, with diverse ways of analyzing and interpreting the data working together to construct a "shared interpretation and understanding of the phenomenon being studied" [27, p. 382]. Once the team reached consensus on the codebook and determined that the codes listed adequately represented all relevant categories, the team moved forward to code each of the 44 transcripts from interviews with professional participants. The final revised codebook is depicted in Table 13.2.

A protocol was developed whereby each transcript was coded by two randomly assigned coders. Coders independently coded each transcript and then met as a pair to discuss and reconcile any disagreements in their coding. Coders coded all participant portions of the transcript, and kept a running log of particularly intriguing quotes, concepts, and analytical questions to be discussed with the research team during coding checks, or potentially used in later stages of analysis. As a final check, the interviewers (two faculty members in the Psychiatry department) met with each pair of coders to review coded transcripts and ensure that assigned codes seemed accurate and consistent.

Table 13.2 Our final revised codebook

Code	Definition	What it is	What it is not
Academic perception	Any reference to opinions about brain research and neurotechnology as experienced by individuals or groups associated with academia; any reference to how academic fields react to research findings within their own field and outside of their own field; any reference to how specific fields of research are perceived or understood by other academics	Any reference to "hype" amongst researchers within a specific field	
Conflict of interest	Any specific reference by the speaker to potential or perceived conflicts of interest within the research process		The coder should not use this to label things they personally think may be a conflict of interest. Instead, those items should be coded using the appropriate overlapping descriptive codes. The coder can also choose to use the "ethical issue" analytic code if they want to highlight that the perceived conflict of interest is ethically problematic
Culture of academia	Any reference to cultural aspects and norms about academia as a whole or about specific fields of study; any reference to career development within academia		References to how some researchers or fields perceive other researchers or field of research should be coded under "academic perception"
Current or past practices	Any reference to the ways that things are generally done, or ways that things have historically been done, in a specific field. Can include criticism or praise. Any reference to how things are typically done in different countries or locales		References to the speaker's current or past work should be coded under "investigator research and findings," not under "current or past practices"

(continued)

Table 13.2 (continued)

Code	Definition	What it is	What it is not
Ethical regulations	Any reference to institutional review boards (IRBs) or other ethics review boards that review and regulate research involving human or animal subjects; any specific reference to overarching ethical principles or regulations that are generally applied within a field of study		
Funding	Any reference to the processes or agencies that fund research; any reference to how access to funding may influence research	Should include a direct mention of funding or resources	Consider using "institutional regulations" for references to funding institutions (e.g., NIH) that don't mention funding or resources
Human subjects considerations	Any reference to research procedures or safeguards that are specific to research projects which involve human subjects Any reference to "burden" on participants in research (e.g., time, effort, travel)	The sub-categories may sometimes overlap and be difficult to separate from one other. If this occurs, code the entire relevant section as "human subjects considerations" and do not worry about labeling the specific sub-categories	
Influences on decision-making	Any reference to reasons why an individual person or a group of people may or may not choose to participate in a research project or treatment. This could include reasons specific to the individual (e.g., desperation, incentives) or reasons specific to the population (e.g., culture of the population, vulnerability of the population)		
Informed consent	Any specific reference to the process of informed consent or assessing an individual's capacity to consent to participate in research; any reference to the use of alternative decision makers		
Privacy and confidentiality	Any reference to concerns about or protections for the privacy or confidentiality of patients or research participants		

Risks and benefits	Any reference to risks and/or benefits of participating in a research project (psychological, physical, legal, social, or economic)		
Subject selection	Any reference to the process of selecting subjects or determining who does or does participate in specific research projects		
Industry	Any reference to the commercialization of technologies or procedures; any reference to how non-academic groups or organizations may be involved in or influence research; any reference to an investigator's relationships (financial, professional, personal) with non-academic groups or organizations		
Innovation	Any specific reference to first-of-its-kind procedures, methods, tools, or technologies; any reference to specific risks and benefits associated with innovative procedures, methods, tools, or technologies; any discussion around the definition of "innovation" or "innovative" procedures		
Institutional regulations	Any reference to institutional regulations, rules, policies, or culture that may impact the process of research	This includes regulations established by academic institutions, funding institutions, and governmental institutions	This does not refer to regulations put in place by private companies or corporations (this would be coded "industry") or regulations put in place by ethics review boards or committees (this would be coded as "ethical regulations")
Investigator research and findings	Any specific reference to the speaker's current or past research, projects, or findings		
Lived experience	Any reference to the experiences of a patient or research participant in the research process or in the daily lives, especially in relation to living with an illness		

(continued)

Table 13.2 (continued)

Code	Definition	What it is	What it is not
Nontherapeutic use	Any reference to augmentation, enhancement, or the use of neurotechnologies or procedures for any reasons that are not therapeutic; any reference to "dual-use," or the possibility of utilizing the same technology or procedure for both beneficial applications (e.g., clinical) and harmful misuse (e.g., bioterrorism); any reference to off-label use, or use by a non-sponsored entity	References to how for-profit clinics may use a tool, technology, or product can also be coded here	
Public perception	Any reference to opinions about brain research and neurotechnology that are experienced on a larger, more public scale; any reference to public confidence in research; any reference about how public perceptions may impact research	Any reference to ideas such as "science fiction" or public "hype" in regard to brain research	
Research vs. treatment	Any reference to the difference between research and treatment or the therapeutic misconception, either from the perspective of the researcher (e.g., the tension they feel between their duties as a clinician vs. as a researcher) or from the perspective of the participant (e.g., the lack of understanding the differences between participating in a research project on their illness and receiving approved treatment for their illness)		
Resources	Any references to resources that are needed for the completion of research (e.g., personnel, equipment, lab space; funds; access)	This code may often overlap with "funding". Be sure to code for both if that is the case	
Sharing of knowledge	Any reference to the sharing of research methods or results, either within the research community or with the public at large. Any reference to the sharing of data	–	
Uniqueness of psychiatry	Any reference to how the fields of psychiatry and/or neuroscience are different than other medical fields; any reference to how the fields of psychiatry and/or neuroscience may be perceived or treated differently than other medical fields	References to how psychiatry as a field is unique from other fields	

Uniqueness of the brain	Any reference to how the brain is different than other organs and how this may impact research; any reference to scientific uncertainty about how the brain functions; any reference to how the brain may or may not contain an individual's personhood or identity	References to how the brain as an organ is unique from other organs	References to precision medicine or n-of-1 studies should be coded as "investigator research and findings" or "current or past practices," not as "uniqueness of the brain"
Validity of the research	Any reference to the validity of specific research in terms of underlying assumptions, methodology, or analysis; any reference to the competence of the researcher to perform valid research; any reference to the robustness of evidence, sample sizes, or amount of data that can be collected	Includes any criticism or praise about the METHODS of research (underlying assumptions, processes of data collection, methods of analysis). Includes mention of incidental findings or p-hacking. This code may often occur in coordination with "current or past practices" or "culture of academia." Be sure to code both/all if this is the case	References that comment on whether or not specific research has value or is useful to society should be coded under "value of research"
Value of the research	Any reference to the potential of specific research to produce meaningful results; any reference to the prioritization of different types of research based on perceived value; any reference to "clinical equipoise" (e.g., there must be some evidence that the new treatment is better than standard therapy), or directing research to serve a clearly defined need	Includes any criticism or praise about the *purpose* of research This code may often occur in coordination with "current or past practices" or "culture of academia." Be sure to code both/all if this is the case	References that comment on whether or not specific research is done correctly or well should be coded under "validity of research"

From Codes to Themes

After initial coding of the data, the team engaged in axial coding to recognize themes emerging within and across interviewee responses [28]. According to Corbin and Strauss [26], axial coding represents a higher level of analysis which enables researchers to identify any connections that may exist between codes. It "further manages, filters, highlights, and focuses the salient features of the qualitative data record for generating categories, themes, and concepts, grasping meaning, and/or building theory" [12, p. 8].

Team members summarized coded passages of data to elucidate similar ideas across participants and find areas of overlap or conceptual connection. By reassessing the data in this way, the research team was better able to grasp the underlying sentiments, concerns, and topics occurring within the narratives of participant responses. These underlying features represented the themes within each code, which Moustakas [5] defined as the essential, invariant structure (or "essence") of a given phenomenon.

The final step of analysis involved organizing the codes and their core themes into broader, overarching categories. These categories, or "buckets," helped to bring meaningful order to the interview data and to relate the story woven throughout the perspectives of our various stakeholders.

The research team utilized a concept-mapping process to assist the categorization of codes and themes. The primary function of this was to re-think and reorganize data in order to highlight the underlying structure of participant responses. Ultimately, the goal was to accurately represent for readers a broad map of the ethical concerns, questions, and issues raised by the diverse set of participants in this study. Our two major categories and their underlying codes are represented in Table 13.3 below. The concept maps in Figs. 13.2 and 13.3 are the final product of

Table 13.3 High level categories and underlying codes derived from participant responses

Category 1: The Human Experience of Research	Category 2: The Structures, Policies, and Practices of Research
Lived experience	Culture of academia
Human subjects considerations	Ethical regulations
Influences on decision-making	Funding
Informed consent	Innovation
Privacy and confidentiality	Industry
Risks and benefits	Resources
Subject selection	Validity of research
Value of research to research participants	Value of research
Public perception	Conflict of interest
Education of research participants ('giving back' to the community)	Improving consent forms
Ethics of innovation	Ethical concerns and risks related to big data
Connecting research to patients/clinical care	Research vs. treatment
Nontherapeutic use	Sharing of knowledge
Funding concerns post-research	Institutional regulations
Patient access to care	Technology's influence on research
Capacity to consent	Innovative methods of subject selection

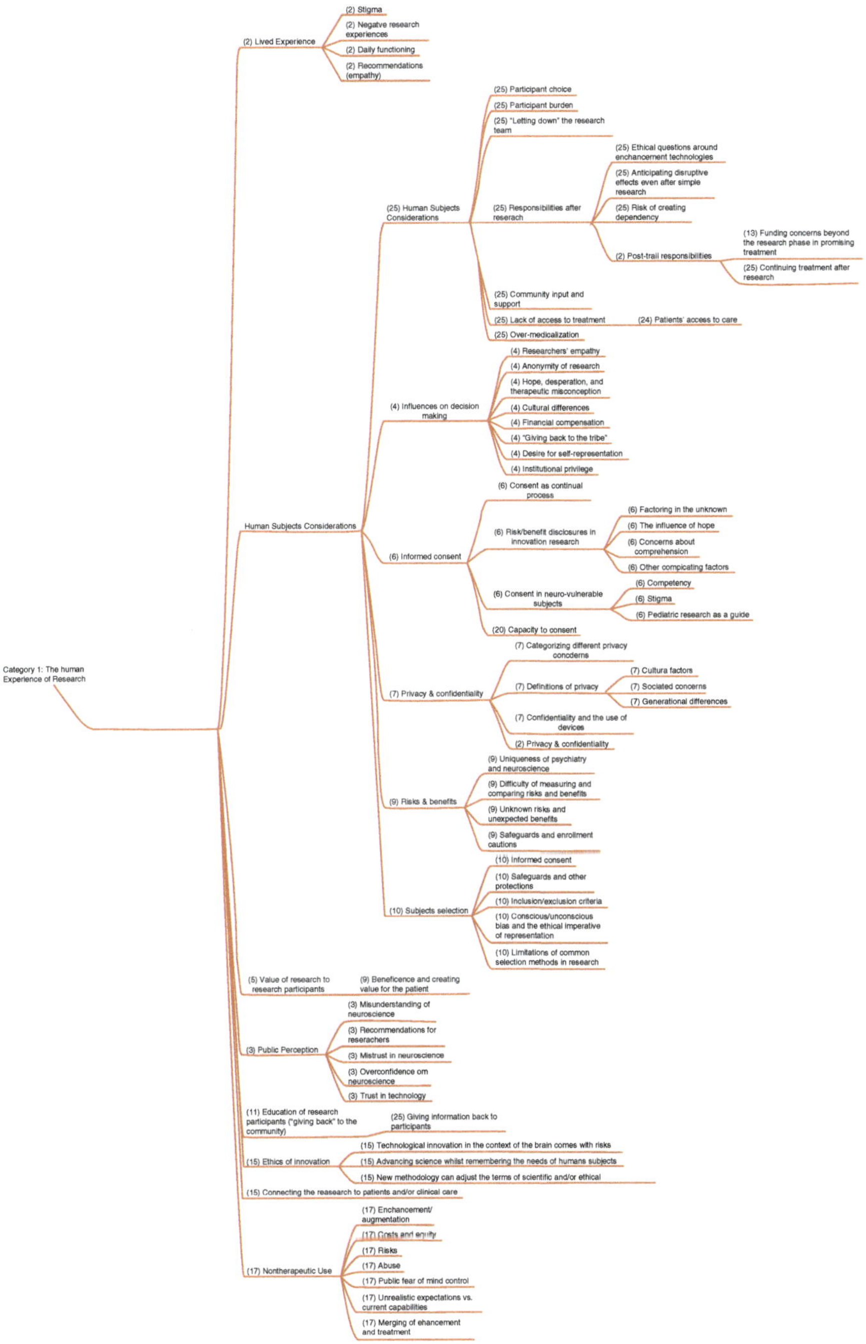

Fig. 13.2 Concept map of Category 1: The Human Experience of Research. Created with MindNode

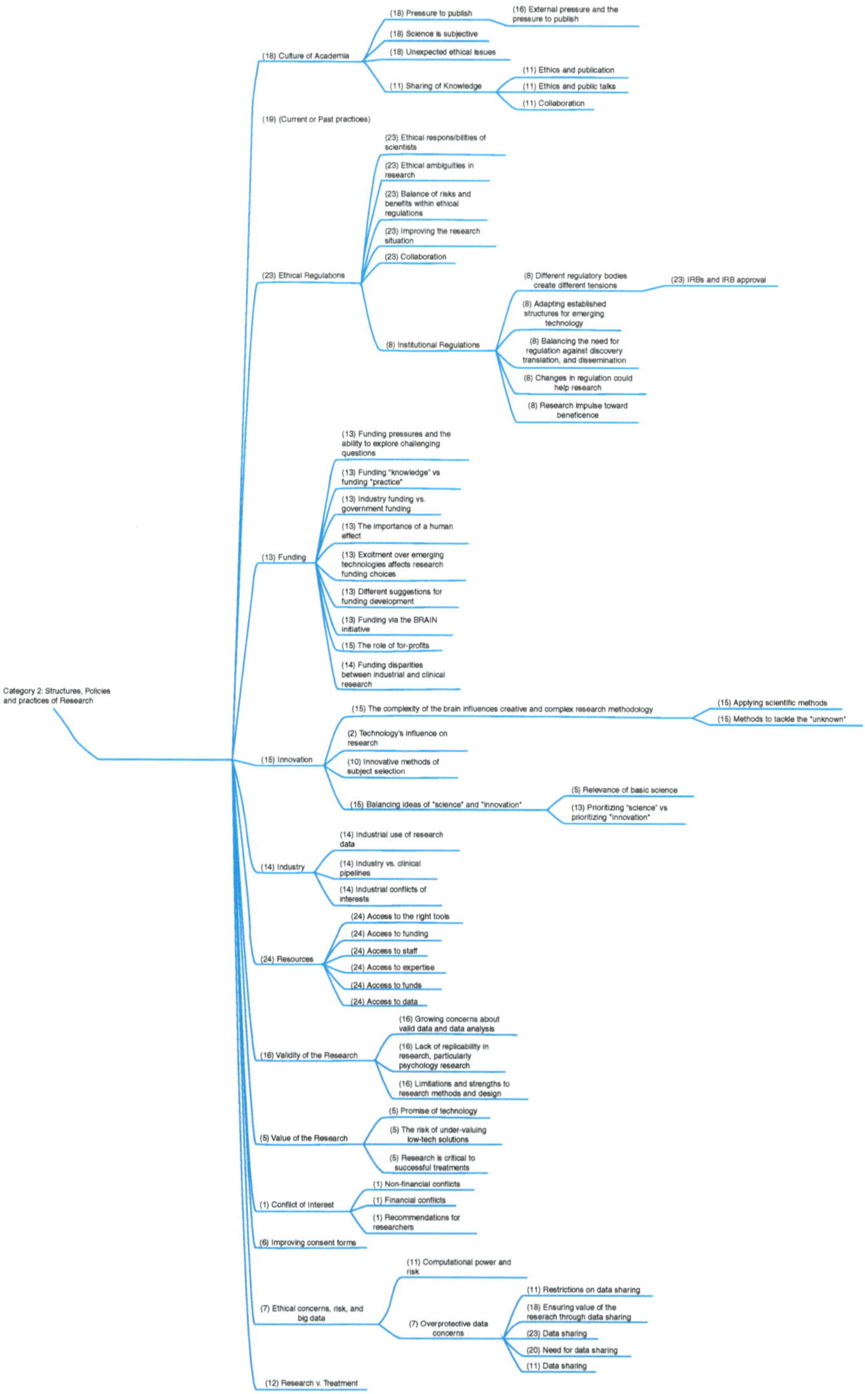

Fig. 13.3 Concept map of Category 2: The Structures, Policies, and Practices of Research. Created with MindNode

our organizing process. Category 1, The Human Experience of Research, is described in depth in Chap. 14. Category 2, The Structures, Policies, and Practices of Research, is described in depth in Chaps. 15 and 16.

Rigor and Trustworthiness

This section will describe steps taken to practice and demonstrate rigor in the conduct of this study and to establish confidence in the findings. The components of qualitative rigor are addressed in turn.

Credibility (Truth Value)

A study is considered credible when it presents an interpretation of an experience in such a way that people sharing that experience immediately recognize it [29]. This study utilized both triangulation of sources (interviews with different populations) and triangulation of analysts (independent and compared analysis by several different members of the research team) to ensure a rich, robust, and complete understanding of the data. Also, quotes from actual participant interviews are used to illustrate codes, categories, and themes as a way to verify that researcher interpretations of the data match the reality of participant experience.

Transferability (Applicability)

Transferability refers to the ability to generalize research findings or methods from one group to another. Transferability is most often a process performed by *readers* of research, noting the descriptions of participants and methods of data collection provided and comparing with their own contexts. This study collected demographic data so that readers may be able to assess the representativeness of the sample in relation to their own. This study also used the same data collection methods with all demographic groups and provides thick descriptions of the coding process.

Dependability (Consistency)

Dependability occurs when fellow researchers can follow the set of decisions made by the researcher in a study. In this project, we ensured dependability from the outset by building peer review into every step of the analysis process. The research team also maintained an audit trail—a complete set of notes on decisions made during the research process, research team meetings, memos and reflective thoughts on the data, sampling choices, the development of instruments, emergence of results and findings, and data management practices. In addition, the inclusion of advisory board members in our research process helped to provide a set of "outside" eyes to hold the research team accountable for all decisions. Finally, a detailed description of the research methods (above) is intended to lend transparency and clarity to our specific goals, steps, and rationale throughout the entire research process.

Confirmability (Neutrality)

Confirmability ensures that the conclusions of a study are based not on the researchers' viewpoints but are instead grounded in the data. The audit trail provides some transparency in regard to the research path and how results were drawn directly from participants' data. It is important to note that utilizing the semi-structured interview protocol, interviewers routinely asked for clarification or elaboration from participants to ensure that the ideas of the participants were being fully understood and accurately captured. Researcher *reflexivity*—self-awareness of assumptions and biases brought to the research process—was facilitated through personal notes and memos wherein research team members noted their subjective responses to the participants and the research process so as not to let these cloud their collection, analysis, and interpretation of the data.

Final Note

Due to the multi-phasic nature of this project, interview data for each stakeholder group was constantly in some stage of collection, coding, analysis, or writing throughout the life of the study. The research team chose to code and analyze data by stakeholder group, once each group's interview transcripts were complete. The data included in the three chapters that follow (Chaps. 14, 15, and 16) is that of the neuroscientists and IRB members, whose interviews were the first to be completed. Responses from ethicists, patients, and family members were analyzed separately and will be reported by other means. An ongoing analysis of all data will surely reveal patterns and differences across and between groups which will be of interest to readers. We continue to explore these rich and varied perspectives for more insight into the ethical issues at stake.

Key Points

1. Qualitative analysis of interview data is performed by a team-based, iterative process including the interviews themselves, codebook development, and consensus-based coding of transcripts.
2. Codes can be, and most often are, both theory- and data-driven.

3. The qualitative rigor and trustworthiness of a coding protocol is determined by assessing its credibility (truth value), transferability (applicability), dependability (consistency), and confirmability (neutrality).
4. The greater the specificity of a codebook, the greater its utility in qualitative research.

Questions to Consider

1. What influence does subject matter (here, neuroscience and the brain) have on our understanding of the components of qualitative rigor (credibility, transferability, dependability, and confirmability)?
2. Does the level of innovation change our understanding of the codebook development process, the creation of themes, or our assessment of qualitative rigor?
3. Is there a difference in codebook development between professional and lived experience interviewee groups? Should there be?

References

1. Hill CE. Qualitative research. In: Norcross JS, Beutler LE, Levant RF, editors. Evidence-based practices in mental health: debate and dialogue on the fundamental questions. Washington, DC: American Psychological Association; 2005. p. 74–80.
2. Morrow SL. Qualitative research in counseling psychology: conceptual foundations. Couns Psychol. 2007;35(2):209–35. https://doi.org/10.1177/0011000006286990.
3. Jaspers K. General psychology. Manchester: Manchester University Press; 1963.
4. Bracken P. Phenomenology and psychiatry. Curr Opin Psychiatry. 1999;12(6):593.
5. Moustakas C. Phenomenological research methods. Thousand Oaks, CA: Sage; 1994.
6. Welton D. The essential Husserl: basic writings in transcendental phenomenology. Bloomington: Indiana University Press; 1999.
7. Moran D. Introduction to phenomenology. Milton Park: Routledge; 2000.
8. Heidegger, M., Macquarrie, J., & Robinson, E. (1962). Being and time.
9. Creswell JW. Qualitative inquiry and research design: Choosing among five approaches. 2nd ed. Thousand Oaks, CA: Sage; 2007.
10. Al-Busaidi ZQ. Qualitative research and its uses in health care. Sultan Qaboos Univ Med J. 2008;8(1):11–9.
11. DeSantis L, Noel Ugarriza D. The concept of theme as used in qualitative nursing research. Western J Nurs Res. 2000;22(3):351–72.
12. Saldaña J. The coding manual for qualitative researchers. Thousand Oaks: Sage; 2015.
13. Miles MB, Huberman AM. Qualitative data analysis: an expanded sourcebook. 2nd ed. Thousand Oaks, CA: Sage; 1994.
14. Ryan GW, Bernard HR. Techniques to identify themes. Field Methods. 2003;15(1):85–109.
15. Emanuel EJ, Wendler D, Grady C. What makes clinical research ethical? JAMA. 2000;283(20):2701–11.
16. Farah MJ. An ethics toolbox for neurotechnology. Neuron. 2015;86(1):34–7.
17. Gardner J, Williams C. Responsible research and innovation: a manifesto for empirical ethics? Clin Ethics. 2015;10(1–2):5–12.
18. Giordano J. A preparatory neuroethical approach to assessing developments in neurotechnology. AMA J Ethics. 2015;17(1):56–61.
19. Goering S, Yuste R. On the necessity of ethical guidelines for novel neurotechnologies. Cell. 2016;167(4):882–5.

20. Ienca M, Jotterand F, Elger BS. From healthcare to warfare and reverse: how should we regulate dual-use Neurotechnology? Neuron. 2018;97(2):269–74.
21. Roberts LW. Ethical dimensions of psychiatric research: a constructive, criterion-based approach to protocol preparation. The research protocol ethics assessment tool (RePEAT). Biol Psychiatry. 1999;46(8):1106–19.
22. Roberts LW. A clinical guide to psychiatric ethics. Washington, DC: American Psychiatric Publications; 2016.
23. United States. The Belmont report: ethical principles and guidelines for the protection of human subjects of research. Bethesda, MD: The Commission; 1978.
24. MacQueen KM, McLellan E, Kay K, Milstein B. Codebook development for team-based qualitative analysis. CAM J. 1998;10(2):31–6.
25. Coffey A, Atkinson P. Making sense of qualitative data: complementary research strategies. Thousand Oaks CA: Sage; 1996.
26. Corbin J, Strauss A. Basics of qualitative research. 3rd ed. Thousand Oaks CA: Sage; 2008.
27. Weston C et. al. Analyzing interview data: the development and evolution of a coding system. Qual Sociol. 2001;24:381–400.
28. Yi E. Themes don't just emerge — coding the qualitative data; 2018. https://medium.com/@projectux/themes-dont-just-emerge-coding-the-qualitative-data-95aff874fdce.
29. Patton MQ. Qualitative evaluation methods. Thousand Oaks, CA: Sage; 2002.

Further Reading

Bracken P. Phenomenology and psychiatry. Curr Opinions Psychiatry. 1999;12(6):593.

Coffey A, Atkinson P. Making sense of qualitative data: complementary research strategies. Thousand Oaks: Sage; 1996.

Emanuel EJ, Wendler D, Grady C. What makes clinical research ethical? JAMA. 2000;283(20):2701–11.

Chapter 14
Qualitative Findings: Diverse Stakeholder Perspectives on Ethical Considerations in Innovative Neuroscience Research Involving Human Volunteers

Laura B. Dunn, Max Kasun, Katie Ryan, Kyle Lane-McKinley, and Laura Weiss Roberts

Introduction

This chapter details key qualitative findings from the first part of the National Institutes of Health's Brain Research through Advancing Innovative Neurotechnologies (BRAIN) Initiative®-funded project, "Enabling ethical participation in innovative neuroscience on mental illness and addiction: Toward a new screening tool enhancing informed consent for transformative research on the human brain." In this first part of the project, members of our research team conducted semi-structured interviews with 44 stakeholders with different professional backgrounds (i.e., neuroscience researchers, Institutional Review Board (IRB) members, and ethicists) in order to better understand stakeholder concerns and perceptions about the ethical issues encountered in innovative neuroscience endeavors. Our research team then undertook an iterative analysis (see Chap. 13) to better delineate and describe the stakeholder perspective. This chapter focuses specifically on our analysis of researchers' and IRB members' perceptions of the ethical concerns related to the participation of human participants in innovative neuroscience and psychiatry research; discussions of ethical issues related to the institutional practices and policies involved in innovative neuroscience research can be found in the following chapters.

The Roberts Valence Model, described in Chap. 3, identifies a number of factors that may influence individuals' decisions regarding participating in research.

L. B. Dunn
Department of Psychiatry, University of Arkansas for Medical Sciences, Little Rock, AR, USA
e-mail: ldunn@uams.edu

M. Kasun · K. Ryan · K. Lane-McKinley (✉) · L. W. Roberts
Department of Psychiatry and Behavioral Sciences, Stanford University School of Medicine, Stanford, CA, USA
e-mail: maxkasun@stanford.edu; kryan2@stanford.edu; kylemck@stanford.edu; robertsl@stanford.edu

L. W. Roberts (ed.), *Ethics and Clinical Neuroinnovation*,
https://doi.org/10.1007/978-3-031-14339-7_14

Whether and to what degree researchers themselves appreciate the range, nature, and impact of factors on participants' decision-making has important implications for the ethical foundations of research. As the interviews with researchers and IRB members were completed, transcribed, and coded, our team began to identify broad topic categories that were found across multiple interviews (see Chap. 13) and a number of common themes related to the ethical completion of innovative neuroscience research involving human participants emerged. The themes and subthemes explored in this chapter demonstrate both the breadth and detail of the questions and observations that researchers and IRB members discussed regarding to the ethical conduct of research involving human volunteers in a rapidly evolving and unique field of study.

Perspectives on Public Understanding and Trust in Neuroscience

In their open-ended responses, a cohort of the interviewed researchers discussed their impressions regarding the general public's understanding of and attitudes toward neuroscience research. As a whole, researchers recognized the draw that neuroscience has on the general public, emphasizing humans' innate fascination with the brain and portrayals of neuroscience in popular media. As one researcher noted:

> Humans I think have this kind of innate dualism where they know that we think with our brain, but still they are somehow amazed if you can show them, "Hey here is how love works in your brain!" Buzzfeed will do a story about it: "This is your brain on X." But still there is this weird disconnect that people have where, even though they understand brains do thinking, they don't think of thinking as a physical phenomenon. (Professional Stakeholder Participant)

Because the brain is widely understood as the home of personality, consciousness, and other elements of the self, there is elevated public interest in research that focuses on it. Researchers emphasized, however, that this heightened interest, paired with the complexity and ongoing mysteriousness of the brain, may lead to flawed understandings and misperceptions about the field of neuroscience overall.

Limits to Understanding

Multiple researchers expressed the view that, in general, the public does not possess a solid understanding of the current capabilities and limitations of psychiatric or neuroscientific research. They articulated that this lack of understanding is likely related to the inherent complexity of the brain and the ever-evolving state of neuroscience research. In some cases, researchers speculated that the public's understanding of neuroscience as a field might also be affected by media hype and misinformation, as media institutions may be tempted to generalize or overstate the likely impact of scientific findings:

> I think what happens in the newspapers and online is that journalists will highlight findings and make them sound more positive than they actually are, because obviously it sells. So like, "New finding! There is a new difference between men and women in the brain," because it's such a hot topic, and then you'll read, scroll down, and see "Well actually the scientist didn't really say this and they're really cautious about this finding." (Professional Stakeholder Participant)

Beyond the impact that media can have on public perception, researchers also discussed how the complexity of the brain itself might lead even the most informed individuals to have incorrect or divergent conceptions of the field of neuroscience. Several researchers interviewed noted that even physicians and scientists whose work is not focused on brain science may possess an incomplete understanding of the goals and possibilities of neuroscience, with one participant stating:

> I would extend the layperson description to other people who are equally researchers, PIs, physicians, who just don't have direct understanding or experience of what neuroscience is doing, what psychiatry is doing. Working on neuroscience you're working on an organ that generates the person in a sense, so there's all these hundreds of year-old mythical ideas we have about souls, and still a lot of people hold — whether they express it in a way that they understand it or not — they hold dualistic views. (Professional Stakeholder Participant)

Some of the researchers commented on how a lack of clarity or understanding regarding science more broadly may encourage portions of the general public to mistrust neuroscientific research. For example, one researcher noted that to someone who has not been educated in the scientific method, the terminology used (e.g., "theory") may lead to the belief that there are no "facts" in science. Others observed that some inherent aspects of science and academia—e.g., debate among researchers about the accuracy or significance of certain findings—may feed into mistrust of science by those less familiar with these processes.

Another factor noted to affect public perceptions of neuroscience—and psychiatry in particular—was that some members of the general public continue to view mental illnesses as not "real" diseases, and therefore less deserving of scientific attention when compared to disorders or injuries that manifest physically. One respondent discussed this issue, noting "Still so much of society views mental disorders as, 'Just toughen up. It's not really a disease. What's the matter with you? Just straighten up, be a man.' So it's a little different from breaking arms." As was noted by more than this one researcher, the perspective that individuals with mental illness simply need to "toughen up" may affect public perspectives on research into these illnesses, furthering stigma or reducing the willingness of people with mental illnesses to participate in research or seek help.

Unrealistic Perceptions about the Capabilities of Neuroscience

Researchers described how the general public's limited understanding of neuroscience research can lead to unrealistic perceptions about the power and possibilities of the field. Such unrealistic perceptions, including the belief that neuroscientists possess capabilities that are widely considered impossible or not possible yet, were discussed by multiple researchers. These researchers noted a tendency for subsets of the population to view certain open epistemological and scientific questions as signifying or even advancing towards dystopian events in the future, a perception that seems to be especially heightened when related to brain research. They commented about the influence of "science fiction" on fear of mind control or manipulation among the public, emphasizing that this fear may represent a mismatch between public perceptions of what neuroscience is currently able to achieve and what the field can realistically do. One researcher noted that "In part, it's giving us way more power than we actually have. If we ever got to that capacity to control brains in that way, that means we've benefited a lot along the way, and that's sort of forgotten."

Other researchers warned about the risks that overconfidence in neuroscience research may pose to potential research participants. One researcher emphasized that, if potential research participants perceive brain research as being more advanced than it actually is, they may in turn underestimate the risks associated with participation, or may be less likely than in other fields to ask questions about the study or to refuse participation:

> They think because there's technology, there's been a lot of safety done. It is sort of like, you go to the doctor and doctor is like god. So you go, "This is this high technology, I don't need to ask questions, this is all ordained." And they don't understand it's not. I think that's what I am trying to put into words what's bothering me. I think that's what it is. They just go into this — "Wow, technology can do all this?!" — so they don't ask questions, they assume it's safe. And because it's so powerful, there is a hubris about it too, and the people using it have a hubris. (Professional Stakeholder Participant)

As is referenced in this quote, researchers observed that when innovative tools are introduced into the research context, participants may be more inclined to place their faith or trust in projects using these tools. Researchers expressed that research participants seem to highly value novelty in technology and may hold assumptions that novel means more powerful or advanced, and that the use of this technology in a research project may result in participants having an unfounded sense of safety or invulnerability.

Ethical Considerations in Recruitment & Participant Selection

Multiple researchers and IRB members discussed their experiences and concerns related to recruiting participants and enrolling them in different types of neuroscience and psychiatry research. As is explored below, recruitment practices—and the resulting research sample—have important ethical implications. Throughout our interviews, researchers and IRB members discussed the importance of recruitment procedures in terms of the identification and enrollment of eligible participants, researchers' ability to fulfill their ethical obligation to complete the project, the need to determine whether the sample of participants is representative of the broader population, and the need for attunement to how recruitment processes may exert subtle or unconscious pressures on research participants.

Racial, Ethnic, and Socioeconomic Biases

Researchers and IRB members agreed that participant selection is susceptible to the conscious and unconscious biases of researchers, as well as self-selection biases that are inherent in the ecological contexts in which the research is conducted—such as the demographic (e.g., racial and socioeconomic) characteristics of a given region. They identified lack of education, unstable income and housing, and other psychosocial and socioeconomic difficulties as conditions that pose barriers for some people to enroll in research. They agreed that these barriers reduce the quality of research by reducing participant diversity, disproportionately incorporating attributes of relatively more affluent, educated, or otherwise homogeneous populations into research findings:

> Another ethical challenge that is more easy to manage is diversity: to make sure that in the program and the projects that we do, we have enough racial diversity, ethnic diversity. Many of the normal range laboratories that we have, or what we consider normal, or the criteria that we have for a diagnosis, are based on studies that have not had that diversity. So when you're seeing an individual that is from a population that is a minority population, you don't know how well or how much you should rely on that criteria, or how much you should rely on this being an abnormal finding. We have a responsibility when we do new research to have that diversity, but sometimes that might be easier than other times. (Professional Stakeholder Participant)

The "Mythical" Psychiatric Patient

When it comes to the recruitment and selection of participants for neuroscience and psychiatry research specifically, additional challenges were identified. Several researchers expressed concern that research in these fields often includes participants who do not accurately represent the populations that clinicians in the field actually serve. Due to the stringent requirements on "clean samples and clean experimental contexts" imposed by funders, researchers often are limited to recruiting single-diagnosis psychiatric patients for inclusion in their studies. This recruitment strategy results in the inclusion of "mythical" patient populations, which do not accurately represent the patients with multiple comorbidities who most often "walk through the door." These researchers noted that, in order to have clean participant groups, "we basically exclude everybody that we typically see in clinic." Researchers viewed comorbid phenomena as valuable objects of study because they are most likely to present in real life cases of mental illness, and argued that a much greater proportion of psychiatric research should be grounded in a pragmatic mindset more focused on the sorts of experiences of disease that the majority of patients actually experience:

> Historically we conduct randomized controlled trials on homogenized samples. We basically exclude everybody that we typically see in clinic. We exclude the people with five different pain conditions who are also depressed, who also have agoraphobia, or who have substance use disorder. All of them get excluded and we end up studying this super homogenized, mythical patient. Then we report on those results and we all pretend that it applies to

> every patient. And it doesn't! This has been a big problem. It's a problem just from a research perspective, it's a problem clinically, but it's a problem ethically that we don't actually know how these treatments work in real-world patients. (Professional Stakeholder Participant)

Other researchers highlighted different issues in the recruitment of participants for psychiatry and neuroscience research that further complicate the conversations around the "mythical patient" and "clean" samples. In the context of research on serious and/or treatment-resistant psychiatric illnesses, one researcher noted that volunteers are often recruited through last resort referrals, after multiple clinical treatments have failed for them, and that these research trials typically attract especially comorbid and vulnerable populations. Disproportionate inclusion of individuals with especially challenging, complex, or difficult-to-treat psychiatric conditions may result in research findings that underestimate a prospective intervention's full strengths on more ecologically valid populations.

Another researcher highlighted that problems can also arise from the inverse case; that is, disproportionate exclusion of individuals with especially challenging conditions. They noted that the high degree of variation in the severity of a given disease can prove a further limitation to research and that potential participants might be excluded from research on the basis of behaviors that are germane to their illnesses and strongly correlated to severity. One researcher described working with a population whose conditions often result in symptoms that prevent participants from being able to remain still enough for the required neuroimaging procedures. As this researcher noted, the exclusion of these individuals from studies may impede advances in the understanding of their severe symptoms, and such research findings might not be able to speak well for these excluded people.

The Researcher's Dilemma: Balancing Roles and Responsibilities

Many of the neuroscience researchers that we interviewed additionally held roles as physicians or clinicians, and throughout the interviews, they reflected deeply on the ethical questions that have emerged during the day-to-day performance of their varied, and sometimes conflicting, roles. These clinician-researchers were cognizant of the aspects in which their multiple professional roles diverged but seemed to have more questions than answers about how to maintain the highest standards of professionalism given their simultaneous roles.

The Inherent Conflicts of the "Dual Role"

In the interviews conducted by our team, researchers discussed their thoughts about how their roles as clinicians may be, or may be perceived as being, in conflict with their roles as researchers, both by patients and in their own self-reflections regarding their work. The ethical challenges inherent in this "dual role" of clinician-researcher were mentioned many times in some of the most emotionally laden statements throughout our set of interviews. One researcher reflected on a comment from their mentor, that "'Most people can't be a clinician and a researcher because you have to brutal to be a researcher," noting that, when assuming a researcher role: "I have to give up the best interest of the patient, which as a clinician I am trained and feel like the autonomy of the individual is the highest ethical standard, so I should be thinking of just the patient, the patient's best interest."

Researchers were candid in admitting that the "lifeblood" of clinical research flows from recruiting participants into trials, and that there is often substantial pressure from funders to complete the enrollment process within a specific time period and to not fall behind. They were cognizant of the ways in which these feelings of pressure could conflict with their best judgment when it came to determining the optimal treatment for their patients:

> We have these regular phone calls every three weeks, but we also get regular emails from the lead investigator: "Okay, how's enrollment going? You're falling a little behind." So there's a lot of pressure. That can conflict with our role as clinicians potentially because we're going to be motivated when a patient comes to the clinic. […] It may be that there would be more flexibility seeing the patient clinically than it would be to participate in the restricted environment that clinical trials often impose, and yet there may be the thinking, "Okay this is somebody who really would meet criteria for this study and they could benefit from that study. Should I consider having them?" And again some of these motivations are unconscious. You may think, "Okay well we're falling behind in enrollment. Maybe rather than suggest another kind of treatment, maybe we can convince ourselves that this is the best option for them." And I think that's something that, as clinicians who also do trials that we all have to try to avoid. (Professional Stakeholder Participant)

Whether or not this tension is exacerbated by enrollment goals, researchers noted that this pressure can lead to a deep internal conflict between the clinician-self and the researcher-self. One researcher described her decision to never enroll her patients into her clinical studies:

> Because once a patient is referred to me as a patient, I'm wearing a clinician hat and that's it. I'm going to do what's best for this patient. I cannot possibly put them in a place where they have 50% chance of not getting the treatment. I just don't feel right to do that…Once you enter my room as a patient, that's one goal. Once you enter my room as a participant, it's another goal. And it's just changes the nature of the relationship. (Professional Stakeholder Participant)

This researcher also noted how the clinician-researcher role could amount to something of a double-edged sword that negatively affects both the doctor–patient relationship and the integrity of the research, stating, "As a clinician you lose the trust of the patient when they become a participant. And as a researcher, I lose the trust that you are really able to be impartial and blind."

The reasons behind the weight accorded to the topic of dual roles and the sensitivity with which it was addressed became more apparent throughout the interviews, for it was clear that the majority of these researchers framed this conflict as a larger philosophical question regarding how they, as trusted healers, should prioritize their time, effort, and care: "It may not be the best thing for her, but it might be best thing for the research. So it's the best thing for the patient versus the best thing for the whole research and understanding and helping this population of the diagnosis. Helping the whole, the group, versus helping the individual."

The "Inheriting" of "Orphan" Populations

Throughout the course of this project, our team interviewed numerous clinician-researchers who were involved in highly innovative research testing novel treatment modalities or applications. These researchers, whose work took place across the spectrum of deep brain stimulation (DBS), transcranial magnetic stimulation (TMS), and virtual reality, openly discussed the challenges that paradoxically emerge when their research is effective and the health of a participant in their project starts to improve as a result of the experimental trial.

As described by these researchers, due to the innovative nature of their treatments and therapies, standard treatments have been ineffective for the participants who typically enroll in research. These treatment-resistant participants are individuals for whom typical interventions are not effective, and who have often been living with severe, life-altering mental illness for decades or more. One researcher referred to these as "orphan" populations, reflecting the fact that these individuals have not yet found a suitable home for themselves within the current realm of mental health care. If the treatment offered in the course of a research project ultimately benefits individuals from these "orphan" populations, researchers emphasized that then "You own these patients. This is a patient population that you essentially will own." With no other avenues to explore for possible relief, these populations become dependent on the treatment or therapy that the researcher is offering, and which is often not commercially accessible outside of their labs. One researcher described this particular conundrum of working with research participants who suffer from treatment-resistant mental illness in this way: "I, without really knowing it, kind of inherited a population of patients where there's no other treatment and that's where the conflict is. The conflict is on the inability to offer them something else that's clinically approved."

Issues that arise from this dependence were discussed by multiple researchers, who grappled with the ethics of who is responsible for the continued care and expenses associated with participants in a variety of innovative trials. These researchers were aware of the possibility that patients may benefit from the experimental treatment being offered, but then be unable to consistently access this treatment after the completion of the study. They questioned what their responsibility to participants should be after the end of a research study, and how this responsibility could effectively be met after the research funding is completed or discontinued:

> The issue is, I've got this group of people that I could keep well but I don't have the financial infrastructure to do so. And so I could elect to just do that, and that's all I do. But then I can't get the method to go any further and I'm not going to get the funding. And so it's very difficult for me because I see them struggling, I have something that can help them, but I have to keep rolling on our end to keep the ball going to get more data and get this thing further out there.... It creates a dilemma in the sense of making a balanced choice of the doctor hat and the scientist hat, and kind of knowing that ultimately getting this approved is the best thing for everybody. (Professional Stakeholder Participant)

This internally experienced tug of responsibility toward these research participants is another example of the researcher's dilemma in balancing their varied roles and responsibilities—to their funders, their research participants, and to themselves and their broader research mission.

Informed Consent

Informed consent was an important concern for the majority of neuroscientist researchers and IRB members interviewed, yielding one of the largest collections of coded responses in our dataset. Respondents valued the process of informed consent as a key tenet of ethical research but reflected on ways the process could be refined to better suit the needs of increasingly innovative lines of inquiry and protect the neuro-vulnerable populations that often participate in neuroscience and psychiatry research.

Consent as an Ongoing Process

Neuroscientist researchers and IRB members reflected on the process of informed consent, endorsing the idea of consent as a dynamic, rather than static, process. Researchers emphasized the importance of understanding consent as a continual process by highlighting two specific instances in neuroscience research where ability to consent might change over time. In one instance, a researcher with experience working with individuals with traumatic brain injuries described cases where a patient or participant may be in a minimally conscious state at the time of enrollment into the research project, necessitating surrogate decision-making by a family member (or other alternative decision maker); the decision-making process must then evolve to include the person, to the extent that they are able to make decisions, as they regain brain functioning.

Another researcher described the reverse instance, in the case of patients or participants who have a degenerative disease such as Alzheimer's, and whose decision-making abilities become impaired as the disease progresses. While a research participant may be healthy enough to make their own decisions upon enrolling in the study, their deteriorating condition may necessitate the involvement of an alternative decision maker over time. In these cases, researchers need to have an established protocol in which they "make sure [they] periodically reassess [the participant's] understanding, and that the caregiver or authorized representative […] continues to have a good understanding of what the study is about and why they are in it and what their expectations are."

Both researchers and IRB members reflected that this concept of consent as a continual process should not be applied solely in situations where the competency of the participant is expected to change dramatically over time. Rather, they emphasized that researchers involved in all types of neuroscience and medical research projects should ensure that the consent is instead an ongoing process where patients and participants can review their participation at all times: "Consent is a continuous process. Clinicians tend to lose sight of that concept. You know, you have your signed form, that's consent right? Well no, it's a continuous process."

Improving Consent Forms

Researchers were in agreement that the current consent form requirements are often cumbersome to researcher and participant alike. The most repeated complaint concerned the length of current consent forms for medical research, which can run up to 40 pages long and often include large portions of text that serve as protections to the funder and institution and are not relevant to the participant's decision-making process. Researchers noted that, due to top-down regulations regarding what must be included in consent forms, these documents have been transformed into protections for the institution instead of for the participant. These researchers emphasized that, in their experience, the way participants view consent forms and process has similarly been transformed, noting that "patients correctly see consent forms as protection for the institution, not for them."

IRB members agreed that consent forms can be overly lengthy and difficult to understand. IRB members in our cohort emphasized the need for the use of clear, easy to understand language within consent forms, especially regarding the potential risks and benefits of a study. However, they also recognized that researchers are often not trained in the development of consent forms and can have difficulty transforming very complex ideas into text understandable to a layperson. As a potential solution, one suggested the establishment of mentoring processes both within and across medical departments. Beyond the text of the consent form, IRB members noted the need for additional training in the process of obtaining informed consent. While research coordinators and other administrative personnel often receive formalized training for this process, physicians and faculty researchers typically do not, which could impact the amount of bias that may be present during the informed consent process.

Researchers and IRB members alike emphasized the need for improved consent processes. Suggestions ranged from post-consent quizzes to assess understanding to using video or even virtual reality to explain the research processes, risks, and benefits. An even simpler suggestion was to send the consent form to potential participants ahead of time, so they could review it without the potentially coercive pressure of being in a research setting.

Risk, Benefits, and Participant Burden

Unique Risks in Psychiatry and Neuroscience Research

Researchers and IRB members we interviewed expressed that, because the brain gives rise to identity and sensory experience, neuroscience research poses risks that raise unique ethical questions and concerns. One researcher discussed in detail how innovative neuroscience and psychiatric interventions can pose unexpected psychosocial risks, including the potential to be profoundly destabilizing to individuals' conceptions of the self's continuity and integrity:

> A patient's self-perception changes when something is implanted into them, implanted into their brain of all intimate organs that one could implant. That is an important part of their identity and so forth, and something changes for those patients. And so we have, again, getting back to this theme of a vulnerable population, that may be different than for a Parkinson's patient who might see it, for example, as an augmenter to their brain that helps control something that's going wrong in their brain. For a depressed patient it may be different in some way. And some of our patients [...][1] would say, "I've got this thing inside me now that has fundamentally changed me, and not for the better." (Professional Stakeholder Participant)

The experience detailed in the above quote additionally revealed to the researcher how difficult it may be to distinguish side effects or problems with the device or treatment from a patient's own psychiatric symptoms. This researcher emphasized that some psychiatric conditions may render patients especially vulnerable to side effects of invasive device trials, and even positive effects may cause anxiety in patients, who may wonder if parameters of the device will be changed (resulting in symptom recurrence), or if the experimental treatment that has been successful for a while could suddenly stop working.

Beyond the distinct psychological impacts patients may experience when participating in innovative neuroscience research, researchers that we interviewed also emphasized the importance of recognizing how a neurological dysfunction resulting from side effects of an experimental treatment can alter a person's quality of life irreversibly and permanently:

> I think that trials or protocols that require some sort of either invasive testing or surgery, the potential side effects can often be a little bit more serious because it alters the mind, and obviously the mind is who we are and how we think and how we perceive the world, and so it can be a little bit more serious and challenging than diabetes research for example... I think if you're talking about implanting electrodes into the brain for functional disorders, or different medications that alter the neurotransmitters, or what we're doing now is potentially injecting viruses into the brain for cancer, I just think that if you have a complication — whether it's a bleed, infection, inflammatory response, or a bad side effect of a certain medication — that can lead to severe neurologic dysfunction. And so often, if you have neurologic dysfunction, it can't be reversed and now you've altered that person's life, quality of life, forever. (Professional Stakeholder Participant)

Furthermore, researchers noted that, because many psychiatric disorders may not be perceived as fatal in and of themselves (despite the heightened risk of suicide in many disorders), the institutional willingness to accept risk in psychiatry and neuroscience research is much lower than in physical diseases or disorders. Cancer trials were mentioned several times as an example of medical research that confers many risks and side effects, sometimes with limited potential benefit, with researchers noting that a similar level of risk and side effects would not be permitted in neuroscience research, even if the result could potentially be more beneficial to a person's long-term health, life expectancy, and quality of life:

> Let's assume you have a nineteen-year-old who walks in, has a very severe form of psychosis, doesn't respond to current medication or responds poorly, has an incredibly high risk of

[1] Suspension points indicate omissions.

> suicide versus the general population. And you could give them a drug that would give them another ten years of normal life, but it comes at a price and in ten years they may develop a lethal liver disease. What do you do?...It's interesting because people immediately assume that for psychiatric disorders, that you should not sacrifice. The perception is always that this is a bad choice: "No, I would rather live with the disease." [...] And the reason why I say this is because in cancer, think what it takes to actually find a cancer drug. All it takes is actually to stop those cells from growing and make sure the patient does not melt. Even if it has huge side effects and long-term consequences, we accept that a drug for cancer is okay to have side effects and to be really bad. But do we accept that for psychiatric disease? (Professional Stakeholder Participant)

In discussing such divergence, this researcher and others highlighted differences in how various types of diseases are understood—not only by the general public, but also within medical science—which may shape perspectives regarding what kinds and levels of risk are acceptable:

> More broadly, what type of risk are we going to be willing to take for curing psychiatric disease if we were to have that opportunity? If we were to have drugs that have severe side effects, long-term side effects, but they can cure some of the psychiatric disease, what would we actually do? Would we accept that? That has probably to do a large extent to [...] the way these disorders are perceived: they're something that is very personal, versus something that is going to cure you; something that you fight in cancer, that is external, that a drug helps you fight, versus something that is who you are in terms of a psychiatric disease. I wonder how that's going to inform some of the decisions. (Professional Stakeholder Participant)

Often, some physical illnesses, such as cancer, are conceived as external threats which intrude on the patient, and therefore should be battled against, regardless of the risks or side effects. In contrast, psychiatric disorders are more frequently perceived as intrinsic—tightly connected to the patient's mind or personality. In the view of some researchers, these unique (and somewhat abstract) aspects of neuroscience affect the types of risks which researchers, institutions, and participants themselves are willing to accept in experimental approaches to psychiatric conditions, and thus shape the direction and speed of advances in the field.

Difficulty of Measuring and Comparing Risks and Benefits

Those we interviewed generally agreed with the utility of the concept of "risk slopes" in appraising and mitigating clinical risk. A risk slope provides a basis for demonstrating that procedures that confer greater risk should also have a greater potential for benefit (either on an individual or societal level) or should be directed toward populations living with conditions where the potential risks of an intervention would be less than the risks that population experiences in day-to-day life.

Despite this general agreement, researchers and IRB members from different backgrounds had highly varied opinions on what is and is not risky. Opinions on risk appeared to depend on a given participant's background and on the types of treatments and interventions they were the most familiar with. Participants who were specialized in certain procedures occasionally posited that procedures in their area of expertise were not as risky or were more heterogeneous or qualified in terms of risk, than perceived by patients or experts in other areas. Individuals who worked closely with certain technologies or procedures seemed to have a better understanding of the potential benefits of specific research projects, while outsiders focused on the risk. With this in mind, however, one researcher mentioned how "we tend to think our treatments work before we've proven that they work," a claim that implies that role- and experience-related cognitive bias may be generative of such diverse interpretations of risk and benefit in research scenarios.

The endeavor to arrive at an integrated and "true" understanding of risk is further complicated by the fact that the use of risk categories and thresholds does not totally resolve the difficulty of transmitting socially, emotionally, and intellectually impacted risk perceptions, and that risk evaluations from different risk categories (i.e., privacy risks vs cognitive risks vs physical risks) are hard to compare in providing optimal advice for potential research volunteers. As an example, one researcher emphasized that the brain-invasive aspect of some treatments does not make them inherently riskier than many noninvasive (drug) treatments:

> There are some differences, but they're not nearly as dramatic, between giving a patient a drug and doing a surgery. And the reason I say that is, think about what giving a drug is like. That is a biochemical sledgehammer to the brain…you're giving a drug that is affecting every single neuron in the brain. And so why is that any different than drilling a hole in somebody's skull and putting a thin little electrode that is only going to act on a cubic centimeter of brain tissue? Which is the more coarse, horrendous manipulation to brain activity? [...] So when you start comparing pharmacotherapies with neurocircuit manipulations like DBS or TMS, I would argue they are much closer to each other than you might imagine. (Professional Stakeholder Participant)

Researchers agreed that patients and the general public are not adept at being able to accurately perceive the likelihood of risk or benefit of participating in research. Researchers did not agree on whether potential participants typically overestimate the risks or benefits of research: "In general people understand probabilities really badly, so it's not clear what they even think when you tell them there is a minute chance of something. Do they overblow it or do they think it's impossible? It's hard to really know."

Unknown Risks

Researchers and IRB members questioned how unknown risks should be handled in the research scenario. Even if all the immediate risks of a procedure are known, the long-term risks can be less easy to predict, especially in more innovative procedures. Researchers and IRB members were in agreement that potential participants need to be informed that there are likely unknown risks associated with participating, but it is unclear how much detail or discussion regarding these potential or unknown risks, if any, should be provided. One researcher emphasized the need for humility, referencing the history of medicine:

> [In] the history of medicine, just because it's a human endeavor, mistakes have been made. We have lots of treatment, not lots, but there's a percentage of treatments that we think work. Ethical physicians are using this treatment, suggesting to patients they take this drug or have this surgery, and then twenty years later we learn that it's not actually doing any good. (Professional Stakeholder Participant)

Several researchers emphasized the need to accept the existence of unknown risks in order to address the urgent need to identify treatments which can potentially reduce suffering in the near-term. They argued that the ethical imperative of reducing the suffering of those living with mental illness must be prioritized, and that attempts to predict or control for all unknown risks—or to answer all scientific questions about the workings of a novel therapy—can hinder the progress needed to improve the lives of those suffering today:

> I believe anything that has the potential [...] to help somebody who is suffering needs to be investigated. Yes, as a basic scientist, I would love it as part of this investigation to have some group of investigators really trying to understand at a mechanistic level how these drugs work. Which molecular targets are they acting on? Which synapses and circuits are they modifying that contribute to its therapeutic benefit? But if the field waits for that understanding before they test for the substances or the treatment [...] before they actually test the treatment in human beings, we're never going to have any new treatments. Because like I said, we've been working on drugs that have been around for 50 or 60, 70 years and we still don't understand how they work. (Professional Stakeholder Participant)

Beneficence and Creating Value for the Patient

Researchers from many different backgrounds expressed concerns about the benefits, or lack of benefits, that their participants received as a result of participating in research. These researchers emphasized that, while the risks of a project can be appraised and mitigated, the benefits of participation are often more difficult for researchers to ensure, especially when it comes to randomized trials where treatment is involved. A primary manifestation of this frequently articulated concern is the sense that while the risks of participation may be low, the benefits of participation that may accrue to participants are not commensurate with the value gained by researchers, institutions, and the field at large. This sentiment is neatly summed up by one researcher, who wondered, "How are we capitalizing on what we're able to learn with neuroimaging research and how does that get back to the families and participants that we're working with?"

Some researchers discussed the difficult task of performing randomized trials on populations who are in need of treatment and are eager to be placed in the experimental arm. One researcher discussed how he designed his protocols to include a "high dose" and "low dose" treatment arm, instead of a "no dose" control group, because the participants, who live with a severe disorder that does not yet have a cure or treatment, are less likely to feel bereft if they are randomized to the low-dose arm. Another researcher discussed the personal challenges encountered when enrolling participants in randomized trials:

> Ethically, the major problem in conducting clinical trials with the hands on the pulse the way I do is that you know quite a bit about the person and then you don't know what they're randomized to. But you see a person and you really wish that they would get the active intervention because you know it's going to help them in your heart. And you can't because we are going to randomize them. So I don't know eventually what happens, but that's the piece that's probably the hardest. (Professional Stakeholder Participant)

Another researcher discussed the worry that by participating in a research study on an experimental treatment, a patient may not receive a different treatment that might be more likely to help them. This researcher expressed hesitation about whether they should be enrolling willing, informed participants who, by their estimation, could potentially see better results through a different method of treatment.

In terms of creating value for the participant, researchers discussed the importance of providing feedback and information to participants whenever possible. This type of feedback, it was felt, may be especially desired by participants in neuroscience or psychiatry research due to people's innate curiosity about the brain:

> What are the responsibilities of researchers to participants in their trials once the trial is over, and how much information do you give back, and how do you do that? And I think that my general feeling is that investigators should always at least plan to provide some kind of summary of the results of the trial to people who participated in it when it is over so that it gives them a sense of being able to see the result of what they've participated in, which I think increases the community engagement and the positive feelings that people have about participating in research trials, if they can see the outcomes, see what the actual product is and have a more directed sense of how their participation is helping to advance the science or to improve the treatment for other people—it's not for themselves—so I think that's always a good idea. (Professional Stakeholder Participant)

While many researchers emphasized the benefits of providing personalized feedback to participants, several discussed the potential risks and difficulties associated with this feedback. In some cases, participants may hope or expect to learn more from this information than is realistic, leading to becoming disheartened or frustrated with the limitations. In other cases, providing honest feedback has the potential for psychological consequences, especially when related to the discovery of unexpected, incidental, intractable, or poorly understood findings which can introduce additional anxiety and despair.

Researchers also mentioned some unexpected benefits which participants themselves have mentioned that they experience. These benefits range from simply receiving a picture of their brain after a neuroimaging study, to having a "reason to wake up in the morning":

> Something I've seen directly with interacting with our research participants is that they really look forward to doing this work. An example is our most recent participant who has paralysis [...] and he looks forward to every day that he is scheduled for research with our team. He says he wakes up excited to do the research that day. And in fact he said it's the one thing that's kept him alive over all these years, it has been keeping himself going to be

> able to participate in this research. So even though there's no direct benefit to individuals, there's certainly indirect benefits to society, but also indirect benefits to the individual for having a purpose and a thing that they feel invested in and to be a part of the team. (Professional Stakeholder Participant)

Participant Burden

Numerous researchers commented on the amount of burden that is placed on participants when they participate in research, and raised questions regarding what was appropriate and fair to ask of them:

> How much burden do you put on a participant, a patient going into treatment? How important is that piece of data that you really want to collect? And why do you want to collect it and all these factors, versus what the person giving you the data has to go through for that data to be collected? I mean we don't coerce anybody and so forth, but the four-hour session? A six-hour session? An eight-hour session? Two days, three days, four days? (Professional Stakeholder Participant)

Many of the researchers we interviewed were cognizant of the lived burden participants experience when volunteering for a research project, and several discussed the importance of establishing a "patient comes first" mentality by being flexible and adaptive in the research scenario in order to ensure that participation is manageable for all participants. They mentioned the innumerable burdens that participants may potentially bear in order to participate in research, burdens that frequently go unobserved or unacknowledged by researchers, including taking unpaid time off from work, obtaining childcare, and arranging transportation. One researcher notes how these considerations "may make it more difficult for them, even if they were willing to participate, to participate. So when we're trying to attempt at diversity, we need to be flexible enough and informed enough to know how to manage and how to support those families so that they can be part of the project."

A different researcher continued to emphasize this point, explaining ways in which his research team attempts to accommodate participants by paying for their transportation or hotel rooms, or by working on the weekends or in the evenings so that the participant does not need to take time off of work. This researcher noted that accommodations such as these, which may feel inconvenient to the research team, must instead be viewed from the perspective of "That is on us to do that, because they don't owe us anything. We are privileged to have them participate."

A handful of researchers and IRB members also addressed the burden of follow up in research. They emphasized that obtaining a high follow-up response rate is important in the research scenario, because it ensures that the data collected is being used to its full potential; however, concern was expressed that participants may not understand the extent of this burden when they agree to participate in research. One IRB member commented that researchers should plan to discuss the importance of follow-up with potential participants ahead of time.

Influences on Participant Decision-Making

Researchers and IRB members we interviewed discussed various influences on participant decision-making in the research scenario. Many of their perspectives overlapped with our team's prior thinking and research that informed the development of the Roberts Valence Model (See Chap. 12).

Hope and Desperation

When discussing reasons why participants may volunteer to engage in innovative neuroscience research, the researchers we interviewed overwhelmingly emphasized the intertwined feelings of desperation and hope that potential participants often express. Researchers were quick to clarify that they did not perceive these signs of desperation and hope as necessarily indicative of the therapeutic misconception when evaluating the voluntariness of decision-making, and instead explained that even after a participant has been explicitly told that they will not receive benefit from a study, they still often express hope that participating will help them personally:

> We're so careful, and you're under great obligation during experiments to explain this to people, and they don't believe you. You tell 'em, and tell 'em, and tell 'em. And they think, "Yeah, okay, that's what he's saying, but really it's probably going to help. I'll find some way to help me." (Professional Stakeholder Participant)

Researchers expressed that, while such participants technically understand the difference between research and treatment and can adequately explain these differences to the researcher, the sense of hope nevertheless is a major factor motivating their willingness to participate in innovative neuroscience research:

> Because often people, I think, tend to be very hopeful, especially if they're in a situation where they have a medical illness that has a poor prognosis, then any little glimmer of hope that there's something on the horizon, people can embrace that sometimes maybe more enthusiastically than the real facts would justify. (Professional Stakeholder Participant)

Multiple researchers expressed a view that, for many research participants, especially those who volunteer to participate in proof-of-concept or Phase 1 trials, this "glimmer" of unfounded hope is a result of their desperation at finding respite from a disorder that has likely inflicted them for a majority of their lives. Eligibility criteria for such studies often require that other treatments or therapies have not worked—or had only partial or short-lived benefit—for these individuals. This is particularly likely to be the case among research participants who are considering enrolling in highly invasive procedures such as deep brain stimulation (DBS), which continue to be reserved for conditions that have not responded to lower-risk treatment options. It is clear to researchers that desperation is a major influence on these participants' decision-making, but these researchers emphasized the broader context—i.e., the patients' ongoing suffering and greatly diminished quality of life—when discussing the ethical implications of "desperation": "They're suffering, they understand it. As long as they're capable of understanding it and understanding the risk and they say, 'I've lived with this for twenty years. Life isn't worth living and I'm willing to try it,' I don't have a problem with that."

Researchers noted that many participants enroll in research because they are desperate for some improvement to their health or functioning, and because they are holding on to hope that there is a solution for them or—importantly—for others like them. Throughout our interviews, researchers used the words "desperate," "out of options," and "in despair" to describe research participants who volunteer to be in innovative studies, but not the word "hopeless." When hope was talked about, it was always mentioned in the positive sense, in that participating in research gave patients hope or made them hopeful, interpreting it as a positive reason for why someone would decide to participate in research, as opposed to a negative influence:

> I take care of cancer patients who all die, unfortunately. Quality of life is not just the ability to walk and talk and not be in pain, but I actually think hope is large part of the quality of life. And so if this clinical trial gives them some hope—The quality of life of going to a hospital every day, five days a week, for a clinical trial, people might automatically say "Well that's actually poor quality of life because they're having to go to the hospital and get out of bed." But to me actually, I would say for most patients who are having end stage cancer, that actually gives them a lot of hope because it gives them purpose and gives them something they're doing and some chance that things might improve. I don't think there's any sort of black and white stance on what exactly the right thing is. (Professional Stakeholder Participant)

Despite their emphasis on hope, researchers emphasized that it remains the researcher's responsibility to ensure that research participants are not inappropriately influenced by their desperation: "Desperation is the desperation of the patient. The ethics are on the physician and the investigator. But the desperation is the patient." One researcher, for example, emphasized a strategy they used to counter the potential role of desperation on the part of patients considering clinical trials—namely, making sure that these patients were aware new research opportunities would continue to appear, and/or that if a particular research study was not something the patient would be comfortable with, that there may be a different research opportunity in the future that the patient might feel more comfortable about. A different researcher noted that assessing whether all participants have the same level of clinical need can be an indicator of whether or not it is ethically acceptable to enroll certain patients or populations, elaborating that when all of the potential participants are essentially desperate or out of treatment options, then desperation is likely a quality of the population that is being studied.

Stigma

Another influence on decision-making that was discussed by researchers related to the stigma felt by potential research participants who are living with a mental illness or addiction. Researchers as a whole noted that many individuals who are eligible to participate in neuroscience research have been living with a mental illness for most of their lives, and they have likely experienced first-hand the stigma that goes along with it. Despite this agreement in understanding, conflicting opinions were expressed regarding how stigma may or may not impact an individual's willingness to enroll in a research studies on mental illnesses or other brain disorders. Some researchers noted that potential participants may fear that, by participating in a research project, they may feel further stigmatized:

> It's tricky because I think part of what we come against are a lot of perceptions that are driven by fear and by stigma. And a lot of the patients have had negative experiences in the past. They feel marginalized, they feel stigmatized, they have a history. They didn't arrive at this perception by accident, it's driven by personal experience. (Professional Stakeholder Participant)

Other researchers suggested that potential participants may feel relieved to find that their condition is being studied and taken seriously: "I think that for these patients, they already experience the stigma and then the fact that someone is actually taking it seriously and researching it, I think may make them potentially even more likely to participate."

With regard to stigma, multiple researchers noted that potential research participants may view different types of studies as more or less stigmatizing by nature of the study design. For example, some potential research participants may feel that their family will not approve of them attending psychotherapy or taking medications, since these types of interventions are heavily associated with psychiatry and often have a negative stigma associated with them. Some technology-heavy interventions (such as those leveraging virtual reality), which have associations outside of the realm of mental illness, may be perceived as less stigmatizing, and patients may therefore be more willing to enroll in them:

> Especially technology and mental illness, they are like, "Oh my family will totally get me doing a VR study for my medical movement disorder, but they are not going to understand me going to a group therapy once a week, so I am going to go for the VR." […] I think it is de-stigmatizing and so they can be more open to it and less skeptical than they need to be, rather than [taking] a pill. Maybe because it's cool, it's the new millennium, it's like "somebody thought about us and used technology to help mental illness, so maybe I'm not so crazy and I'm not so different than a medical illness." And I think it's de-stigmatizing in that way. (Professional Stakeholder Participant)

While the de-stigmatizing effects of some lines of research may appear more inclusive and inviting to potential participants, this researcher also worries about how patients may prioritize participation in these trials versus obtaining more traditional forms of treatment that are available and that would likely be more beneficial to the patient.

Daily Functioning

Researchers also discussed what they perceived as a primary motivator for patients who are beginning a new treatment or enroll in a research study. They shared that patients often reveal that their primary hope is not that their symptoms will be fully alleviated, but rather that their ability to function in day-to-day life will improve in its physical, psychological, social, or occupational domains. As one researcher succinctly put it, "They want to function better. They want to do more."

Standardized classifications and assessment methods, however, largely measure successful outcomes in treatment or research as the alleviation of a specific symptom or set of symptoms. Therefore, researchers noted the need to keep in mind that research-defined "successes" (positive outcomes) are rather narrowly defined and do not necessarily translate into improved real-life functioning for patients:

> If you talk to patients, symptoms aren't always a thing that bothers patients. Functioning in life, relationships, social and occupational kind of stuff, is usually what the problem is more so than a particular symptom that can be palliated in some way. And that's largely ignored by our classification system. (Professional Stakeholder Participant)

Privacy and Confidentiality

Due to the stigmatization of mental illness, privacy and confidentiality take on a very important role in participant decision-making. Researchers noted that participants are sometimes concerned or fearful that their records, medications, or diagnosis would be shared. They described the importance of recognizing these concerns, including when participants themselves may not. This was especially noted in cases where innovative neuroscience research produces "first responders," who, in other medical fields, traditionally have their disease details published: "The patients that are first responders to these early interventions, what about them and their mental health issues being released? [Publishers] are like, what's the big deal? I am like, it's a different deal. It's not like you had cancer, it's something else."

Researchers also described the potential influence of confidentiality protections attained by virtue of being in research on decisions to enroll in research, and the corollary concern that participants desiring such confidentiality might forego established treatment options for their conditions:

> Part of the motivation of being in a study for some people is that there's some anonymity associated with it. What we tell patients is that no records are completely secure, that with the right legal piece of paper, any record could be obtained. But one of the advantages of participating in a trial if they're concerned about that is that […] the basic information that's being obtained about their psychiatric history doesn't necessarily go in Epic, for example, doesn't necessarily go in the medical record. And so patients are often concerned that employers, that others involved in some legal issues like divorces and so forth, that those records might be easily accessible. And even though the general medical records should not be easily accessible, these [research] records may have another layer of anonymity associated with them, because for the most part we're talking about numbers as opposed to names and that sort of thing. (Professional Stakeholder Participant)

As noted by this researcher, some participants might be influenced by the idea that their involvement in the study would not appear in their official medical record or in documents which could impact other aspects of their life. To our knowledge, this finding regarding the role of confidentiality protections as an influence on decision-making for research participants has been minimally previously discussed.

Institutional Prestige

Multiple researchers noted the impact that institutions themselves can potentially have on participant decision-making. Several noted that when research is completed at or associated with an institution that the potential participant views as prestigious or highly trustworthy, the participant may be more likely to expect benefit and overlook risk: "People who are desperate say 'Well, it's worth it and [this University] wouldn't be sponsoring it if it weren't okay'."

This effect may be compounded when the potential participant perceives a large educational or socioeconomic imbalance between themselves and the researcher or institution, for these participants may be overly trusting of the information being presented to them or uncomfortable asking questions of the researcher:

> In our work we are working with a lot of families from underserved, low-socioeconomic communities that don't come in with the best understanding or previous education or background. […] It can be kind of fancy and intriguing for them to say "Oh, we're going to go to [this University] and be in this study! There's so much we can learn from this! How great for my kid!" But then I don't want there to be a disappointment for them to walk out feeling that they got a bad deal or something like that because they didn't get the sort of information that they wanted. (Professional Stakeholder Participant)

Another researcher noted how this effect could be further heightened in studies where participants are being recruited from distant locations. The opportunity to participate in a study at a prestigious institution with name recognition in a desirable location can function as an influence on decision-making, even as the difficulty of travel for those prospective participants represents a barrier and an influence in itself.

"Letting Down" the Research Team

One researcher addressed the pressure that potential participants may feel to "succeed" in a research project. Once the research team has committed time to them, they may feel as if they are already committed to participating and that they will be letting the research team down if they decide not to enroll or to withdraw from the study. While this was only directly discussed by one researcher, indirect references to this issue (including in combination with the "institutional prestige" issue described above) emerged in many of the interviews.

> Some parents may well be overly anxious coming to [here for our study] because it's a very prestigious institution and they want their child to be able to perform adequately. And so if, for example, they turn up and we decide that their child really couldn't go into the scanner and shouldn't because they are going to move too much or they might become too anxious and it's best for them not to do it, sometimes the parents feel anxious because they've let us down. Because they said that they thought their child could do it and they've been prepping for their child to do it and they feel disappointed and upset, either with themselves or with their child, and it's not a good feeling for them because they feel like they failed. (Professional Stakeholder Participant)

Another researcher noted the importance of recognizing cultural differences in the research scenario. People from different cultures and background may have varying levels of comfort asking questions of individuals whom they perceive as authority figures, saying "no" after someone has spent time informing them of the study details, or simply engaging in the research process at all.

"Giving Back"

The impulse to contribute to scientific knowledge, and particularly to participate in a study which may lead to future treatment options for people with the same disorder or disease, was discussed by researchers as an influence on decision-making. It was summed up by one researcher who referred to this desire as "giving back" to their community. To the extent that research participants feel part of a community of patients living with the same or similar psychiatric conditions, an important influence on the decision to participate in a research study may be the sense that, in doing so, one is making a contribution to that community.

"Allure" of Self-Imagery or Self-Representation

Researchers described another possible factor in participant decision-making as the desire, on the part of at least some participants, to receive some type of representation, picture, or image derived from their participation in research, such as a brain scan. This desire, researchers noted, seemed to be disconnected from the actual utility of these representations at the level of individual treatment or diagnosis. For instance, one researcher stated bluntly, "I think the seductive allure for them is often they get a picture of their brain," noting that "being able to see inside of yourself," and particularly into your own brain seems to act as an influential factor in some participants' decision-making. To the extent that receiving images or other forms of self-representation is an influence on research decision-making, it may act outside of participants' and researchers' awareness, and is likely not considered a "benefit" of participation in the same way that potential direct (e.g., medical) benefits or other forms of compensation or payment are.

Financial Compensation

Financial compensation as an influence on decision-making was discussed by numerous IRB members, but by very few researchers. IRB members noted the importance of making sure that the financial compensation provided was not an undue influence, especially in the case of healthy participants, whereas some researchers wondered if the amounts that they were providing were really sufficient to compensate for the amount of burden introduced into the participants' lives by participation in their study.

Conclusion

In reviewing discussions that dealt with the topics addressed in this chapter, one noteworthy takeaway was the sympathy and empathy that commonly permeated each discussion as researchers expressed their thoughts, concerns, and experiences in working with research participants. This emotional content circumscribes the specific topics of discussion, and reflects the sense of care and compassion that researchers we spoke with feel and demonstrate for the participants with whom they work, regardless of their understanding of the healing capabilities of the current state of their field.

Throughout these discussions, the researchers we interviewed frequently conveyed great concern and respect for the participants they engage. In addition, they were particularly attuned to, and concerned about, the possibility that participants could feel (or really be) unintentionally exploited or manipulated. We believe that this sense of respect and empathy in the research culture is important to highlight in broader discussions of neuroethics. Of course, as with any study that relies on individual self-report, there is the potential for "social desirability bias" in the statements these professionals made when describing their own approaches and attitudes, so these findings should be viewed with some caution.

Nevertheless, expressed qualities such as empathy and concern are worth pursuing further as potential influences on how research is conducted; they are not typically discussed in funding discussions or IRB review processes. In fact, these qualities may tend to be viewed as "soft" (whether from a personal, psychological, or professional perspective) or—in some instances—less desirable for a competitive neuroscience career. Yet, it may be such qualities that provide some kind of unseen buffer or safeguard against a variety of potential harms (including unintended consequences) that may result from research participation. Scientific researchers are provided with enormous trust and respect by our society; perhaps empathy is the least we should ask in return. As one of our interviewees put it,

> I think it's something that we really have to keep in mind and to really think, "Well what if we were that patient?" always, rather than just looking at it from an IRB committee or a principal investigator point of view. If you just have that sort of overarching thought process, I would say all the other little things will fall into that. (Professional Stakeholder Participant)

Key Points

1. Stakeholders expressed concern about public perception of neuroinnovative research.
2. Stakeholders discussed the nuances of ethical recruitment of participants in neuroinnovative research.
3. Stakeholders were concerned about the balance between their dual roles as clinicians and researchers.

4. Stakeholders held different opinions about risk, but generally agreed with the concept of a "risk slope," where procedures that confer greater risk should also have a greater potential for benefit.
5. Stakeholders' perspectives aligned with the concepts described in the Roberts Valence Model for participant decision-making.
6. Stakeholders expressed noteworthy sympathy and empathy for participants.

Questions to Consider

1. What level of understanding of the emerging brain neuroscience is necessary for members of the medical community, in order to enhance awareness of the range of ethical issues associated with this research? What level of understanding of this science is needed among the general public?
2. Does the exclusion of research participants with comorbid conditions present a problem of justice? For example, a large proportion of patients with psychiatric illnesses have more than one diagnosis; do strict exclusion criteria affect the scientific value or generalizability of research findings? Why or why not?
3. Clinicians who are also researchers endorsed the concern that, by virtue of their dual roles, they might, even unintentionally, have too much influence on their patients' research decision-making. Other clinician-researchers expressed the conviction that they have a duty to inform their patients about any study that could help with the patients' condition. What are some ways that clinician-researchers could address their concurrent ethical duties as clinicians and researchers?
4. Some researchers described a tension between the commitment they feel to a group of research participants who have benefited from experimental treatments and the necessity of using limited resources to generate new knowledge and new treatment options. Such tensions come into stark relief in instances where ongoing procedures available via research context are not available to participants in clinical settings. What ethical principles are in tension when it comes to offering post-trial treatment? How can researchers and sponsors address concerns about "abandoning" a patient population who may have benefited from an experimental treatment? Is this concern unique to neuroscience research?
5. Research procedures can sometimes create burdens for participants and their families, including travel, loss of work, and emotional strain. Because these burdens are more acutely felt among lower income populations, they tend to disproportionately impact participants of color, which can make the task of ensuring a diverse participant pool more difficult. How does the principle of justice thereby inform decisions about how to accommodate the needs of research participants?

Further Reading

Ballard ED, Waldman L, Yarrington JS, Gerlus N, Newman LE, Lee L, Sparks M, Liberty V, Pao M, Park L, Zarate CA Jr. Neurobiological research with suicidal participants: a framework for investigators. Gen Hosp Psychiatry. 2020;62:43–8. https://doi.org/10.1016/j.genhosppsych.2019.11.007. Epub 2019 Nov 19. PMID: 31775068; PMCID: PMC6980757.

Bell E, Racine E. Ethics guidance for neurological and psychiatric deep brain stimulation. Handb Clin Neurol. 2013;116:313–25. https://doi.org/10.1016/B978-0-444-53497-2.00026-7.
Bell E, Racine E, Chiasson P, Dufourcq-Brana M, Dunn LB, Fins JJ, Ford PJ, Glannon W, Lipsman N, Macdonald ME, Mathews DJ, McAndrews MP. Beyond consent in research. Revisiting vulnerability in deep brain stimulation for psychiatric disorders. Camb Q Healthc Ethics. 2014;23(3):361–8. https://doi.org/10.1017/S0963180113000984.
Borgelt EL, Buchman DZ, Weiss M, Illes J. In search of "anything that would help": parent perspectives on emerging neurotechnologies. J Atten Disord. 2014;18(5):395–401. https://doi.org/10.1177/1087054712445781. Epub 2012 May 22.
Bracken-Roche D, Bell E, Karpowicz L, Racine E. Disclosure, consent, and the exercise of patient autonomy in surgical innovation: a systematic content analysis of the conceptual literature. Account Res. 2014;21(6):331–52. https://doi.org/10.1080/08989621.2013.866045.
Bracken-Roche D, Bell E, Racine E. The "vulnerability" of psychiatric research participants: why this research ethics concept needs to be revisited. Can J Psychiatry. 2016;61(6):335–9. https://doi.org/10.1177/0706743716633422. Epub 2016 Apr 5. PMID: 27254842; PMCID: PMC4872238.
Buchman DZ, Borgelt EL, Whiteley L, Illes J. Neurobiological narratives: experiences of mood disorder through the lens of neuroimaging. Sociol Health Illn. 2013;35(1):66–81. https://doi.org/10.1111/j.1467-9566.2012.01478.x. Epub 2012 May 3. PMID: 22554090; PMCID: PMC3414674.
Cabrera LY, Bittlinger M, Lou H, Müller S, Illes J. Reader comments to media reports on psychiatric neurosurgery: past history casts shadows on the future. Acta Neurochir (Wien). 2018a;160(12):2501–7. https://doi.org/10.1007/s00701-018-3696-4. Epub 2018 Oct 24.
Cabrera LY, Bittlinger M, Lou H, Müller S, Illes J. The re-emergence of psychiatric neurosurgery: insights from a cross-national study of newspaper and magazine coverage. Acta Neurochir (Wien). 2018b;160(3):625–35. https://doi.org/10.1007/s00701-017-3428-1. Epub 2017 Dec 21.
Cabrera LY, Courchesne C, Bittlinger M, Müller S, Martinez R, Racine E, Illes J. Authentic self and last resort: international perceptions of psychiatric neurosurgery. Cult Med Psychiatry. 2021;45(1):141–61. https://doi.org/10.1007/s11013-020-09679-1.
Cascio A, Racine E, editors. Research involving participants with cognitive disability & difference: ethics, autonomy, inclusion, and innovation. Oxford: Oxford University Press; 2019.
Christopher PP, Appelbaum PS, Truong D, Albert K, Maranda L, Lidz C. Reducing therapeutic misconception: a randomized intervention trial in hypothetical clinical trials. PLoS One. 2017;12(9):e0184224. https://doi.org/10.1371/journal.pone.0184224. PMID: 28931031; PMCID: PMC5607126.
Christopher PP, Leykin Y, Appelbaum PS, Holtzheimer PE 3rd, Mayberg HS, Dunn LB. Enrolling in deep brain stimulation research for depression: influences on potential subjects' decision making. Depress Anxiety. 2012 Feb;29(2):139–146. https://doi.org/10.1002/da.20916. Epub 2011 Nov 17.
King NM. Defining and describing benefit appropriately in clinical trials. J Law Med Ethics. 2000 Winter;28(4):332–43. https://doi.org/10.1111/j.1748-720x.2000.tb00685.x.
Michaels TI, Purdon J, Collins A, Williams MT. Inclusion of people of color in psychedelic-assisted psychotherapy: a review of the literature. BMC Psychiatry. 2018;18(1):245. https://doi.org/10.1186/s12888-018-1824-6. PMID: 30064392; PMCID: PMC6069717.
Nebeker C, Leow AD, Moore RC. From return of information to return of value: ethical considerations when sharing individual-level research data. J Alzheimers Dis. 2019;71(4):1081–8. https://doi.org/10.3233/JAD-190589.
Sierra-Mercado D, Zuk P, Beauchamp MS, Sheth SA, Yoshor D, Goodman WK, McGuire AL, Lázaro-Muñoz G. Device removal following brain implant research. Neuron. 2019;103(5):759–61. https://doi.org/10.1016/j.neuron.2019.08.024.

Chapter 15
Qualitative Findings: A Focus on Professional Stakeholder Perspectives on the Environments and Challenges of Innovative Neuroscience Research

Max Kasun, Jodi Paik, Katie Ryan, and Laura Weiss Roberts

Introduction

This chapter, like Chaps. 14 and 16, details key qualitative findings from the first part of our National Institutes of Health's (NIH's) Brain Research through Advancing Innovative Neurotechnologies (BRAIN) Initiative®-funded project, "Enabling ethical participation in innovative neuroscience on mental illness and addiction: Toward a new screening tool enhancing informed consent for transformative research on the human brain." As part of this project, members of our research team conducted semi-structured interviews with 44 "professional" stakeholders (neuroscientists, Institutional Review Board (IRB) members, and ethicists) in order to better understand stakeholder concerns and perceptions about the ethical issues encountered through their work in innovative neuroscience endeavors. Our research team then undertook an iterative analysis (see Chap. 13 for a description of methods) to better capture and describe these stakeholders' perspectives. This chapter focuses specifically on our analysis of stakeholder perceptions regarding the environments in which research occurs, and how these environments uniquely impact the fields of psychiatry and neuroscience research.

The Varied Environments in which Research Takes Place

Neuroscientist respondents and our IRB cohort expressed strong positive views about the purpose and possibilities of brain research (see Chap. 16). In describing contemporary brain research, respondents spoke generally of three research

M. Kasun · J. Paik · K. Ryan (✉) · L. W. Roberts
Department of Psychiatry and Behavioral Sciences, Stanford University School of Medicine, Stanford, CA, USA
e-mail: maxkasun@stanford.edu; jodi.paik@stanford.edu; kryan2@stanford.edu; robertsl@stanford.edu

L. W. Roberts (ed.), *Ethics and Clinical Neuroinnovation*,
https://doi.org/10.1007/978-3-031-14339-7_15

environments—the academic, industrial, and regulatory—that exert a powerful influence on the conduct of science, shaping what research does and finds, and how (or whether) its findings are valued and shared. Topics ranged from local influences, such as the publication pressures that drive short-term productivity; to constraints in the broader research environment, such as resource scarcity, regulatory effects, and common ethical quandaries in academic-industrial cooperation. Neuroscientists and IRB members shared a concern for ensuring mutualistic interrelationships between research and its academic, industrial, and regulatory environments in order to maximize brain research's benefits to humanity.

The Academic Environment

In our interviews, the academic environment of research stood out as an important influence, with particular respect to the pressure to publish and related concerns over career advancement. Respondents expressed that publishing work is a very socially and ethically complex act, involving deep questions about a researcher's competing responsibilities to share, protect, and take accountability for their work, attend to enduring ethical principles such as beneficence, non-maleficence, and justice, and advance a career that is necessary to support the maturation of their work.

Research Funding and the Pressure to Publish

Respondents shared that professional advancement generally depends on continuous research funding, whether from federal sources, universities, or industry. Some expressed ambivalence that published work and its underlying data are more and more often resembling commercial products. The increasing commercialization of contemporary research can make researchers feel compelled to take on additional roles not necessary in the past—for example, the role of the entrepreneur or marketing agent. Some respondents perceived pressures to publish to be at odds with what they viewed as a free-ranging, exploratory ideal of scientific investigation that may have occasionally proven more effective in the past:

> It's bad in that the goal is not a scientific goal — the goal is to publish. It's interesting when I need to justify to my students, your skills nowadays are not to be an amazing scientist but be a good writer because you will have to sell your grants and your data. (Professional Stakeholder Participant)

The same participant noted that the explicit breakdown of many academics' salaries into discrete research funding sources reinforces a pressure to publish. Many funding mechanisms, including grants, provide relatively short-term funding, and successive funding tends to stem from the continual production of data and published papers. Continuing this line of agreement, one participant reflected that major scientists of the past benefitted from fewer economic constraints, which allowed them more time for scientific exploration, rather than short-term funding from large entities:

> It's not like, Newton or whatever had the NIH […][1] All those major scientists, they had position in universities […] Those were lifetime positions. With this hard money comes stability, with stability comes challenging questions. Of course if you say that to a subset of the community here, they will say "Oh but we're tackling challenging questions." They don't realize that they may be able to take paths that are way more original or difficult. (Professional Stakeholder Participant)

Some respondents expressed concern that the cultural pressure to "publish or perish" may not only curtail and foreclose the exploration of many potentially fruitful paths but can also incentivize hastily conceived work. Researchers who prioritize the pressure to publish against the conflicting professional duty to conduct rigorous research run the risk of oversimplifying their objects of study. One researcher suggested that undue publication pressure may ultimately delay scientific advancement by opening or reinforcing demonstrably weak lines of inquiry. They viewed this problem as especially vexing because publication volume is an important index of scholarly achievement that is reviewed by promotional committees in determining whether a candidate is competitive for promotion. In turn, the structure of academic promotion can increase the pressures to publish that may result in ill-conceived work, with one respondent noting "Somebody counts how many publications you have, so it's in your best career interest to publish things that might not replicate. And you do. And that's where we're getting into trouble."

The Effect of Bias on Research

Researchers expressed concerns that publication and funding pressures can pose an enormous threat to the intellectual trust invested in research given their risk of introducing bias into the processes of knowledge discovery. While conflicts of interest in research and medicine are often understood as referring to *financial* conflicts of interest, respondents were not heavily concerned with these types of conflicts, and more often discussed concerns surrounding conflicts that are much more nuanced and difficult to measure. One researcher viewed publication pressure as a potential conflict of interest:

> Most of the scientific misconduct and fraud has nothing to do with financial conflicts of interests. It's due to the fact that people want to get their paper into better journals, so they falsify their research. I think that those levels of conflict of interest have always been more interesting to me than financial conflicts of interest, but financial conflicts of interest seem to be more things that people are concerned about: one, because it's easily measured, and, two, because bias toward thinking that money drives people to do bad things, which it probably does, but I think there are other conflicts of interest that we don't measure that may have more ethical consequences. (Professional Stakeholder Participant)

Respondents agreed that the pressure to publish can cultivate bias and other threats to scientific rigor. A frequently identified source of bias was the tendency of publishers to favor studies that find significant differences between groups. An

[1] Suspension points indicate omissions.

ethically problematic "p-value culture" can cause researchers to perceive as necessary and engage in data manipulation to secure grants and promotions, as one researcher described:

> Your salary depends on your grants, and your grants depend on your data, so for me in terms of bias and ethics, you have your problem right there. (Professional Stakeholder Participant)

Respondents identified several other manifestations of bias that factor into the academic environment. A researcher's own self-interest may make it more difficult for them to elevate other work that they perceive to be in competition with their own or accept limitations in their own work, and one's training and background in a field may sway them to recommend certain therapies or treatments. There is not a clear way that these types of conflicts can be reported, even if the researcher is able to recognize them. To mitigate these subtle but serious conflicts of interest, respondents were generally of the view that researchers presuppose the natural existence of some degree of intellectual conflict of interest, and work to discern and resolve it to the greatest possible extent. One researcher noted the importance of self-analysis as a key aspect of this practice:

> The very first step is just to recognize that the conflicts exist and tell yourself that you are conflicted. Once you call that out, once you've named it, once you've though about it, you can then take steps, inadequate as they always will be, but at least you can do your best to try to mitigate those things. (Professional Stakeholder Participant)

There was an overall sentiment that intellectual conflicts of interest, while potentially ethically problematic, are not always unethical. Most individuals encounter intellectual conflicts of interest in their own work and research. Our interviewees felt that it is important, however, that researchers are willing and able to identify and disclose these conflicts.

The Industrial Environment

Interviewees discussed the sometimes generative, sometimes conflicting interactions between academia and industry, frequently referencing industry's powerful role in shaping contemporary scientific advancement and expressing a variety of opinions on the ethical acceptability of some emerging patterns and norms. They expressed many concerns related to power asymmetries arising from the greater financial capital and control over research directions generally held by industry firms. They explored differences between industry's aims and scope and those of academic research, drawing particular attention to the effects of industry's different set of incentives.

Federal Versus Industry Funding

Neuroscientist and IRB member respondents reflected on the differences that exist between federal and industry funding sources. Some respondents felt that industry enables clinical trials that would not otherwise be possible due to limited federal and other sources of academic funding, with one participant noting that "Much of the work ends up ultimately being funded not by federal grants but by industry, because they have the resources to do those kinds of trials and the NIMH doesn't."

This participant emphasized that industry is especially adept at developing efficient translational pipelines and conducting systematic analyses of small projects' potential. In developing these pipelines, big pharmaceutical and big device firms use the vast amount of resources at their disposal to perform impact analyses on smaller companies' products and, using this knowledge, attempt to acquire the ones that seem most promising. One respondent noted that many for-profit companies are moving away from funding innovative science internally, due to increased regulatory hurdles and a stronger financial rationale to simply provide venture capital:

> Twenty or twenty-five years ago, many of the big companies [...] did innovative stuff. As the regulatory hurdles increased and the competition increased, they ended up kind of pulling away [...] I think that certainly industry is interested in innovation, but big industry, big pharma, big device companies… they're less interested in taking the chance, increasingly, of funding things that will not pan out. They've become sort of the venture capital for incubator companies that will take the chance of doing the bigger proof of concept studies that might lead to a registration and then scooping them up [...] somebody else has taken the chance and it looks promising, and they now have the resources to bring it to the finish line. (Professional Stakeholder Participant)

Many respondents were skeptical about industry's ability to align with the moral and intellectual goals of scientific research conducted in academia, emphasizing how the different interests, including self-interests, of industrial and federal funding entities may lead to differences in how their respective research projects are funded and advanced. One researcher noted that promising clinical trials have the benefit of potentially obtaining neutral funding that has no bearing on the outcome of the trial and is therefore protected against some kinds of conflicts of interest. Yet, while NIH funds a number of these proof-of-concept trials, many large registration and device trials end up being funded by industry. This difference highlights a potential limitation of industry, in that it tends to fund work that offers clear returns on investment and tends to ignore work which may greatly benefit humanity but offers fewer or no clear short-term economic returns.

Additionally, where industry excels at developing new products, it may lack certain crucial stakeholder perspectives, especially those of vulnerable and special populations and people living with serious illnesses. One researcher expressed that, in their experience, companies tend to lack experts in some areas who could help develop ethically and scientifically robust research, from trial design to responding to concerns that arise in real time and in the longer term:

> I think that if I was going to conduct a trial like this, I would hire experts in the field, not only to help design the trial [...] [but also] who really could respond to these kinds of concerns. Because when questions would go back to the [industry] sponsor, they really didn't have anybody. (Professional Stakeholder Participant)

Downstream Effects

Respondents were attentive to the complexity of relationships between research institutions, patient needs, and the large companies who might potentially fund such projects. In one example, a researcher focused on conflicts of interest in such relationships, referencing an instance when a larger company pitched a promising treatment to them, and successfully convinced them to use it in their own clinical research. As noted by this researcher, these overlaps between industry and academic research involve ethically problematic dual role situations that can include conflicts of interest.

Researchers also noted how the influence of industry in clinical research can indirectly impact labs and projects that do not receive industry funding. As one researcher noted, it is sometimes difficult for investigators to find or retain qualified research staff because talented staff can find a higher salary in industry; this may place researchers in a predicament where they are required to frequently train, hire and retrain new staff throughout a study:

> It's actually difficult to find research assistants who have those qualifications, because they're not paid an enormous amount of money here [in academia]. And so that's a problem, to find really qualified staff who are willing to forgo a salary that they could probably get higher in industry, because they can make a lot more money. [...] So then it requires more retraining of the new staff. That's an issue obviously, and I think it's true with any lab. (Professional Stakeholder Participant)

Respondents also discussed how financial concerns may persist even after research yields important findings. Researchers noted that a serious ethical quandary arises in cases where well-funded research yields a potential treatment, but limited funding outside of the study itself threatens or discontinues access to the treatment for the research participants who benefited. This problem prompted more than one researcher to ask, "If it turns out that these people are benefited by this, how do we pay for it to continue?"

Several neuroscientists discussed the ethical problem of withdrawing a treatment or therapy for patients who show symptom reduction or remission during the research stage, noting that continued treatment is typically not possible due to funding limitations. One researcher described how their trial intervention was able to remit most participants for days to months after treatment, at which point the participant would need continuing treatment. Because of the research protocol and limited funding, however, the researcher was unable to provide this care and instead refocused on enrolling new participants:

> The issue is, I've got this group of people that I could keep well but I don't have the financial infrastructure to do so. I could elect to just do [the research], and that's all I do. But then I can't get the method to go any further and I'm not going to get the funding [for that]. (Professional Stakeholder Participant)

The Regulatory Environment

Research is funded by both private and public institutions, with accompanying foci and concerns. Likewise, institutional regulations derive from many places—IRBs, the NIH, or the U.S. Food and Drug Administration (FDA) directly. In academic settings, the IRB is paramount in the formal assessment of ethics in research design, responsible for approving, rejecting, monitoring, and reviewing research involving human volunteers; the NIH or FDA may have additional requirements. Neuroscientists were sensitive to the number of institutions that issue regulatory standards, as well as the competing motivations informing them. As one IRB member noted:

> There are a lot of just little minor rules that the FDA and OHRP are pretty insistent on, so you have to kind of guide the investigator into checking the right boxes and answering the right questions. [...] If it's an NIH funded study then we generally presume that that's been done by the review committee. If it's a multi-center drug study, then typically those are going to be watched by the FDA. [...] There are, as you may know, reporting requirements for adverse events that go either to a central review committee or to the IRB or somewhere, so that there is a fairly robust review of what we consider the data safety plan. (Professional Stakeholder Participant)

A handful of researchers expressed somewhat jaded attitudes toward IRBs. Some believed that the quality and integrity of IRBs differed based on the institution. Some felt that it can be excessively easy to obtain IRB approval if one knows how to "work the system." Others noted that investigators may perceive an IRB as the final ethical litmus test for their project and neglect asking enough ethical questions on their own. Some researchers felt that IRB regulations have led to some undesirable effects. One noted that increasingly stringent animal ethics regulations in some countries are forcing researchers to outsource animal research. With regard to human trials, one respondent noted a seeming contradiction that a surgical innovation can be tested in vivo with minimal IRB approval, while a pharmaceutical or technological intervention with similar risk requires a disproportionately large amount of paperwork and long approval process. Respondents perceived many current ethical ambiguities in research and felt that data safety monitoring boards, ethics consultants, clinicians, and research volunteers can conflict in their ethical judgments, leading to contentious outcomes. One respondent perceived considerable polarization around about IRBs in the research community, suggesting that many think that IRBs do more to block up the research and translation pipeline than to support it, while others hold the opposite view.

Regulations on Financial Interests

Regulatory bodies and the public generally focus their attention on financial conflicts of interest when it comes to the fields of medicine and research. Noting that rules are set in place requiring the reporting of financial conflicts of interest in these fields, many researchers felt that these types of conflicts are adequately covered and accounted for.

Some researchers shared concerns that regulations regarding financial conflict of interest may have gone too far and could negatively impact sharing of scientific knowledge and tools. Top-down regulations that prohibit doctors or researchers from participating in activities associated with for-profit entities (e.g., attending or giving talks about a new treatment or therapy), though usually thoughtfully applied, could also discourage collaboration and negatively impact the distribution of useful treatments and therapies to patients. One respondent noted that institutional regulations can paradoxically undermine the ethical imperative to serve patients and advance science by suppressing data about promising new treatments that have yet to pass through lengthy regulatory processes.

Researchers overwhelmingly agreed that financial conflicts can potentially be problematic and should be publicly disclosed. However, several researchers expressed beliefs that public disclosure of these financial interests should be sufficient, for this action allows patients and research participants to take this into consideration when making a treatment- or participation-related decision.

Regulations Concerning Confidentiality and Data Sharing

When discussing the influences that regulatory bodies have on the completion of innovative research, several respondents expressed concerns about how increased regulations on reporting have led to the underutilization of valuable data. Respondents felt that many current regulations—citing the Health Insurance Portability and Accountability Act of 1996 (HIPAA) rules or IRB requirements, for example—did a good job of protecting research participants. Other regulations, however, were seen as slowing the progress of neuroscience research. If data collected from a study lacks existing IRB approval for sharing at study initiation, for example, it may be limited to analysis by only the initial research team, and the benefit of sharing the data with other researchers can be lost:

> You can have rules that limit the ability for people to learn things from the data, and that slows the pace of scientific progress because this group collected this data but they didn't get the IRB approval at the original time to be able to share the data with the group that has better technique. So as a result the data sit there and they're analyzed poorly, and the other group that could, in principle, have come in and done something useful with it is prevented from doing it because of IRB rules. I consider that to be an ethical problem. (Professional Stakeholder Participant)

Researchers described increasingly complex regulations on confidentiality in the research approval process, and expressed concern that increasing complexity can limit or delay research without sufficiently good reason. Some members of our cohort identified privacy and confidentiality regulations as examples where the ethical principle of non-maleficence and the ideals of scientific advancement may conflict. In one researchers's words: "Of all the things I'm worried about ethically, I'm worried about the patient privacy as interfering with the ability for doctors to share." For this respondent, such concerns were two-fold: on the one hand, well-intentioned regulations governing the sharing of patient or participant data can make it difficult for researchers to gain access to meaningfully large datasets, and on the other hand, risk-averse institutions may shy away from the collection of many forms of data to avoid lawsuits:

> There are other aspects of the measurements of the patients that you might not realize that are correlated with the outcome for the patient. You might only have tried a small set of

> things, whereas the guy down the road might've tried something else [...] People are terrified to accumulate this information because the lawyers will sue them if that guy had a better outcome and you had a worse outcome. (Professional Stakeholder Participant)

Respondents felt that data sharing is critical to accelerate neuroscience research, and generally perceived the sharing of de-identified data as safe. Several respondents were skeptical of the application of standardized risk categories to data, suggesting that risk assessment based on a priori categories may not provide necessary protection for research subjects. One participant expressed their opinion that current data risk assessments are highly burdensome for researchers, and therefore may induce pressures to underestimate the risk of certain data in order to move research along more quickly.

Several respondents shared frustration that ethical regulations may not always properly balance risks and benefits. Many respondents perceived regulations on secondary analyses of blinded data to be too strict, as risks are understood to be very low and potential benefits are perceived to be relatively high. Overall, scientists and IRB members expected very little risk to patients' privacy if data were to be more openly shared. Some suggested that there the research community is too risk-averse regarding the use of personal data. To relieve some of this fear, one researcher advocated for informed consent language that is less aggressive in its presentation of risk, arguing that current consent disclosures are too lengthy and overly "risk-weighted." Another researcher used the human genetics community as a model example of data sharing, as it crosses national borders and its members perceive great benefits to sharing their data and research and are frequently willing to share it.

Challenges Specific to Psychiatry and Neuroscience Research

The perspectives regarding the academic, industrial, and regulatory environments described above are ones that—though discussed here in regard to the fields of psychiatry and neuroscience—are likely held in many, if not all, fields of academic research. While these environments influence many different disciplines, respondents to our interviews also discussed the ways in which their confluence uniquely impacts research activity in the fields of psychiatry and neuroscience.

Although neuroscience research is a subset of medical research, the current physical inaccessibility and complexity of the brain relative to other organs create challenges for researchers that are not found elsewhere in medicine. In the present day, the study of brain-based diseases is often conducted through indirect means: either though molecular and cellular level science with animal models and organoids, or through non-invasive tools which measure and map connectivity in the brain at lower than optimal resolution. These indirect methods lead to unique concerns regarding the validity and value of research within psychiatry and neuroscience, which were expressed numerous times throughout the interviews.

Overall, the researchers we interviewed were highly motivated by the urgency and severity of symptom burdens experienced by those living with mental illness and were eager to discover more real-world applications of psychiatry and neuroscience research. With this urgency, however, a large segment of respondents expressed questions about the ultimate value of different types of neuroscience research, noting a seeming mismatch between tha magnitude of financial investment in new technologies and techniques and the translational returns that have thus far been derived from such investments. It can seem, even to researchers themselves, that intellectual resources are too often pointed in research directions that already have demonstrated limited promise, and that research is too often not primarily motivated by medicine's imperative to reduce suffering and related hardship. This concern was typically reflected in one of two ways—first, in beliefs that translational research overemphasizes the study of non-human biological models (e.g., animal models) which tend to have limited relevance to human brains and disorders, and second, in perceived limitations of the research tools and methodologies currently used to attempt to understand the brain and develop treatments.

The Conflict between Basic Research and Translation

The use of animal models in basic neuroscience research generated much discussion related to the potential value it has for understanding the human brain and developing meaningful treatments that could potentially benefit patients. One researcher provided a succinct overview of what he saw as the current state of neuroscience research, in which the use of animal models may be too deeply entrenched:

> I would say that there's kind of a default set of operations that neuroscience engages in, and people don't think about it and challenge it all that much… mostly we're just locked into this world of do something in a mouse and claiming that it is going to matter for humans [...] We're kind of in a certain frozen state where in the human stuff you can't get the same spatial resolution and molecular resolution as you can in the animal models, so that makes it seem a little less science-y. And the animal stuff, you can do quite remarkable genomic, molecular, cellular manipulations, but it may not have any relevance to the human. (Professional Stakeholder Participant)

This concern regarding the ability of animal model research to provide relevance to addressing human hardship was brought up by multiple respondents. Researchers expressed wariness about the large amount of funding and attention that is directed toward animal model research, despite the low likelihood of most of it effectively translating to humans in the form of new treatments or therapies. In particular, there was skepticism regarding whether the majority of basic science that is explored under the umbrella of psychiatry and neuroscience will directly benefit humans suffering from psychiatric and neurological disorders given that the circuits responsible for such disorders can be unique to the human brain and many diagnostic criteria are currently predicated on subjective reports of suffering. One researcher described this concern:

> I think one of the bigger issues for me is where understanding will really come from. There's a lot of value placed on basic science as generating that understanding. And basic science, being in animals, has value. There's no doubt about that, and you can understand a lot of biology from looking at basic science. Whether that will ever help psychiatry, I don't know. (Professional Stakeholder Participant)

Several respondents described how this issue has been complicated by the institutionalization of "translation" within basic research. One researcher described how funding sources will often require researchers to explain how their basic research will apply to a human population, creating a false impression of translational promise, even if the proposed research, as yet, "has no real meaning for coupling any human disease." The pressure that researchers may sometimes feel to assert a link to human illness or treatments in even the most preliminary of studies was further contextualized by other respondents, who referenced how the aforementioned "publish or perish" research culture can influence researchers to ascribe greater translational potential to their work than is likely to be found:

> Right now, what you see in the field is a huge emphasis on technology in animals, and all that gets rewarded at high impact journals is more and more technology. And what you actually have is a situation that that field has diverged from anything that could help it translate to patients. You're rewarded for more and more elaborate experiments, doing things that will never have any relevance to people because it is not intended to. The actual hard work of translating between animals and humans and invalidating or validating models is being done less and less. (Professional Stakeholder Participant)

The above observation highlights the distinction between the types of research that are incentivized and rewarded by the field in the form of publications, career advancement, and grant funding, and the approaches that the participant perceived as urgently needed. This researcher continued to describe how, in his time in the field of psychiatry research, he has noticed that even the language used within the basic science literature seems to have evolved to accommodate the belief that all basic science should have clear implications for translation:

> For a long time, people would use words to describe the behavioral phenotypes of "depression-like" or "anxiety-like," realizing that they don't know if the animal is depressed or anxious or psychotic or whatever. They infer that by looking at behavior and they would make that clear in how they refer to things. And that's gone. That word "like" has disappeared. There is this pretending that you are actually studying something clinical but really you are not. What you are doing is advancing technologies that push you into this fancier experimentation. It might be great for understanding how the brain works, but not at all relevant to understanding the patients. (Professional Stakeholder Participant)

This shift in language was concerning to this researcher, who understood it as a symptom of a larger problem—that the field of psychiatry as a whole tends to overestimate the weight that should be accorded to animal model research. In turn, other types of research within the field—including promising translational studies which attempt to bring tools and therapies into human trials—may be viewed as more of an afterthought, leaving these efforts struggling for limited resources and funding.

Multiple researchers proposed variations of the question, "What are we are trying to achieve here?" in regard to the use of animal models. They questioned how findings from animal model research can be best incorporated into translational research and supported further inquiry into what translational research should really mean:

> You can't just look at something to give you a story that gives you a patina of translation. Translational work has to be done seriously and in a directed way. I think that the animal literature, in a sense, need to have a reckoning in terms of what the purpose is. Is the purpose to understand the brain? That could be a fine goal by itself. Or is the purpose to really do translation? (Professional Stakeholder Participant)

More context for this issue was provided by researchers who work on basic animal model research in psychiatry. These researchers shared concerns that the institutionalized focus on "translation" detracts from the quality of basic research, and that it pushes researchers too far, too fast. As one respondent explained, "I think psychiatrists are put in this spot where they need to cure patients, and this is the medical field for which we have the least basic knowledge. It's scary because you are put in a spot where you need to find solutions without it." Many of the researchers interviewed repeatedly highlighted how little is known about the functioning of the human brain, and how critical it is for researchers to continue pursuing basic research in order to obtain a more advanced level of understanding.

These researchers provided additional commentary regarding how academic and funding norms may not optimally support the progress of necessary basic psychiatry research with less clear impact. When funding and publications give the most attention to animal model studies that involve new technologies which provide a hint of possibility for translation to human studies, research that aims to develop a better basic level of understanding of human brain mechanisms is more likely to be ignored. As one respondent noted, "Do you know how many times you have a finding and you know it's going to be translational? Very rarely. It's just like, electricity was not discovered in an attempt to improve the candle," emphasizing that through extensive methodological work and steady progress, scientific breakthroughs can take place. By compelling basic researchers to focus on finding cures or translational returns, however, attention may be subtracted from basic discovery, reducing the likelihood of major breakthroughs.

Technological and Methodological Limitations

As discussed earlier, much of the human studies undertaken in neuroscience currently utilize tools that measure brain activity through non-invasive means, such as functional magnetic resonance imaging (fMRI) and electroencephalography (EEG). These tools, which can measure brain activity in the cortex, are the two most widely used tools used in the field of neuroscience research. While researchers still viewed these tools as highly useful, many drew attention to their limitations and the lack of other effective tools available to address their limitations. One expressed this directly:

> The other barrier is in terms of just having the right tools to look at the human brain […] because you can't open up the skull and start to do things inside the brain, and so you have to go with non-invasive techniques and they have their methodological limitations. So somewhat of the barrier to treatment and getting deep understanding becomes, what kinds of neuroscience signals can you get [through EEG and fMRI]? They have their limitations. (Professional Stakeholder Participant)

This point was re-emphasized by multiple researchers, who expressed concerns that these technologies, which were developed to measure connectivity within the cortex only, are often extrapolated to research that they are not designed for, simply because the field does not have appropriate alternatives to study much of the brain.

Researchers who use EEG and fMRI as their primary methods of research also reflected on the limitations of the output of these tools, noting that there is often an over-interpretation of the data that is obtained from these methods: "a lot of the analysis methods aren't worked out yet, it's very easy for the data to be looked at in many, many different ways." Several of the researchers expressed concerns regarding how extensive p-hacking and data mining have become a common practice in studies that involve neuroimaging and questioned the ethical implications that this had for the field.

Conclusion

Our interviews illustrated the many ways in which neuroscientists and IRB members think deeply and critically about ethics in the context of their work. Members of the research community clearly valued regulations that protect against unethical behavior and recognized that the process of making ethical choices requires a sustained endeavor as circumstances, concerns, and tools, and scientific objects change and newly emerge. As one respondent expressed, ethics are not about arriving at unassailable or universal theoretical rules, but about reaching decisions pragmatically, given environmental constraints, with attention to relevant relationships and conflicts of interest.

Clearly, our participants felt strongly that differences in the research milieu—whether a project is deployed in academia or in industry; how close a research topic might be to translational integration in healthcare; specifics of the project funding mechanism, for example—had great bearing on the particular concerns that a researcher must consider in order to perform ethically sound investigations. While there were some differences in their perspectives, our participants were unmistakably committed to navigating the features of various research environments in ways that best honored their responsibilities to their research participants and to humanity. Increased awareness and active consideration of the range of ethically salient factors that shape the conduct of neuroscience research can help researchers and other stakeholders sharpen their sense of how research should continue to evolve.

Key Points

1. Our professional stakeholder participants discussed three broad environments that shape neuroscience research: the industrial environment, the academic environment, and the regulatory environment.
2. Participants identified areas where academic and industrial research was not aligned in intellectual interests, often due to economic pressures (or lack thereof).
3. Participants frequently perceived that some regulations, especially with respect to data sharing, limited collaboration between stakeholders and the robustness and rate of research discoveries.
4. Participants noted some limitations regarding the applicability of animal model and organoid research to the human brain and its disorders as well as limitations of contemporary research tools and methodologies.

Questions to Consider

1. How might insights from stakeholders be used to help foster a more active and robust culture of knowledge sharing in psychiatry and neuroscience?
2. Academic, industrial, and regulatory forces influence how neuroscience research is conducted, but they also exert influence on each other. In what ways do they exert a mutualistic influence? Are there ways in which their influences may be ethically problematic or misaligned with the ethical ideals of medicine?

Further Reading

Abreu R, Leal A, Figueiredo P. EEG-informed fMRI: a review of data analysis methods. Frontiers in human neuroscience. 2018;12:29.

Goering S, Klein E. Fostering neuroethics integration with neuroscience in the BRAIN initiative: comments on the NIH neuroethics roadmap. AJOB neuroscience. 2020;11(3):184–8.

Hanson KA, Almeida N, Traylor JI, Rajagopalan D, Johnson J. Profile of data sharing in the clinical neurosciences. Cureus. 2020;12(8):e9927.

Chapter 16
Qualitative Findings: A Focus on Professional Stakeholder Perspectives on Additional Issues in Research and Clinical Innovation in the Brain

Max Kasun, Jodi Paik, Katie Ryan, and Laura Weiss Roberts

Introduction

This chapter, like Chaps. 14 and 15 before it, details key qualitative findings from the first part of the National Institutes of Health's (NIH's) Brain Research through Advancing Innovative Neurotechnologies (BRAIN) Initiative®-funded project, "Enabling ethical participation in innovative neuroscience on mental illness and addiction: Toward a new screening tool enhancing informed consent for transformative research on the human brain." Members of our research team conducted semi-structured interviews with 44 "professional" stakeholders (neuroscientists, Institutional Review Board (IRB) members, and ethicists) in order to better understand stakeholder concerns and perceptions about the ethical issues encountered through their work in innovative neuroscience. Using an iterative approach (see Chap. 13 for further description of methods), our research team then analyzed the interviews to better delineate and describe the stakeholders' perspectives. This chapter focuses specifically on our analysis of stakeholder perspectives related to clinical innovation in the specific context of neuroscience and psychiatry research.

Innovation, defined as the process of making *changes in something established, especially by introducing new methods, ideas, or products*, has always been a part of medicine. Historically, new understanding of disease causes and mechanisms, changes in population-level disease burdens, and new models of care provision have caused health care systems to evolve frequently and rapidly. More recently, the development of previously aspirational and theoretical technologies—and the speed at which these technologies are advanced—has changed the landscape of medical research, influencing how we understand health and disease and our sense of what treatment directions may look like in the not-too-distant future.

M. Kasun · J. Paik · K. Ryan (✉) · L. W. Roberts
Department of Psychiatry and Behavioral Sciences, Stanford University School of Medicine, Stanford, CA, USA
e-mail: maxkasun@stanford.edu; jodi.paik@stanford.edu; kryan2@stanford.edu

L. W. Roberts (ed.), *Ethics and Clinical Neuroinnovation*,
https://doi.org/10.1007/978-3-031-14339-7_16

Since 2013, when the NIH launched its BRAIN Initiative, focus on innovation in neuroscience has increased sharply, with innovative neuroscience research endeavors ongoing in many different areas. Members of our interviewee cohort highlighted projects of all types, ranging from big data analysis to informatics, stem cell research to innovative device development, and drug-related randomized-control trials. Their excitement about the possibilities of innovative neuroscience was clear, and respondents reported looking forward to a better understanding of how different mechanisms connect to individual patients, a better characterization of the physiological basis of mental illness, and, most importantly for many, new ways to translate their research into clinical care and honed more precisely to the needs of individual patients.

Distinguishing Between "Science" and "Innovation"

Previous chapters detailed many of the administrative- and infrastructure-related concerns discussed by participants as they reflected on the practical realities of undertaking neuroscience research. Interviewees also conveyed philosophical concerns. Resonating with discussions about the role of scientific discovery versus engineering in the wider technology industry, for many researchers, neuroinnovation provoked questions about its ideal relationship with neuroscience. Neuroscientists and IRB members in our subject pool viewed neuroscience and neuroinnovation as dynamic endeavors, and sometimes diverged in their views about whether they described essentially the same processes or were functionally harmonized. One respondent compared science to an artistic endeavor, noting that "There are huge disagreements about what is the best path forward, what is good science, what is bad science. […][1] Science is much closer to art than any scientist would ever want the lay public to know about."

The idea of innovation inspired our respondents and was often reflected in how they described innovative projects. One respondent extolled the culture of neuroinnovation's pursuit of solutions that "nobody has thought of before," and are "ambitious" and "cutting edge." Respondents often referenced the potential for innovative techniques to open up paths towards previously unattainable kinds and degrees of knowledge, citing emerging examples such as the development of manipulable brain organoids, the fields of optogenetics and artificial intelligence, and progressive advances in computing. They felt that the historical lack of techniques able to explore and manipulate the brain at high enough levels of resolution and precision represented frustrating limits of brain research that can now be pushed further. In the words of one IRB member, "we're really excited about all the studies [psychiatry and neuroscience] are doing because [understanding of] the brain to me seem[ed] to be lagging."

This excitement carried on through many interviews, as respondents spoke about the desire to tackle historically fuzzy and intractable problems, and about how best to approach them. They wanted to know more about the nature and functional roles of

[1] Suspension points indicate omissions.

neurological circuits (among many objects of study) and hoped that better understanding of the brain would assist discovery and reduce morbidity and mortality in neurobiology and psychiatry.

Respondents spoke frequently about "new models," "new conceptual structures," and "new treatment modalities" possible via neuroinnovation. One researcher, characterizing his work, explained that "the kind of neuroscience I do is a new form." Many were motivated by their understanding that existing knowledge of the brain is miniscule in comparison to the level of knowledge that could be achieved, such as one researcher who reflected:

> I think the creativity [interests me], because I can all of a sudden work with industry and with other people in the department and in other disciplines and where nobody knows anything. We're exploring. I am an adventurer at heart, so I like the exploration process. (Professional Stakeholder Participant)

Several interviewees tempered their excitement, sharing concern that innovative techniques might not lead to great success without very careful development and implementation. One respondent expressed worry that the strong affective and economic pull of "innovation" could diminish interest in scientific rigor, sidelining "good" traditional science, as they defined it:

> It should not be about innovation and it drives me crazy. It should be about good science, and sometimes the best science is doing an old-fashioned methodology. We are treading down a very dangerous path of not doing the best science anymore and just promoting "innovation," in quotes. (Professional Stakeholder Participant)

Because the term "innovation" refers to novelty and does not necessarily include a value claim or judgment, some respondents were concerned that the idea might currently exert an undue influence on research funding. For them, intrinsic value was one attribute that distinguished biomedical research from innovation. The respondent above continued:

> It drives me crazy that on every NIH grant now you get scored on innovation. That's not what should be the criteria. It should be the best science to answer the question you're addressing. And certainly innovation is important and should be encouraged, but it should not be encouraged at the expense of the best science. (Professional Stakeholder Participant)

The Role of Innovation in Changing Established Scientific Frameworks

New tools enable new types of research, and the introduction of innovative research tools into the field of neuroscience can alter established protocols and generate new ethical and methodological concerns. Interviewees reflected on how caution is required in these burgeoning inquiries, in that the innovative methods developed may have adverse consequences or necessitate reevaluation of the scientific frameworks that shape the conduct of research.

Effects on Life Science Research Methodology

Respondents explored ways that innovative neuroscience could mutualistically support other biological research. Central to many of these examples was the idea that new research techniques might enable kinds of biological knowledge previously thought unattainable. Several respondents described biological methods they felt were especially promising and expressed excitement at the ways usage of these methods might allow for a new understanding of the brain. One participant pointed to the "amazing breakthrough" of optogenetics as one example of a game-changing technique that allowed for a new way to understand neuronal function, pointing to optogenetics as "this huge breakthrough that for once, in contrast with fMRI, can allow us to address the function of a neuron."

As another example, several respondents cited the creation of brain organoids as a new scientific advancement that could fundamentally enhance neuroscience research. Derived from pluripotent stem cells and maintained in vitro, these organoids function as developmentally conserved, three-dimensional cultures, facilitating the discovery of new models of neuropsychiatric disorders and treatments. Several respondents discussed the profound scientific impact of this new technique, in that it can surmount previously intractable ethical and experimental barriers in the modeling of complex neurological and neuropsychiatric conditions. As one researcher explained, the advancement in the use of brain organoids "solves the problem of not having access to human tissue and it releases some of the need, for instance, for using human primary tissue."

Different biological approaches impact established research frameworks in different ways. Respondents described the possibility that new approaches such as optogenetics and brain organoids will alter not only the methods of study but also the established pipeline of scientific inquiry. One researcher described ways that new biological techniques may influence how the quality or usefulness of different types of research is perceived: "Very soon people are going to say, 'Well unless you make a little bit of a brain tissue and test it, then we're not going to believe this data'."

While excitement for these new biological methods was strong, respondents considered their utility in solving clinical problems for actual patients an important part of their value. One researcher summed up this point, explicitly mentioning the importance of "coming back to, 'How do we fix us?'"

Technological Changes

Innovation in neuroscience ran in parallel with innovation in technology for many of our respondents, and concerns described initially in the tech space were revisited in the context of clinical neuroinnovation. Respondents expressed agreement that as interest in computational neuroscience has grown, big data and artificial intelligence have increasingly found a foothold in the research endeavor. One researcher described how big data platforms can generate insights even at the level of an individual patient, and the hope it implies for the future of patient care:

> [Our department] has developed an informatics platform [...] it gives people the ability to conduct these pragmatic trials. [...] It actually has smaller standard error, so you get precision with lower burden to the patient, and you get a beautiful, collated, real time, and longitudinal data display of how your patient is doing. (Professional Stakeholder Participant)

While patient care was one major area of interest, respondents also described ways in which technological changes have affected the path of research. Aware of neuroinnovation's close relationship with advanced computing, many respondents shared a desire to incorporate new computational techniques into their work, but also expressed concerns about its effectiveness, and awareness of potential threats to ethical commitments, such as patient privacy. Some respondents pointed to the potential for machine learning and artificial intelligence applications to describe, prevent, predict, and help manage disease as a reason to pursue its development. There was agreement among several respondents that these tools could prove generative, although some respondents tempered this view with observations that artificial intelligence and machine learning are still in their infancy. One researcher expressed some ambivalence about the current usability of machine learning in neuroscience, expressing the opinion that it could be more powerful "maybe in the future." Others described the success of computational techniques in other disciplines and hoped to bring them into the field of neuroscience. As one respondent explained, "There is a whole field of computational neuroscience that's working on that that I am at the edge of, but we want to kind of steal conceptual stuff from them and bring it to bear it on neuroimaging work."

Emerging Ethical Concerns in Innovative Neuroscience

While highly valuing the goals of science, respondents also recognized the needs of human subjects and affirmed the importance of human subject protections in safeguarding the ability to perform innovative, and potentially risky, research studies. Their excitement about neuroinnovation was moderated by the desire for thoughtful and robust ethnical frameworks. One researcher detailed a concern that not addressing the ethical dimensions of innovative projects could potentially slow the advancement of the field:

> [It worries me that] because everything is so promising and so exciting, in terms of what is happening in neuroscience, that anything that you put in the way of neuroscience progressing could be [seen as] an ethical dilemma. You're preventing science from going forward [...] You get into your own bubble, if you will. I think that's a danger. I love addressing curiosity, but I feel that it cannot be that alone. It has to advance science and it has to advance how human beings live a better life. (Professional Stakeholder Participant)

The Complexity of the Brain and Unique Risks of Neuroscience

The brain is a unique organ with complex biology and a distinct role as the root of consciousness, emotion, self-governance, and personhood. The uniqueness of the brain entails that brain research is also unique, carrying specific risks and consequences by virtue of its distinctive and fundamental role in the mind and body—an understanding our interviewees shared. Many reflected on the ways that this unique status creates novel tensions in the lab and in the clinic.

Researchers reported employing a variety of novel technical methodologies to help science make progress, and many respondents were moved to describe those techniques in detail, highlighting the ways that the complexity of the brain can motivate the use or creation of innovative tools. Respondents expressed concern that research involving cutting-edge tools, especially in the context of the brain, carries unique risks. The risks they identified were both actual (e.g., surgical risks from implantation of a novel deep brain stimulation device) and theoretical (e.g., the future possibility of brain organoids advanced enough to have some experiential awareness that could be violated). They suggested that some concerns could be originate in an innovative technology itself (e.g., psychosocial or privacy threats related to wearable devices that collect and analyze biometric data). Participants also expressed concerns about the societal context of use (e.g., "biohacking"). Reflecting on the uniqueness of the brain, some members of our cohort described a need to balance excitement with caution, especially in research endeavors with novel interventions, as expressed by one respondent:

> There's just the need to be really cautious. We may not have a full way of understanding or a full way of measuring or capturing what the impact is of either a measurement tool or a neurological intervention. [...] There's just a lot that feels kind of unknown, so with innovation you have to be very cautious about that. [...] I think maybe initially when starting a project, particularly when it is an innovative neuroscience study, really asking participants what they're expecting to get out of it and why. (Professional Stakeholder Participant)

Innovative Devices and Technology

There is a wide range of methods and tools that can be applied in the context of neuroinnovation, encompassing everything from novel biologically based methods such as optogenetics and brain organoids; to neuromodulating interventions like deep brain stimulation (DBS) and transcranial magnetic stimulation (TMS); to ex vivo computational methods, including artificial intelligence and machine learning. Regardless of whether the technology considered was in vivo, in vitro, or in silico, however, respondents described an awareness of technology-specific concerns and valued considering the pros and cons of new technologies' use. Like the many demonstrated implications of other transformative technologies (e.g., nuclear energy, DNA manipulation, deepfakes), respondents found both promise and peril.

Recognizing Limitations

Respondents reflected on the status of innovative devices and methodologies like DBS and repetitive transcranial magnetic stimulation (rTMS), expressing both excitement about their utility and concern about some of the details of their deployment. Recognizing that "we're getting more and more advanced technologies, but some of those are more and more invasive," one researcher cautioned about the need to evaluate research rationales: "are we doing things just for the fact that we can do it and it's just new technology and we want to try it? Or do we really think it's going to help these patients or further the field?" Another researcher echoed the importance of having an explainable scientific rationale for utilizing tools that could be explored across a potentially wide problem space, perceiving a problematic tendency: "When there is some new treatment modality [...] one of the first things that happens is that people try to generalize it." While acknowledging that generalization might be safe in lower risk studies, this respondent stressed that the risks of some neuroinnovations could be considerably higher than those of usual medical interventions and thus deserved more careful consideration:

> You need to be really careful not to just sort of say "Well if it worked there let's try it here," without really having a very sound pre-clinical scientific basis for why you think it might work in that setting. It's a lot easier if you've got a new antibiotic to try it on 53 different bacteria to see if it works, but if you're doing some invasive — either an electrical stimulation or something that requires that you place electrodes or some physical sensor of some kind in somebody's brain — then just because it worked for Parkinson's disease doesn't mean it's going to work for ALS, or however you might define that. Stimulating nerves with an electrical pulse might be very helpful in one disease, but you need to make sure you have a very sound scientific rationale before you say "Well it worked for that, let's try it for this and that." (Professional Stakeholder Participant)

Many respondents described their wish for researchers to maintain a critical attitude when assessing the limitations and ethical considerations of new technologies. One respondent warned against getting "wowed again by the technology driving, kind of the tail wagging the dog in a way." Some respondents expressed the view that innovative methods can unduly inspire public trust and sway prospective research volunteers, particularly in the case of innovative technologies, because of their novelty and advanced appearance, as described by one researcher:

> Because [subjects] think because there's technology, there's been a lot of safety done. You know it's sort of like, you go to the doctor and the doctor is like God. So you go, "This is the high technology. I don't need to ask questions, this is all ordained." And they don't understand it's not. I think that's what I am trying to put into words what's bothering me. I think that's what it is. It's like you just go into this, "Wow, technology can do all this!" so they don't ask questions, they assume it's safe. And because it's so powerful, there is a hubris about it too, and the people using it have a hubris. (Professional Stakeholder Participant)

Several respondents expressed similar concerns related to how perceived innovativeness may influence research participation, feeling that an overabundance of trust could lead research subjects, and researchers themselves, to underestimate the risk and overestimate the potential benefits of their involvement. As one researcher noted, "There's a hubris about it, technology and the brain, and that we know things and we underestimate risk."

Privacy and Confidentiality Concerns

Many concerns arising from the use of innovative devices and technology in neuroscience and psychiatry were specifically related to privacy. Respondents pointed to the increasing use of devices that collect patient data as a growing region of ethical concern in terms of patient confidentiality:

> A lot of the protocols that are coming along in psychiatry use devices, and so a shift away from medication to devices has called for a different understanding of what the ethical challenges are of doing it because of issues regarding privacy and sharing data with outside companies. (Professional Stakeholder Participant)

Concerns about privacy, data-sharing, and the use of big data were repeatedly brought up in conversations regarding devices, which are frequently owned by "companies that hospitals might contract with" and that might collect many types of data. Interviewees expressed apprehension about the ownership and security of that data, which could be stolen and mined for information, and had questions about "what happens if those [data] get out into the open." Researchers expressed concerns that, in the hands of actors outside the clinical setting, health data could be applied in ways that could negatively impact patients' lives, with examples being that such data may be used to deny health insurance or increase premiums, or even prevent certain individuals from being hired into specific jobs or fields. One respondent hinted at these concerns, asking "If people get hold of those kinds of data and they can mine it to see who might have depression or clinical symptoms, how might they actually use it?" while another addressed them explicitly:

> There are unknown potential risks of that re-identification and we don't know what those are. You can imagine health insurance risks, and unlike genomic data, you're not protected legally from the negative consequences of somebody using these kinds of data. (Professional Stakeholder Participant)

Specific attention was given to the inability to clearly anticipate future risks entailed by new technologies. Respondents repeatedly expressed that many of their concerns surrounding the collection and sharing of neural or biological data stem from the fact that the future capabilities of technologies could re-identify individuals more easily. Neuroscientists interviewed by our team were wary of the still-hypothetical but increasingly likely potential for anonymized neuroimaging data (such as MRI scans) to be re-associated with an individual, with its own new set of risks:

> The question is, if you knew that about someone, if you had their just raw neural data, what could you learn? Is it violating their privacy in any way? Right now, I don't necessarily think so. We've maintained that we'd rather err on the side of caution and not release that data if we can avoid it because who knows in the future what sort of deep learning techniques—, you might be able to identify a person's neural signature from their data. So I think this is a thing as we're starting to get bigger and bigger data sets with more and more of a subset of the neurons that make up a person's brain, that we may be able to find out more about them through that. (Professional Stakeholder Participant)

Enhancement and Augmentation

The potential for unintended downstream effects of some innovative neuroscience technologies weighed on the minds of many of our respondents. One respondent discussed the high likelihood of science aiming increasingly towards augmentation of the brain, expressing that "augmentation is one of the topics that people are really interested in because our history as humans is, we find something that makes us able to do something better, we're going to try to do that." As neuroscience research has become more innovative in design and scope, the augmentative potential of much neuroscience research has become more obvious and its ethical limits less clear, introducing the question, "Under what circumstances would we ever consider augmenting someone?"

Treatment Versus Enhancement

Respondents noted that a new host of ethical issues arises when we begin moving past treatment and into enhancement or augmentation, which people may perceive as undue alteration of an individual's apparently healthy state or to increase their mental abilities beyond their "natural" level of functioning. The limits of appropriate use of hypothetical cognition-enhancing therapies, and their risks, are undefined and difficult to predict. Researchers discussed the distinctions that society tends to draw between treatment and enhancement. One respondent felt that "anything that puts you back to your baseline, I think people would be really comfortable with that," in the context of potential enhancing technologies. When similar approaches are instead utilized purely for purposes of self-enhancement or improvement, however, researchers suggested that society may not be as open or accepting:

> You could do something that would make someone the best that they are, to make their brain function at its highest level. If you could make someone think better and be smarter, I think most people would think that was a good thing, but that would be changing what people perceive as a basic thing about someone. So even if you make someone their "best" with some of these changes, I think people are going to be uncomfortable with any change, just as a society [...] But I think most people are comfortable if you're treating something that people are suffering from so that they feel better or so that they are where they were before. But when you start doing things that change them from baseline and things that may be an enhanced intelligence or mood, I don't know. I think that's where the ethical issues are going to come in. (Professional Stakeholder Participant)

One respondent noted that conceptions of "therapy" and "enhancement" tend to become merged in hypothetical advanced technologies, because some could potentially serve both purposes. Hypothetical enhancement of our mental functioning through neural interfaces (already being developed) and other prostheses could result from devices supposedly originally developed as therapies. One researcher expressed concern that such enhancing technologies will be prematurely applied in clinical contexts:

> There is this superman kind of trend, which is like, "Oh I don't sleep. I sleep four hours. I take a cold shower in the morning, fifteen minute nap in the afternoon, and I'm just fine." Yes, for a couple of months. And the trend to me is like, whatever we do that would show that we can shorten sleep, that will be taken too rapidly to the clinic without evaluating the very long-term effects. (Professional Stakeholder Participant)

These concerns extended to outside of the clinic, with some researchers emphasizing that novel risks and vulnerabilities may emerge in uncontrolled and non-therapeutic contexts, and that these implications are intrinsically difficult to predict. One researcher noted that techniques that target neural circuitry pose special risks to autonomy and risk being misused outside the therapeutic setting:

> We're taking people and driving their brain transiently into a much more highly receptive state and then having them enter into a therapeutic environment where they get suggestions that normally their brain would clamp down or reject because the conflict circuity is intact, but in this case after TMS, after shutting down the conflict circuitry, after allowing information flow to happen more readily, they just take it in. [...] Say this got approved [...] and then Joe Blow in the community does this. (Professional Stakeholder Participant)

These concerns are linked by the idea that some individuals and entities outside the therapeutic setting have norms or incentives that conflict with the medical community's principles and safeguards, such as beneficence and informed consent. At the same time, other researchers noted that much of individual non-therapeutic use is driven by a desire for optimal well-being, and that healthy people have and likely will always find ways to use innovation for personal enhancement:

> We forget that we actually do those things all the time in our daily lives. Like, I just had a cup of coffee. I am going for neurostimulation pharmacologically, and I do that every day, multiple times a day. We do these things anyway in our daily lives. [...] People are so afraid of what might happen that they forget that they're doing some of that every day, that humans for millennia have been trying to neuromodulate themselves to seek advantages, to get better test scores, to run faster, to do something better. That's kind of what people have been doing always. (Professional Stakeholder Participant)

Justice and Autonomy

Multiple researchers addressed the ethical concerns surrounding justice (e.g., equity, access) and autonomy that may emerge when technologies are used for enhancement purposes. One researcher noted that the likely immense costs of new technologies may pose a justice issue as it relates to enhancement, as only those with the financial resources will be able to access such tools for improved functioning. Differential access to cognitive enhancement could become yet another frustrating device for reinforcing existing systemic inequality or inequity, with one respondent noting that "I think that if you start doing things, then I think people are going to start to worry about these inequalities. It may be only the people who are the wealthiest who can afford to do some of these treatments to make them better and smarter."

One respondent proposed that enhancements may be one day perceived as necessary for certain social roles and labor. Concerns arose in regard to specific jobs or careers that may encourage the use of enhancement, and how the availability and use of these technologies could have lasting implications on both the individuals and the field. One participant cited work on a deep brain stimulation system that was shown to support increases in focus and concentration and hypothetically could unjustly advantage certain individuals. This researcher questioned the impact that such tools could have if they were made available outside of a medical context:

> Is there either overt or subtle societal coercion? Is there individual coercion? What about equitability? Can everybody get it? Or can only a certain few people get it? […] What are the risks? What are the societal risks? What about others in [the] squadron? Are they being coerced now because [someone else] has an advantage? Is there subtle pressure that they should do it? (Professional Stakeholder Participant)

Such questions regarding coercion and societal pressures emphasize the effect that augmentative technologies could potentially have on an individual's bodily autonomy, further demonstrating how wide-ranging and unpredictable the downstream effects of innovative neurotechnology may be once outside of the controlled environment of a research laboratory.

Conclusion

Innovative research can be transformative, and nowhere is this truer than in the brain. Clearly, neuroscientists and IRB members in our interview cohort gave critical attention to the specific context of neuroscience as a field. Our interviewees shared specific concerns arising from the brain's unique role as the seat of human consciousness. Guided by a shared motivation to ensure research is conducted with great care in this particular field of study, our participants explored the unique concerns arising from brain-based innovations. They carefully considered the line between "science" and "innovation" in the brain, explored the unique protections required in neuroinnovation due to specific privacy and security issues specific to brain-based study, and expressed concern about the line between treatment and enhancement. Our respondents also examined issues related to new technologies utilized in investigative neuroscience. As in previous chapters, these interviews highlight a shared desire to protect the ethical foundation of research focused on the brain. With a shared sense of the promise neuroinnovation holds for improving human health, our interviewees reflected a common desire to advance sophisticated and stakeholder-informed approaches to addressing accompanying ethical issues.

Key Points

1. Some of our professional stakeholder participants viewed innovative qualities as important in evaluating the usefulness of research, while others believed it was overemphasized for reasons such as its value-agnostic nature.
2. Participants were optimistic about the potential for new tools to achieve greater biological resolution and control of the brain (and derived organoids) and ulti-

mately to reduce human suffering. Still, they pointed out unresolved ethical issues surrounding these tools, including that some tools' promise has yet to be widely demonstrated, that related perceptions may involve a degree of hubris, and that they likely entail new biological, psychiatric, and societal risks.

Questions to Consider

1. Are neuroscience and neuroinnovation distinct professional endeavors? To what extent (if any) do they overlap or align (e.g., in values, motivations, or functional aspects)?
2. Some neuroinnovations, such as machine learning-derived tools for modeling features of mental illness, have been critiqued for potential overreliance on biological frameworks for understanding illness. Is this critique valid? In what way(s) could such biological essentialism matter ethically?
3. Besides biological and justice risks (e.g., invasive nature of a brain implant procedure, social inequality), what other risks might be posited for brain-enhancing technologies?

Further Reading

Haselager DR, Boers SN, Jongsma KR, Vinkers CH, Broekman ML, Bredenoord AL. Breeding brains? Patients' and laymen's perspectives on cerebral organoids. Regenerat Med. 2020;15(12):2351–60.

Lipsman N, Zener R, Bernstein M. Personal identity, enhancement and neurosurgery: a qualitative study in applied neuroethics. Bioethics. 2009;23(6):375–83.

Chapter 17
An Innovation in Neuroscience and Neuroethics Survey Research: Amazon MTurk

Tenzin Tsungmey, Jane Paik Kim, Henry Termuehlen, Jodi Paik, and Laura Weiss Roberts

Introduction

Our project, "Enabling ethical participation in innovative neuroscience on mental illness and addiction: towards a new screening tool enhancing informed consent for transformative research on the human brain," utilizes a unique online crowdsourcing marketplace to recruit participants for important parts of our research. Amazon, the world's largest internet company, launched Amazon Mechanical Turk (MTurk) in 2005 as a tool to recruit remote workers to complete tasks that could not be performed by artificial intelligence. The online service was inspired by the Turk, a famed chess-playing machine hoax developed in the eighteenth century. Challengers who encountered the Turk believed they were playing chess against a sophisticated machine; rather than being automatic, however, the Turk was controlled by a hidden human chess master enclosed within a small cabinet (see Fig. 17.1). Amazon's MTurk allows for the deployment of human intelligence under a technological interface and was explicitly designed as a platform to crowdsource workers who could perform tasks that a computer could not. Initially, these crowdsourced workers were used to remove duplicate products from Amazon's own website.

The MTurk platform launched in November of 2005 and its user base grew rapidly. Realizing the potential of the platform, and recognizing that "while technology continues to improve, there are still many things that human beings can do much more effectively than computers," Amazon extended the services of MTurk to other companies also looking for workers to complete small jobs or projects known as

T. Tsungmey · J. P. Kim · J. Paik · L. W. Roberts
Department of Psychiatry and Behavioral Sciences, Stanford University School of Medicine, Stanford, CA, USA
e-mail: janepkim@stanford.edu; jodi.paik@stanford.edu; robertsl@stanford.edu

H. Termuehlen (✉)
Medical College of Wisconsin, Milwaukee, WI, USA
e-mail: htermuehlen@mcw.edu

L. W. Roberts (ed.), *Ethics and Clinical Neuroinnovation*,
https://doi.org/10.1007/978-3-031-14339-7_17

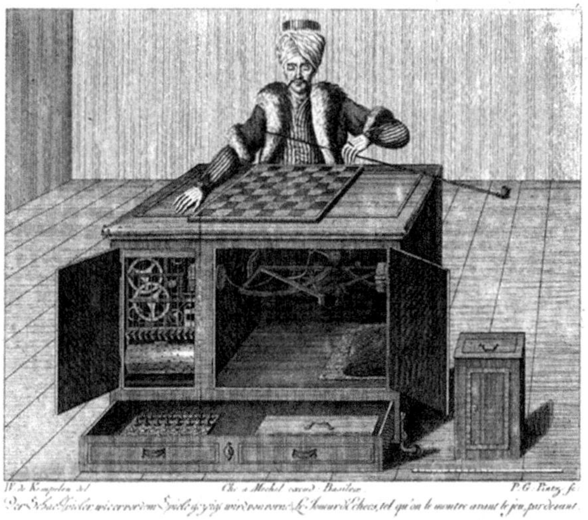

Fig. 17.1 Copper engraving from the book: Karl Gottlieb von Windisch, Briefe über den Schachspieler des Hrn. von Kempelen, nebst drei Kupferstichen die diese berühmte Maschine vorstellen. 1783. Public Domain, https://commons.wikimedia.org/w/index.php?curid=424092

Human Intelligence Tasks (HIT)—single, self-contained, virtual tasks, such as moderating content, removing duplications, or participating in survey research—that require human participation to complete [1]. Amazon describes MTurk as follows:

> Amazon Mechanical Turk (MTurk) is a crowdsourcing marketplace that makes it easier for individuals and businesses to outsource their processes and jobs to a distributed workforce who can perform these tasks virtually. This could include anything from conducting simple data validation and research to more subjective tasks like survey participation, content moderation, and more. MTurk enables companies to harness the collective intelligence, skills, and insights from a global workforce to streamline business processes, augment data collection and analysis, and accelerate machine learning development [1].

The process of assigning tasks to be completed by anyone who willingly takes up the job has become known as crowdsourcing. Crowdsourcing involves gathering labor or information from a large and open group of participants. Work that is crowdsourced is typically not assigned directly to any specific individual. It is instead willingly taken up by an individual that chooses to complete the task for the reward that is agreed upon. Crowdsourcing differs from outsourcing because the task is completed via a less specified and more public group (whereas in outsourcing a task is typically assigned to a specific group). Because of the nature of crowdsourcing, the workforce composition is fluid, often gaining new workers and losing past workers over the course of time. It is estimated that between 100,000 and 200,000 workers were on the MTurk site as of 2018, and that half of all MTurk workers will perform tasks on the site for about 400 days before leaving the platform [2].

The MTurk platform has been explored and adopted as a survey platform in many fields. In this chapter, we discuss the use of MTurk in the conduct of neuroscience and neuroethics survey research. We begin by providing background on MTurk as a survey platform and then describe some early studies in the domains of neuroscience and neuroethics conducted via MTurk.

Amazon MTurk and Academia

Survey participant recruitment increasingly utilizes internet platforms instead of traditional sampling techniques. Amazon's MTurk has been a tool for recruitment and data gathering for social science research since 2010, and approximately one-third of all jobs on MTurk are from individuals in academic sectors [3]. Researchers in the social sciences—including marketing, psychology, and political science—rely on MTurk as a source of participant recruitment. Crowdsourced research is increasingly popular; as an example, over 40% of behavioral studies in volume 43 of the Journal of Consumer Research (June 2015–April 2016) utilized data from surveys deployed through the MTurk marketplace [4]. Other academic fields, e.g., sociology and communication, have also adopted crowdsourcing method of collecting data.

Increasingly, MTurk also has been used to improve training data for machine learning artificial intelligence systems. For example, YouTube uses MTurk workers to review videos to determine if they contain any crude or inappropriate content that is not allowed on the site. Workers watch videos and then answer some questions about what was contained in the video; their answers improve the data used to develop YouTube's machine learning tools for automated content detection [5]. Machine comprehension of text also drives some MTurk projects. The Stanford Question Answering Dataset, or SQuAD, utilized MTurk to answer 100,000+ questions posed by workers on a set of Wikipedia articles; the goal of this project was to provide a dataset to improve the reading comprehension of artificial intelligence systems [6]. More recently, academic medicine researchers have used MTurk to assess clinical populations [7].

How Does MTurk Work?

MTurk is simple and relatively easy to use. Participants register as "Requestors" (task creators) or "Workers" (contractors or paid task completers). Requestors can post any task or HIT and define how many workers are needed to finish the task (i.e., 500 workers to complete the task). Requesters have the option to make sure that the best workers complete the HITs by setting qualifications for workers in terms of the following parameters:

- Location (i.e., only workers in the U.S. to complete the HIT),
- The number of HITs a worker has completed (i.e., choosing workers who have completed at least 200 HITs),

- The percent of completed HITs accepted (i.e., choosing workers who have had 95% or more of their HITs accepted by other requesters).

MTurk workers who meet the above qualifications can then finish HITs in the time frame selected by the requester (e.g., within 2 hours) and get paid for their work within hours or days of completing the task. Payments for tasks that are completed are transferred directly from a requester's account to a worker's account. For international MTurk workers, payment might be in Amazon credit.

Registration, Sign Up, and Payment

Workers register for MTurk using their email address. After registering, workers can view a list of available HITs and can opt to participate in HITs for which they are eligible. Workers can link their MTurk account to their bank account for the purposes of payment. The HITs that the workers complete are linked to online survey platforms such as Survey Monkey and Qualtrics [7].

Fee Structure

Amazon's MTurk profit is derived from the fees charged to requesters (on top of payments to any worker). Amazon adds a 20% fee on top of the reward (i.e., payment) and any bonus or tip amount payed to workers; HITs with 10 or more assignments are charged an additional 20% fee [8].

Requesters are responsible for payment, and they can decide what work to pay for and what work to reject. In the example of a survey, requesters can reject a survey completed by a MTurk worker on the basis of the requester's subjective criteria, e.g., if the survey respondent misses one too many attention-check questions or if the time to completion was deemed insufficient. Amazon is not responsible if a MTurk worker believes that their work was rejected unfairly [8]. Requesters looking to conduct longitudinal studies and repeated measures experiments can do so by issuing follow-up invitations to their previous MTurk workers, providing payment on more than one occasion. Requesters can also exclude MTurk workers from one study to the next study if conducting longitudinal studies by collecting user IDs from the first study [8].

Who Are the MTurk Workers?

The Amazon MTurk workforce is global, although most workers are from the United States (75%), followed by India (16%), Canada (1.1%), Great Britain (0.7%), Philippines (0.35%), and Germany (0.27%) [2]. MTurk workers are younger than the overall population of the United States (88% of workers under the age of 50 compared to 66% of U.S. working adults) and are more educated (51% of respondents have college degrees, compared to 36% of working adults in the U.S.) [7]. Initial examinations of the worker population revealed that MTurk workers were comprised of significantly more females (64.85%) than males (35.15%) [9], but more recent studies show an almost equal representation of males (49%) and females (51%) [10].

Many workers report that they join MTurk because of the financial opportunity it represents. Although most workers use MTurk to supplement their income, about 25% of workers make most or all their income from MTurk [7]. Other reported motivations for participating on MTurk are diverse, including entertainment, "killing time," personal growth, skill building, and contributing to knowledge and society [7, 9]. Even though some MTurk workers are motivated by the money they make from taking part in MTurk, studies show that compensation alone is not the only reason for participation [9, 11].

Are MTurk Data Reliable and Valid?

Concerns about the quality of MTurk data generally reflect an awareness that the anonymized, virtual-only connection between requesters and workers may result in a lack of monitoring that could cause problems with data integrity. To ensure that workers are incentivized to provide data of good quality, MTurk allows requesters to reject work and not pay workers or even to block workers from future work after completion. This encourages workers to be motivated to follow instructions and pay attention to the research survey, particularly research surveys with screening methods and attention checks to assess their concentration on the task at hand. Most of the time requesters require workers to have a high approval rating for HITs. Approval ratings are ratings made by MTurk requesters and are updated each time an HIT is completed. MTurk workers are motivated to keep up their high approval ratings in order to have access to more HITs in the future [7, 12].

A genuine concern is whether workers are diligent and honest. Issues that can affect data integrity in traditional research settings (cognitive biases, logical fallacies, and other behaviors among research participants) remain in the crowdsourced setting [12]. These issues may be compounded with additional concerns related to the unique dynamics of crowdsourced, internet-based participation, although this has not necessarily been confirmed in the MTurk literature. Comparing MTurk workers with traditional research participants, Paolacci and Chandler found few differences in the successful replication of attention-sensitive tasks, and direct assessment of attentiveness revealed few differences [12]. Additionally, although overall demographic characteristics of workers vary regularly by time of day, the location and demographic information provided by individual workers is consistent over time, suggesting that data collected through MTurk is reliable [10].

Bunge et al. compared the recruitment process and participant characteristics between two similar randomized controlled trials of interventions for mood disorders [11]. The first trial utilized MTurk to recruit 795 participants while a second trial, which recruited via unpaid internet resources (UIR), ultimately included 329 participants, indicating not only a higher recruitment rate per month via MTurk but higher retention rates as well. Importantly, however, participants recruited via MTurk may not represent the general online population of interested in clinical intervention, and data from clinical studies resulting from MTurk should be interpreted with caution.

To specifically check for data quality, Kees et al., conducted a study with five different participant groups, including two study samples (one group took the study in a lab and the other group took the study online) and three online platforms (Lightspeed, Qualtrics, and MTurk) [13]. When comparing across the five sample groups for data quality, MTurk data outperformed the other four groups. MTurk workers finished the survey before the other four groups, but also provided longer answers to the open-ended questions. Thus, MTurk is a practical platform for academic data collection. Similarly, Buhrmester compared MTurk participants with typical American college samples, concluding that "data obtained are at least as reliable as those obtained via traditional methods" and confirming that requesters can feel confident using MTurk for most study designs, as "Workers are diligent because of their intrinsic motivations and the incentive structure of MTurk: Requesters are not forced to approve submissions and can screen workers on the basis of past approval rates," respectively [14]. MTurk samples have continued to be shown to be representative of empirical, offline research [15]. Even so, due to the virtual nature of the data, inquiry into its validity remains a focus of study [16] and quality control is increasingly sophisticated. Agley performed a randomized controlled study, assigning workers to one of four study arms with different quality control measures. The utility of quality control measures was confirmed in this sample, although the authors note that care must be placed in their deployment, as "use, or lack thereof, of quality control questions in crowdsourced research may substantively affect findings, as might the types of quality control items" [17].

Ethical Issues in MTurk Crowdsourced Data

Ethical issues exist with the MTurk platform. As MTurk is an unregulated workforce with researchers essentially hiring an MTurk worker to complete a task in a certain amount of time in exchange for payment, a power imbalance exists between MTurk workers and requesters. Workers depend on the approval of the requester for both payment for the study they are part of and eligibility for future tasks which can result in a high risk for coercion. Researchers must mitigate these concerns and remain mindful of providing fair pay to workers [18]. As more researchers use MTurk, concerns have been raised regarding the ethics of using MTurk workers for research experiments, on the basis of the low wages that workers are given for their participation in research [2].

MTurk allows researchers to conduct studies for low cost, but it is important to be mindful of ethical considerations in research and to compensate workers fairly [13]. A recent task-level analysis of MTurk workers recorded 2676 MTurk workers performing 3.8 million tasks and calculated a median hourly wage of only ~$2 an hour, with only 4% of workers earning more than $7.25 an hour [19]. Ninety-six percent of MTurk workers earn below the U.S. federal minimum wage. On average, requesters are paying $11.58 an hour, but dominant requesters who post many low-wage HITs (such as content creation tasks) bring down the overall wage distribution.

In terms of safeguards for MTurk workers, at present Amazon has no rating system for workers to rate requesters. Independent sites like the Turkopticon (https://turkopticon.info; dedicated to workers rating requesters) as well as Turkopticon2 (workers can rate specific HITs) allow workers to share numerical ratings as well as comments. The Turkopticon allows workers to rate requesters on four characteristics; each characteristic is rated on a scale from 1 to 5, with 1 being low and 5 being high [7]. The rated characteristics include:

1. Communication: the responsiveness of the requester to emails expressing concerns. Workers can email requesters with questions or comments directly from MTurk if they are having a problem with the HIT itself.
2. Generosity: how well workers are compensated for the amount of time it takes to complete the HIT.
3. Fairness: the degree to which the requester is fair in approving or rejecting work.
4. Promptness: how quickly work is approved and paid for.

Additionally, workers write comments to provide context for their ratings. Many MTurk workers frequently check the Turkopticon before agreeing to sign up for a HIT [7].

Ideally, requesters are transparent and honest with their potential MTurk workers. In addition to supporting ethical principles such as autonomy and beneficence, transparency also yields positive effects on sites like Turkopticon. In their HIT instructions, requesters can use real names and their institutional affiliations. Informed consent (a paramount requirement for ethical research supported by all Institutional Review Board (IRB) protocols) is also integral to protect research integrity [7]. To better understand workers' experiences on MTurk, requesters are ideally encouraged to complete a few HITs as MTurk workers to get a better sense of the ethical issues of participation in this platform [20]. Improving communication channels could make it easier for requesters to identify and fix broken HITs [19].

Neuroscience and Neuroethics-Related Studies Using MTurk

Our team learned from several studies relevant to neuroscience and neuroethics conducted using MTurk. Here, we feature several projects that illustrate the platform's utility across various types of research design, including different lines of research inquiry and differing levels of participant vulnerability, including those projects that engage individuals living with mental health conditions or studies that relate to scientific, clinical, or ethically important issues regarding the treatment of mental illness.

MTurk Is a Reliable and Valid Research Method

The use of MTurk for data collection in the behavioral sciences has increased for several reasons [21, 22]. The speed of data collection, the ability to recruit large samples at relatively lower cost, and growing evidence that the crowdsourced data is of equal or better quality to that collected with traditional populations and methods supports MTurk's growing usage. Shapiro was one of the first studies to look at the unique aspects of utilizing the MTurk workplace and examined the feasibility of crowdsourcing programs to conduct research on psychopathology [22]. This project reviewed the frequency of several psychiatric disorders and related problems using MTurk, additionally checking the reliability and validity of participant reports. The findings suggest that MTurk offers many advantages for clinical research when collecting data online, but that potential problems, such as misrepresentation, could impact research and must be addressed.

Schleider extended the assessment of the feasibility of MTurk to examine its utility for longitudinal clinical research [23]. Traditional longitudinal research is generally expensive and demands a lot of resources. Through MTurk, researchers were able to recruit 177 participants to link youth mental health to family functioning, highlighting the need to document causal pathways. Parents who participated in the study provided "high-quality data (e.g., passed consistency checks); measures showed acceptable psychometrics at each time-point; and correlations among study measures paralleled those observed in prior research" [23]. When compared to previous longitudinal study methods, the MTurk method needed less resources, had relatively low cost, was comparable in participant attrition, and similar to traditional studies in measures of attrition bias, participant race/ethicality, and enrollment of single parents. The study suggests that "MTurk is a viable tool with its own strengths and limitations, and a potentially useful complement to traditional longitudinal methods" [23].

Research into issues like stigma can be performed effectively using the MTurk platform. Corrigan utilized MTurk to examine the ways that stigmatizing attitudes might pose a threat to open-mindedness in compared to the endorsement of difference [24]. To examine the psychometrics of different assessments of perceived difference from a person with mental illness, 460 participants were recruited using MTurk, and four measures of difference, the Likert Scale of Difference, Semantic Differential: Similar-Different Scale, Semantic Differential: Mental Illness versus Other Illness Scale, and Cause of Perceived Difference Scale were compared to measures of stereotypes, affirming attitude and care seeking. Measures of difference produced significantly higher endorsements than measures of stereotypes, with the Semantic Differential: Similar-Different scale endorsed at a higher rate than other difference scales.

MTurk also appears an effective tool for randomized controlled trials. Cunningham conducted two randomized controlled trials (RCTs) to explore the use of MTurk to evaluate rapid online interventions for unhealthy alcohol use [25]. In the first trial, password-protected, online interventions accessible by the study portal were randomized to condition. In the second trial, participants were directed to free-of-charge interventions, with proof of engagement provided by participant submission of a screenshot of intervention usage. Although neither trial demonstrated that access to online interventions could lead to reduction in alcohol usage, the effectiveness of a RCT model that was effective and quick to deploy was confirmed.

MTurk Can Be Effectively Utilized for Research Involving Vulnerable Participants and Subjects

Studies recruiting more vulnerable individuals, including individuals with substance use disorders or addictive concerns (e.g., gambling) have been effectively deployed via the MTurk platform. Kim [27] used MTurk to recruit individuals with substance use disorders and gambling addiction. After informed consent, 208 drinkers, 200 cannabis users, and 200 gamblers completed measures of alcohol, cannabis and gambling severity, psychological constructs (e.g., measures of impulsivity) related to their disorders, overt and subtle measures of valid responding, and assessment of their motivations for MTurk study participation [26]. Measures of addictive behavior were significantly correlated with each other and 80–85% of study participants provided responses that appeared consistent and valid. Participants reported answering questions honestly, with financial motives being the most popular motivation. Regarding the use of MTurk, self-reported data collected from participants with alcohol use disorder or gambling addiction were assessed to be of high quality, but this differed from the data collected from individuals with a cannabis use disorder. Thus, MTurk is a valid tool to recruit participants with some forms of addictive behaviors.

Similarly, MTurk has been effectively and ethically deployed in studies with other vulnerable populations, including adults with attention deficit/hyperactivity disorder (ADHD) and with personality disorders. Wymbs [26] used the MTurk platform to investigate ADHD in adulthood. 6526 workers completed an online screening survey to assess their diagnostic histories and symptoms of ADHD [26]. MTurk results showed that the percentage of workers with ADHD that continued from childhood to adulthood was consistent with those observed from offline samples. Comparing workers with ADHD diagnosed as adults with those diagnosed as children, those who were diagnosed in early stages of life were more likely to be male and without college degrees, and less likely to have comorbid depression or anxiety disorders. The MTurk platform was effectively used as a recruitment tool to study adults with ADHD.

Participants with other mental health, neurological, or behavioral disorders can also participate in research using the MTurk platform. Miller evaluated the strengths and limitations of data collection through MTurk, examined how MTurk has been used in personality disorder research, and compared MTurk research with personality disorder research done in other settings, concluding that MTurk is an effective tool for studying personality disorders, especially for those studies collecting large data [21].

The increasing evidence that MTurk is an effective, valid, and ethically deployable tool for research across populations has been instructive for our own project, which examines different stakeholder views on innovative neuroscience research ethics, comparing across healthy individuals and individuals who self-identify as living with mental health or physical health disorders.

Conclusion

Machine learning and artificial intelligence programs are developing at an increasingly fast rate, but some tasks, such as content curation, remain best performed by humans. Initially developed to support internal needs to crowdsource iterative, human intelligence-requiring tasks, Amazon's MTurk workplace has expanded outside Amazon itself to become an important tool for businesses and social science research. Using the MTurk website, MTurk requesters engage workers, who function as contractors to complete described tasks at a relatively lower cost and with greater ease. MTurk has become an established method to recruit participants for human subject research, including our own, but attention to ethical issues remains paramount.

Key Points

1. Despite the increasing sophistication of machine learning and artificial intelligence, some tasks remain better performed by human intelligence.
2. Amazon's Mechanical Turk (MTurk), initially designed as an internal crowdsourcing platform for human intelligence-requiring tasks (HITs), has expanded to function as a popular marketplace, where requesters can recruit workers to complete carefully defined tasks for monetary compensation.
3. The utility of MTurk crowdsourcing has made inroads in the research sphere, but as with all human subject research, attention to both traditional and unique ethical issues is paramount.

Questions to Consider

1. What ethical principles are highlighted when MTurk workers function as moderators or content-creators (e.g., when they assess YouTube videos for company standards) vs. when they participate in IRB-approved, university-supported academic research?
2. What additional concerns arise in the expansion of MTurk workers to include a more global workforce?
3. What unique concerns are introduced by inviting MTurk workers to participate in research addressing mental health needs?

References

1. Amazon. Amazon Mechanical Turk: Access a global, on-demand, 24x7 workforce; 2020. https://www.MTurk.com. Accessed 30 Jul 2020.
2. Difallah D, et al. Demographics and dynamics of mechanical Turk workers. In: Proceedings of the eleventh ACM international conference on web search and data mining - WSDM 18; 2018, https://doi.org/10.1145/3159652.3159661.
3. Chandler J, Shapiro D. Conducting clinical research using crowdsourced convenience samples. Annu Rev Clin Psychol. 2016;12(1):53–81. https://doi.org/10.1146/annurev-clinpsy-021815-093623.
4. Goodman JK, Paolacci G. Crowdsourcing consumer research. J Consum Res. 2017;44(1):196–210. https://doi.org/10.1093/jcr/ucx047. 22 February 2017
5. Matsakis L. How YouTube uses mechanical Turk tasks to help train its AI. Wired, Conde Nast, 22 Mar. 2018. www.wired.com/story/youtube-mechanical-turk-content-moderation-ai/
6. Rajpurkar P, et al. SQuAD: 100,000 questions for machine comprehension of text. In: Proceedings of the 2016 conference on empirical methods in natural language processing, Nov. 2016, https://doi.org/10.18653/v1/d16-1264.
7. Sheehan KB. Crowdsourcing research: data collection with Amazon's mechanical Turk. Commun Monogr. 2018;85(1):140–56. https://doi.org/10.1080/03637751.2017.1342043.
8. Amazon MTurk Requester. Amazon mechanical turk pricing; 2022. https://requester.MTurk.com/pricing
9. Paolacci G, Chandler J, Ipeirotis PG, Stern LN. Running experiments on Amazon mechanical Turk. Judg Decision Making. 2010;5:411–9.
10. Casey LS, Chandler J, Levine AS, Proctor A, Strolovitch DZ. Intertemporal differences among MTurk workers: time-based sample variations and implications for online data collection. SAGE Open. 2017;2:1–15.
11. Bunge E, Cook HM, Bond M, et al. Comparing Amazon mechanical Turk with unpaid internet resources in online clinical trials. Internet Interv. 2018;12:68–73.
12. Paolacci G, Chandler J. Inside the Turk: understanding mechanical Turk as a participant Pool. Curr Dir Psychol Sci. 2014;23(3):184–8. https://doi.org/10.1177/0963721414531598.
13. Kees J, Berry C, Burton S, et al. An analysis of data quality: professional panels, student subject pools, and Amazon's mechanical Turk. J Advert. 2017;46(10):141–55. https://doi.org/10.1080/00913367.2016.1269304.
14. Buhrmester M, Kwang T, Gosling SD. Amazon's mechanical Turk: a new source of inexpensive, yet high quality, data? Perspect Psychol Sci. 2011;6(1):3–5.
15. Merz ZC, Lace JW, Einstein AM. Examining broad intellectual abilities obtained within an MTurk internet sample. Curr Psychol. 2020;41:2241. https://doi.org/10.1007/s12144-020-00741-0.

16. Kim HS, Hodgins DC. Are you for real? Maximizing participant eligibility on Amazon's mechanical Turk. Addiction. 2020;115(10):1969–71. https://doi.org/10.1111/add.15065.
17. Agley J, Xiao Y, Nolan R, Golzarri-Arroyo L. Quality control questions on Amazon's mechanical Turk (MTurk): a randomized trial of impact on the USAUDIT, PHQ-9, and GAD-7. Behav Res Methods. 54:885. https://doi.org/10.3758/s13428-021-01665-8.
18. Summerville A, Chartier CR. Pseudo-dyadic "interaction" on Amazon's mechanical Turk. Behav Res Methods. 2013;45(1):116–24.
19. Hara K, Adams A, Milland K, Savage S, Callison-Burch C, Bigham JP. A data-driven analysis of workers' earnings on Amazon mechanical Turk. In: Proceedings of the 2018 CHI conference on human factors in computing systems; 2018, pp. 1–14. https://doi.org/10.1145/3173574.3174023.
20. Buhrmester MD, Talaifar S, Gosling SD. An evaluation of Amazon's Mechanical Turk, its rapid rise, and its effective use. Perspect Psychol Sci. 2018;13(2):149–54.
21. Miller JD, Crowe M, Weiss B, Maples-Keller JL, Lynam DR. Using online, crowdsourcing platforms for data collection in personality disorder research: the example of Amazon's mechanical Turk. Personal Disord Theory Res Treat. 2017;8(1):26–34.
22. Shapiro DN, Chandler J, Mueller PA. Using mechanical Turk to study clinical populations. Clin Psychol Sci. 2013;1(2):213–20.
23. Schleider JL, Weisz JR. Using mechanical Turk to study family processes and youth mental health: a test of feasibility. J Child Fam Stud. 2015;24:3235–46.
24. Corrigan PW, Bink AB, Konadu Fokuo J, Schmidt A. The public stigma of mental illness means the difference between you and me. Psychiatry Res. 2015;226:186–91.
25. Cunningham JA, Godinho A, Bertholet N. Outcomes of two randomized controlled trials, employing participants recruited through mechanical Turk, of internet interventions targeting unhealthy alcohol use. BMC Med Res Methodol. 2019;19:124. https://doi.org/10.1186/s12874-019-0770-4.
26. Wymbs BT, Dawson AE. Screening Amazon's mechanical Turk for adults with ADHD. J Atten Disord. 2019;23(10):1178–87.
27. Kim HS, Hodgins DC. Reliability and validity of data obtained from alcohol, cannabis, and gambling populations on Amazon's mechanical Turk. Psychol Addict Behav. 2017;31(1):85–94. https://doi.org/10.1037/adb0000219.

Further Reading

Chandler J, Shapiro D. Conducting clinical research using crowdsourced convenience samples. Ann Rev Clin Psychol. 2016;12(1):53–81. https://doi.org/10.1146/annurev-clinpsy-021815-093623.

Difallah D, et al. Demographics and dynamics of mechanical Turk workers. In: Proceedings of the eleventh ACM international conference on web search and data mining - WSDM 18, Feb. 2018, https://doi.org/10.1145/3159652.3159661.

O'Neal C, Schutt R. Doing data science: straight talk from the front line. Sebastopol, CA: O'Reilly; 2014.

Appendix 1: Pilot Quantitative Phase: Initial Results

Introduction

Our National Institutes of Health (NIH)-funded, Brain Research through Advancing Innovative Neurotechnologies (BRAIN) Initiative® project, "Enabling ethical participation in innovative neuroscience on mental illness and addiction: toward a new screening tool enhancing informed consent for transformative research on the human brain," is guided by the understanding that the ethical engagement of research volunteers in novel human studies requires a rigorous, nuanced informed consent process. Informed by stakeholder interviews from an earlier aim of the project, we used Amazon MTurk to pilot test an online survey assessing positive and negative valence factors influencing research participation over a large population, further refining the Roberts Valence Model for ethical research participation (see Chap. 12). These pilot findings, presented in this Appendix, add to a limited empirical literature on the ethically salient factors influencing research voluntarism and may contribute assurance as to the decisional capacity of prospective volunteers, especially those living with mental illness. In presenting the analyses in this Appendix, we hope to illustrate how empirical analyses can be applied in the field of neuroethics to provide evidence demonstrating the role of valence factors—often and widely not considered in practice—in enrollment decisions made by prospective research participants.

In this pilot survey, we collected information pertaining to demographics, psychosocial factors, attitudinal beliefs regarding medical research, and perceptions of risk associated with specific research protocols provided as examples in the survey. With these measures, we were able to test hypotheses around the Roberts Ethical Valence model (Box A.1), which posits that ethically acceptable research requires an appropriate balance of positive and valence factors that bring a potential research participant to a decision. With this in mind, we have considered factors, such as attitudinal beliefs about research and perceptions of risk, in contexts that are truly cutting edge, and in others that are more traditional.

L. W. Roberts (ed.), *Ethics and Clinical Neuroinnovation*,
https://doi.org/10.1007/978-3-031-14339-7

Box A.1 Pilot Hypotheses Informed by Overall Grant Hypotheses

1. Valence Factor Assessment: Measures of valence factors will vary by study attributes (e.g., innovative project type) and/or personal attributes (e.g., self-reported illness). We will consider valence factors such as perceived stigma, perceptions towards medical research, optimism, perceived protectiveness, perceived risk, and perceived helpfulness.
2. Participation Willingness and Valence Factors: Respondents will express greater willingness to participate when they express higher levels of *positive valence factors* **such as** perceived helpfulness of research, optimism, and trust in medical research.

The subsections in this Appendix are aligned with central questions of the Roberts Valence Model and understanding the key attitudes, psychosocial factors, and salient ethical perceptions that enable people to make decisions about research participation. Readers are encouraged to follow the subsections of this Appendix for details on implementation, methodology, and analytic approaches.

Overall Methods

Survey Instrument

A 175-question online survey instrument was developed for the pilot project based on prior work in empirical ethics, and included questions on psychological symptoms, personality traits, research attitudes, and perspectives on ethically salient aspects of research, as well as four attention check questions [1–4]. This instrument was hosted online on Qualtrics and distributed electronically through Amazon MTurk (See Chap. 17 for details on the process of recruiting participant populations via MTurk).

The individual measures that were used in our pilot survey instrument included:

Demographics

Demographic variables, such as health status, gender, race, ethnicity, and education level, were collected.

Trust in Medical Research

Study respondents completed 10 items from the Trust in Medical Research scale [5]. Each item was rated on a 5-point Likert scale. Scores could range from 10 to 50, with higher scores representing more trust in medical research. In addition to examining scores as a continuous variable, we also created a dichotomous variable based on exploratory data analysis, i.e., “weak to moderate trust in medical research” (<40) and “strong trust in medical research” (≥40).

Medical Research Attitudes

The Research Attitudes Questionnaire (RAQ) evaluated respondents’ attitudes toward biomedical research [6]. Respondents were asked to rate each of the seven statements on a 5-point Likert scale (i.e., 1 = “strongly disagree” to 5 = “strongly agree”), and a total score was calculated (possible range: 7–35; higher scores reflect more positive attitudes toward research).

Optimism

Study respondents completed the 10-item Life Orientation Test-Revised (LOT-R) [7]. Each item on the LOT-R is rated on a 5-point Likert scale. Three items measure optimism, three items measure pessimism, and four are filler items. For this analysis, we used the three items assessing optimism (possible range: 3–15, with higher scores indicating higher levels of optimism). Similar to Trust in Medical Research, in addition to examining scores as a continuous variable, we also created a dichotomous variable, i.e., “weak to moderate optimism” (<12) and “strong optimism” (≥12), based on exploratory data analysis examining the distribution of scores.

Stigma

An internally generated stigma scale assessed participants’ perceptions of stigma for several types of health problems. Participants rated nine items on a 5-point Likert scale (i.e., 1 = None, 2 = A little bit, 3 = Some, 4 = Quite a bit, 5 = A lot) and a total score was calculated (possible range: 9–45; higher scores reflect higher level of stigma). For interpretability reasons, we created a dichotomized variable based on exploratory data analysis, i.e., “weak to moderate stigma” (≤27) and “high stigma” (>27).

Perceived Risk, Helpfulness, Likelihood to Participate in Research Projects in the Presence of Potential Influences, and Participation Willingness

This internally generated component measures items related to risk, helpfulness, likelihood to participate, and participation willingness regarding two research projects. A more detailed description of this component can be found in the “Valence factors and participation willingness” section later in this Appendix. See Table A.1 for item descriptions and Table A.2 for questions.

Table A.1 Excerpts of text from our survey vignettes describing two innovative research projects

Survey section: Perceived risk, helpfulness, likelihood to participate in research projects in the presence of potential influences, and participation willingness	
Innovative project	**Survey text**
Wearable device	This research project involves wearing a device on your wrist that collects digital data about your activity levels, exercise patterns, sleep, and your geographical location. The aim of this project is to study differences and similarities among people with certain illnesses or other health traits. Your digital data and health information will be stored for future studies. The risks involved in this study include the possibility that someone outside of the research team could learn personal information about you and your health. Every precaution will be taken to minimize this risk.
Ketamine infusion	This research project involves receiving a medication called ketamine through an intravenous (IV) tube to treat symptoms of specific brain-based illnesses. Participants in this study will receive three IV infusions (each of which takes about 100 min) over the course of 2 weeks. The potential side effects of ketamine infusion include dissociation (feeling that things are not real, or feeling of not being in one's body), dizziness, numbness or tingling in the hands or feet, sleepiness, increased emotionality or tearfulness, and facial numbness. These side effects, if they occur, usually resolve within 2 hours after the infusion.

Table A.2 Survey questions measuring risk, helpfulness, likelihood to participate, and participation willingness regarding two innovative research projects

Survey section: Perceived risk, helpfulness, likelihood to participate in research projects in the presence of potential influences, and participation willingness	
Factors	**Survey questions**
Risk	How risky is this project?
	How do the risks of this project compare with the usual risks you live with every day?
Helpfulness	How helpful is this project to society?
	How helpful is this project to the person who takes part in it?
Other influences	If you were offered $100 one time at the beginning of the study, how much more likely would you be to participate?
	If you were offered $500 one time at the beginning of the study, how much more likely would you be to participate?
	If you had an illness being studied in this research project, how much more likely would you be to participate?
	If someone important in your life wanted you to, how much more likely would you be to participate?

Safeguard Procedures: Perceived Protectiveness and Influence on Willingness to Participate

This internally generated component measures items related to five different human subjects safeguard procedures. A more detailed description of this component can be found in the "Perceived protectiveness and participation willingness in context of five safeguard procedures" section later in this Appendix. See Table A.3 for safeguard descriptions and Table A.4 for questions.

Table A.3 Excerpts of text from our survey vignettes describing five different human subjects safeguard procedures

Survey section: Safeguard procedures: perceived protectiveness and influence on willingness to participate	
Safeguard procedure	**Survey text**
Institutional review board (IRB) review	Before a research study starts, a group of people discuss the project carefully to decide whether it is safe and helpful to do. This group is made up of doctors, researchers, and people from the community. It is usually called an **IRB** or **institutional review board (IRB review)**.
Data safety monitoring board (DSMB) review	While a research study is being performed, an independent group of experts may watch over it. This group of people, a **data safety monitoring board (DSMB oversight)**, makes sure that the risks of the research are not too serious. They also make sure that the project is being performed properly. This group sometimes will stop projects that are too risky or are causing problems that the researchers did not expect.
Informed consent	Before a person even begins their participation in a research project, the researchers will talk to them about the purpose of the project and its risks and benefits. The researchers will also explain that participation is voluntary and will answer any questions that people may have before they choose to join the project. Almost always, the researchers will review this information with each person in writing and get the person's signature. This process is called getting **informed consent**.
Designated decision maker (alternative decision maker)	Sometimes patients who could take part in research have difficulty understanding information about the project, or they are too upset or too ill to make a decision for themselves. In these situations, another person such as a family member can make decisions for the patient about their research participation. The job of the **designated decision maker (alternative decision maker)** is to make the choice that the patient would want, if the patient was feeling better or better able to make decisions themselves.
Special code number (confidential coding)	When information is gathered from a person in a research project, the person is assigned a **special code number (confidential coding)**. That number is used to store personal information from the project separately from a person's name. In this way, a person's confidentiality will be protected.

Table A.4 Survey questions related to perceptions of five different human subjects safeguard procedures

Survey section: Safeguard procedures: perceived protectiveness and influence on willingness to participate	
Factors	**Survey questions**
Protectiveness	How much does this protect people who participate in research?
Willingness to participate—IRB review	If you knew that a project was approved by an institutional review board (IRB) how would it influence your willingness to participate in the project?
Willingness to participate—DSMB oversight	If you knew that a project was being watched over by a data safety monitoring board (DSMB), how would it influence your willingness to participate?
Willingness to participate—Informed consent	If you knew that a project involved this informed consent process, how would it influence your willingness to participate in the project?
Willingness to participate—Alternative decision maker	If you knew that a project involved this designated decision maker, how would it influence your willingness to participate in the project?
Willingness to participate—Confidential coding	If you knew researchers were using a special code to protect your confidentiality, how would it influence your willingness to participate?

General Analytic Methods

Several analytical methods were used to model relationships between the outcome (e.g., participation willingness) and valence factors of interest (e.g., perceived protectiveness of safeguards). Many of these methods are used across our analysis and are useful in empirical studies. The application of these methods as they pertain to our hypotheses are described in their respective analysis sections.

The statistical methods we have used to conduct these analyses include:

1. *Generalized Estimating Equations (GEE)*: GEE is a statistical method to model marginal means for repeated measures data. Each GEE model was used with a Gaussian link and unstructured correlation structure.
2. *Least Absolute Selection and Shrinkage Operator (LASSO)* [8]: LASSO is a method used for variable selection in the context of high dimensional data (i.e., many variables), where the residual sum of squares is subject to a linear equality constraint. For the analyses, we used linear regression-based LASSO. All predictor variables were standardized. Mean squared error based on three-fold cross validation was used to select the tuning parameter, which controls the amount of shrinkage that is applied to the estimates of the regression coefficients. The penalty parameter that was chosen corresponded to the model with minimum mean square error.
3. *Backward selection procedures for linear regression*: Backward stepwise regression is a method which involves starting with all candidate variables, testing the deletion of each variable using a chosen model fit criterion and deleting the variable whose loss gives the most statistically insignificant deterioration of the

model fit. This process is repeated until no further variables can be deleted without a statistically insignificant loss of fit.

All statistical analyses were performed using SPSS Statistics (version 25) and R version 1.1.453 (IBM SPSS Statistics for Windows; R Core Team).

Participants

Pilot Sample

For this pilot project, we surveyed three groups of respondents: (1) individuals who self-reported as having a mental illness and/or substance use disorder, (2) individuals who self-reported as having a physical illness, and (3) individuals who self-reported as being in good health.

Participants were selected for inclusion based on their responses to screening items. Eligible participants had *one of* the following:

(a) A self-reported mental illness or substance use disorder (individuals who responded "Yes" to having a mental illness or substance use disorder and without any self-reported physical disorders); or
(b) A self-reported physical illness (individuals who responded "Yes" to having a physical illness and without any self-reported mental illness or substance use disorders); or
(c) Self-reported "Very good" or "Excellent" health (i.e., response of ≥4 to the question, "In general, would you say your health is: 1 = Poor, 2 = Fair, 3 = Good, 4 = Very good, 5 = Excellent"), and without any self-reported mental or physical illnesses.

Eligible participants were given a short description and a web link to the survey, hosted on Qualtrics. Before taking the survey, respondents were required to read and agree to an electronic consent. 151 individuals consented to participate and completed the 175-item survey (mental illness or substance use disorder, $n = 50$; physical illness, $n = 51$; in good health, $n = 50$). Participants were paid $8.00 for completing the survey.

Background Characteristics of Study Respondents

Demographic characteristics are reported in Table A.5. Descriptive statistics were generated for continuous and categorical variables. Chi-square tests and analysis of variance (ANOVA) tests were used to examine differences by health status.

The respondents to our pilot survey reflected a young sample with the majority having at least a college education. The mean age of respondents was 38.3 years (SD = 13.6 years). Respondents with mental illness and/or substance use disorder were the youngest group, followed by respondents in good health; respondents with

Table A.5 Background characteristics of respondents for pilot project

–	Mental illness and/ or substance use disorder ($n = 50$)	Physical illness ($n = 51$)	Healthy ($n = 50$)	Overall ($n = 151$)	P value
Age[a]					
Mean years (SD)	[xyz]31.8 (7.4)	[xyz]44.6 (16.0)	[xyz]38.4 (12.7)	38.3 (13.6)	****<0.001**
Gender (%, *n*)[b]					*0.786
Women Men	44.0 (22) 56.0 (28)	41.2 (21) 54.9 (28)	36.0 (18) 60.0 (30)	40.4 (61) 57.0 (86)	
Ethnicity (%, *n*)					*0.451
Not Hispanic or Latino Hispanic or Latino	90.0 (45) 10.0 (5)	90.2 (46) 9.8 (5)	96.0 (48) 4.0 (2)	92.1 (139) 7.9 (12)	
Race (%, *n*)[a]					*0.989
Non-white White	30.0 (15) 68.0 (34)	31.4 (16) 68.6 (35)	30.0 (15) 70.0 (35)	30.5 (46) 68.9 (104)	
Education level (%, *n*)					*0.268
High school Some college College Graduate or professional school	16.0 (8) 36.0 (18) 38.0 (19) 10.0 (5)	13.7 (7) 27.5 (14) 41.2 (21) 17.6 (9)	16.0 (8) 16.0 (8) 42.0 (21) 26.0 (13)	15.2 (23) 26.5 (40) 40.4 (61) 17.9 (27)	
Working in the health field (%, *n*)[a]					*0.864
No Yes	88.0 (44) 10.0 (5)	86.3 (44) 13.7 (7)	88.0 (44) 12.0 (6)	87.4 (132) 11.9 (18)	
Optimism[c]					
Optimism (mean, SD) Strong optimistic (%, *n*) Weak to moderate optimistic (%, *n*)	[xy]8.3 (3.3) 24.0 (12) 76.0 (38)	[x]11.1 (2.3) 47.1 (24) 52.9 (27)	[y]11.4 (2.3) 52.0 (20) 48.0 (30)	10.3 (3.0) 41.1 (62) 58.9 (89)	**** < 0.001** *** < 0.001**
Trust in Medical Research[d]					
Trust in Medical Research (mean, SD) Strong trust in MR (%, *n*) Weak to moderate trust in MR (%, *n*)	35.1 (7.5) 28 (14) 72 (36)	36.5 (8.0) 39.2 (20) 60.8 (31)	35.5 (7.0) 32 (16) 68 (34)	35.5 (7.5) 33.1 (50) 66.9 (101)	**0.608 *** < 0.001**

x & x, y & y, z & z: *P* values less than 0.05 and corresponds to Fisher's Least Significant Difference (LSD) post-hoc test

**P* value correspond to Chi-square test

***P* value correspond to ANOVA test

[a] 1 participant failed to respond

[b] 4 participants failed to respond

[c] *Optimism: Strong optimistic ≥12, Weak to Moderate optimistic* <12

[d] Trust in Medical Research (MR): Strong trust in MR ≥40, Weak to Moderate trust in MR <40

physical illness were the oldest group (mean [SD]: 31.8 [7.4] vs. 38.4 [12.7] vs. 44.6 [16.0], *P* value <0.001). Other than age, there were no significant differences between the three respondent groups in any of the demographic variables.

Interestingly, individuals with and without illness were quite different in terms of their personality traits, and in particular, optimism. On average, respondents with physical illness and healthy respondents were more optimistic than respondents with mental illness and/or substance use disorder (mean [SD]: 11.1 [2.3], 11.4 [2.3] and 8.3 [3.3] out of 15, respectively, *P* value <0.001). Moreover, a greater proportion of respondents with self-reported mental illness and/or substance use disorder reported "weak to moderate" levels of optimism (76%, *P* value <0.001). On average, respondents expressed "weak to moderate" trust in medical research (mean [SD]: 35.5 [7.5] out of 50) and a greater proportion of respondents with self-reported mental illness and/or substance use disorder reported "weak to moderate" levels of trust in medical research (72%, *P* value <0.001).

Valence Factors and Participation Willingness

Methods

Variables

The survey instrument was used as described in the Overall Methods section. In particular, this analysis uses the following set of survey items: (1) *demographics*; (2) *perceived risk, helpfulness, likelihood to participate in research projects in the presence of potential influences, and participation willingness;* (3) *trust in medical research; (4) medical research attitudes;* (5) *optimism*; (6) *personality*; and (7) *stigma.*

For this analysis, we analyzed responses assessing respondents' perspectives *on perceived risk, helpfulness, likelihood to participate in the presence of potential influences, and participation willingness* regarding two research projects: (1) a research project involving a *ketamine infusion* and (2) a research project involving a *wearable* device (see Table A.1 for study descriptions). After a description of each of the two research projects, respondents were asked about their perceptions of the (a) *risks* of the projects, (b) *helpfulness* of the projects, (c) *influences* on their likelihood to participate in the projects, and (d) *willingness to participate* in of the projects (i.e., participation willingness).

Respondents were asked to rate two items related to *risks*: (1) "How risky is this project?" (1 = Not at all risky, 2 = A little risky, 3 = Somewhat risky, 4 = Quite risky, 5 = Very risky) and (2) "How do the risks of this project compare with the usual risks you take every day?" (1 = Not at all more risky, 2 = A little more risky, 3 = Somewhat more risky, 4 = Quite a bit more risky, 5 = Very much more risky). Respondents were asked to rate two items related to *helpfulness*: (1) "How helpful is this project to the person who takes part in it?" and (2) "How helpful is this project to society?" (both rated on a 5-point Likert scale, 1 = Not at all helpful, 2 = A little helpful, 3 = Somewhat helpful, 4 = Quite helpful, 5 = Very helpful).

Respondents were asked to rate four questions related to *influences* on their likelihood to participate: (1) "If you were offered \$100 one time at the beginning of the study, how much more likely would you be to participate," (2) "If someone important in your life wanted you to, how much more likely would you be to participate,"' (3) "If you had an illness being studied in this research project, how much more likely would you be to participate," and (4) "If you were offered \$500 one time at the beginning of the study, how much more likely would you be to participate?" (all rated on a 5-point Likert scale, 1 = Not at all more likely, 2 = A little bit more likely, 3 = Somewhat more likely, 4 = Quite a bit more likely, 5 = Very much more likely). See Table A.4 for risk, helpfulness, and influence questions.

Respondents were asked to rate one item related to *participation willingness*: "Would you participate in this project?" (rated on a 5-point Likert scale; 1 = Absolutely not, 2 = Probably not, 3 = Indifferent, 4 = Probably yes; 5 = Absolutely).

Furthermore, a different but overlapping set of variables were used for the LASSO models (see "LASSO Results" in the Results section further below). These include the initial set of covariates as described above and other measures such as select Patient-Reported Outcomes Measurement Information System (PROMIS) measures and the 10 Item Personality Index (TIPI). Refer to Table A.6 for a description of this full set of covariates.

Table A.6 Additional Variables Used for LASSO analysis

Measure	Description of measure
Patient-reported outcomes measurement information system (PROMIS) scale v1.2 Global Health	The PROMIS Global Health measure was used to produce two scores: a physical health and mental health raw score. Higher scores indicate better physical and mental health, respectively. Both the global physical health and mental health raw score were composed of four questions each (eight questions total). Two questions were filler. Each composite ranged from 5 (low health) to 20 (high health).
PROMIS item Bank v1.0—meaning and purpose—short form 4a	This item bank was developed to evaluate an individual's sense of having a purposeful life and substantial reason for living. The four items on the short form were summed to obtain a meaning and purpose score (minimum: 4, maximum: 20). Higher scores indicate more hopefulness, optimism, and goal-directedness.
PROMIS item Bank v1.0—general self-efficacy—short form 4a	This scale measures an individual's confidence to successfully perform general and health-related tasks or behaviors in various settings. The four items were summed to obtain a self-efficacy score (minimum: 4, maximum: 20). Higher scores indicate higher confidence in self-efficacy.
PROMIS item Bank v1.0—cognitive function abilities subset—short form 4a	This scale measures an individual's perceived cognitive deficits. The four items on the short form are summed to obtain a raw score between 4 (minimum) and 20 (maximum). Higher scores indicate higher cognitive functioning.

Table A.6 (continued)

Measure	Description of measure
PROMIS item Bank v1.0—emotional distress-anxiety—short form 4a	This scale assesses an individual's self-reported fear (i.e., fearfulness, panic), apprehension (i.e., worry, dread), hyperarousal (i.e., tension, nervousness, and restlessness), and somatic symptoms related to arousal including racing heart and dizziness. The four items on the short form are summed to obtain a raw scare between 4 (minimum) and 20 (maximum). Higher scores indicate higher levels of anxiety.
PROMIS item Bank v1.0—emotional distress-depression—short form 4a	This scale assesses an individual's self-reported negative mood (i.e., sadness, guilt), negative views of self (i.e., self- criticism, worthlessness), negative social cognition (i.e., loneliness, interpersonal alienation), and decreased positive affect (i.e., loss of interest, meaning, and purpose). The four items were summed to obtain a raw score between 4 (minimum) and 20 (maximum). Higher scores indicate higher levels of depression.
Altruism scale	A 13-item scale used to measure an individual's level of altruism with higher scores indicating less altruism. Scores ranged between 13 (minimum) and 65 (maximum).
Davis empathy scale	A 7-item scale used to measure an individual's sense of empathy. Higher scores indicate less empathy.
Altruism value scale	A 4-item scale used to measure an individual's sense of altruistic values. Scores run from 4 (someone giving the most altruistic response to all items) to 20 (someone giving the least altruistic response to all items).
10-item personality inventory (TIPI)	A 10-item scale used to assess five different personality domains: Agreeableness, conscientiousness, emotional stability, extraversion, and openness to experience. Each personality domain was a mean of two items with mean score ranging from 1 (having less of the personality trait) to 7 (having more of the personality trait).
Internally generated research project questionnaire (extension of perceived risk, helpfulness, likelihood to participate in research projects in the presence of potential influences, and participation willingness section described above)	Questions were as follows for each research project: 1. How important to your decision is the possibility that you might personally be helped by participating? 2. How important to your decision is the possibility that others might be helped by your participation? 3. After weighing the risks and benefits of this research, do you find it to be more risky or more beneficial?

Statistical Analysis

To assess within- and between-group differences in responses regarding valence factors, we used paired t-tests, analysis of variance (ANOVA) tests, Fisher's Least Significant Difference (LSD) post-hoc tests, and repeated measures ANOVA tests as appropriate.

To assess the association between participation willingness (primary outcome) and valence factors, we performed generalized estimating equations (GEE). See General Analytic Methods section for more details about GEE. Participation likelihood (or participation willingness) was treated as a repeated measures outcome. Confounders such as health status, gender, race, optimism (dichotomized), and trust in medical research (dichotomized) were included in the GEE model.

To examine important valence factors in predicting participation willingness, a method called least absolute shrinkage and selection operator (LASSO) was used. LASSO was performed for each study type to select valence factors that were significant. Refer to the Description of General Analytic Methods section for more information about the LASSO procedure.

Descriptive Results

Perceived Risk of Research Projects by Project Type (Table A.7)

Among the entire sample, respondents perceived the ketamine infusion project as significantly riskier than the wearable device project; i.e., the ketamine infusion project was viewed as "quite risky," whereas the wearable device project was viewed as "a little risky" (mean [SD]: 3.8 [1.1] vs. 1.8 [0.9], *P* value <0.001). Similarly, when asked how the risks of the projects compared with everyday risks, the ketamine infusion project was perceived to be significantly riskier than the wearable device project (mean [SD]: 3.8 [1.0] vs. 1.7 [0.9], *P* value <0.001).

Responses differed significantly depending on the phrasing of the two risk items only for the wearable device project. Respondents perceived significantly greater risk when asked "How risky is this [wearable device] project?" compared to when asked about the "risks of this [wearable device] project compared with the usual risks you take every day" (mean [SD]: 1.8 [0.9] vs. 1.7 [0.9], *P* value = 0.026).

Table A.7 Perceived risk of wearable device and ketamine infusion research projects

	Wearable device (*n* = 151)	Ketamine infusion (*n* = 151)	
–	Mean (SD)	Mean (SD)	*P* value[a]
How risky is this project?	1.8 (0.9)	3.8 (1.1)	**<0.001**
How do the risks of this project compare with the usual risks you live with every day?	1.7 (0.9)	3.8 (1.0)	**<0.001**
P value[a]	**0.026**	0.913	

Participants were asked:

Q.1 How risky is this project?

Scale: 1 = Not at all risky, 2 = A little risky, 3 = Somewhat risky, 4 = Quite risky, 5 = Very risky

Q.2 How do the risks of this project compare with the usual risks you live with every day?

Scale: 1 = Not at all more risky, 2 = A little more risky, 3 = Somewhat more risky, 4 = Quite a bit more risky, 5 = Very much more risky

[a] *P* values correspond to paired t-test

Perceived Risk of Research Projects by Health Status (Table A.8)

Respondents with mental illness and/or substance use disorder, with physical illness, and in good health did not differ in their perceived risks of the wearable device project and ketamine infusion project, regardless of how the questions were asked.

For the wearable device project only, however, respondents with physical illness perceived significantly greater risk when asked "How risky is this [wearable device] project," compared to when asked about the "risks of this [wearable device] project compared with the usual risks you take every day" (mean [SD]: 1.7 [0.9] vs. 1.5 [0.6], *P* value = 0.015).

Table A.8 Perceived risk of wearable device and ketamine infusion research projects by health status group

	Wearable device				Ketamine infusion			
	Mental illness and/ or substance use disorder ($n = 50$)	Physical illness ($n = 51$)	Healthy ($n = 50$)		Mental illness and/ or substance use disorder ($n = 50$)	Physical illness ($n = 51$)	Healthy ($n = 50$)	
–	Mean (SD)	Mean (SD)	Mean (SD)	*P* value[a]	Mean (SD)	Mean (SD)	Mean (SD)	*P* value[a]
How risky is this project?	2.0 (1.1)	1.7 (0.9)	1.8 (0.9)	0.223	3.8 (1.1)	3.8 (1.1)	3.7 (1.0)	0.801
How do the risks of this project compare with thc usual risks you live with every day?	1.9 (1.0)	1.5 (0.6)	1.7 (1.0)	0.075	3.8 (1.0)	3.8 (1.1)	3.7 (1.1)	0.920
P value[b]	0.322	**0.015**	0.485		1.000	0.837	0.709	

Participants were asked:

Q.1 How risky is this project?

Scale: 1 = Not at all risky, 2 = A little risky, 3 = Somewhat risky, 4 = Quite risky, 5 = Very risky

Q.2 How do the risks of this project compare with the usual risks you live with every day?

Scale: 1 = Not at all more risky, 2 = A little more risky, 3 = Somewhat more risky, 4 = Quite a bit more risky, 5 = Very much more risky

[a] *P* values correspond to ANOVA t-test

[b] *P* values correspond to paired t-test

Table A.9 Perceived helpfulness of wearable device and ketamine research projects

–	Wearable device (n = 151) Mean (SD)	Ketamine infusion (n = 151) Mean (SD)	P value[a]
How helpful is this project to the person who takes part in it?	3.0 (1.2)	2.7 (1.2)	**0.002**
How helpful is this project to society?	3.6 (1.1)	3.4 (1.2)	0.084
P value[a]	**<0.001**	**<0.001**	

Participants were asked:

Q.1 How helpful is this project to society?

Q.2 How helpful is this project to the person who takes part in it?

Scale: 1 = Not at all helpful, 2 = A little helpful, 3 = Somewhat helpful, 4 = Quite helpful, 5 = Very helpful

[a] P values correspond to paired t-test

Perceived Helpfulness of Research Projects by Project Type (Table A.9)

Among the entire sample, respondents perceived the wearable device project as significantly more helpful "to the person who takes part in it" than the ketamine infusion project (mean [SD]: 3.0 [1.2] vs. 2.7 [1.2], P value <0.001). When asked about helpfulness of the wearable device project and ketamine infusion project "to society," respondents shared similar views (mean [SD]: 3.6 [1.1] vs. 3.4 [1.2], P value = 0.084).

Respondents rated both the wearable device project and ketamine infusion project as more helpful "to society" than "to the person who takes part in it" (wearable device project, mean [SD]: 3.6 [1.1] vs. 3.0 [1.2], P value <0.001 and ketamine infusion project, mean [SD]: 3.4 [1.2] vs. 2.7 [1.2], P value <0.001).

Perceived Helpfulness of Research Projects by Health Status (Table A.10)

Respondents with mental illness and/or substance use disorder perceived research involving the wearable device to be less helpful "to the person who takes part in it" compared to respondents with physical illness and in good health (mean [SD]: 2.7 [1.2] vs. 3.2 [1.2] vs. 3.2 [1.1], P value = 0.045). Similarly, respondents with mental illness and/or substance use disorder perceived lower levels of helpfulness "to society" than respondents with physical illness and in good health (mean [SD]: 3.2 [1.0] vs. 3.7 [1.1] vs. 3.8 [1.0], P value = 0.022).

Healthy respondents perceived the ketamine infusion project to be more helpful "to society" than respondents with mental illness and/or substance use disorder and physical illness (mean [SD]: 3.8 [1.1] vs. 3.2 [1.1] vs. 3.3 [1.2], P value = 0.027).

For both the wearable device and ketamine infusion project, respondent groups, regardless of their health status, rated helpfulness "to society" as higher compared to helpfulness "to the person who takes part in it" (wearable device project: mental illness and/or substance use disorder, mean [SD]: 3.2 [1.0] vs. 2.7 [1.2], P

Table A.10 Perceived helpfulness of wearable device and ketamine research projects by health status group

	Wearable device				Ketamine infusion			
	Mental illness and/or substance use disorder ($n = 50$)	Physical illness ($n = 51$)	Healthy ($n = 50$)		Mental illness and/or substance use disorder ($n = 50$)	Physical illness ($n = 51$)	Healthy ($n = 50$)	
–	Mean (SD)	Mean (SD)	Mean (SD)	*P* value[a]	Mean (SD)	Mean (SD)	Mean (SD)	*P* value[a]
How helpful is this project to the person who takes part in it?	[xy]2.7 (1.2)	[x]3.2 (1.2)	[y]3.2 (1.1)	**0.045**	2.5 (1.1)	2.7 (1.2)	3.0 (1.2)	0.109
How helpful is this project to society?	[xy]3.2 (1.0)	[x]3.7 (1.1)	[y]3.8 (1.0)	**0.022**	[x]3.2 (1.1)	[y]3.3 (1.2)	[xy]3.8 (1.1)	**0.027**
P value[b]	**0.005**	**0.001**	**<0.001**		**<0.001**	**<0.001**	**<0.001**	

Participants were asked:

Q.1 How helpful is this project to society?

Q.2 How helpful is this project to the person who takes part in it?

Scale: 1 = Not at all helpful, 2 = A little helpful, 3 = Somewhat helpful, 4 = Quite helpful, 5 = Very helpful

x & x, y & y: *P* values less than 0.05 and corresponds to Fisher's Least Significant Difference (LSD) post-hoc test

[a] *P* values correspond to ANOVA t-test

[b] *P* values correspond to paired t-test

value = 0.005; physical illness, mean [SD]: 3.7 [1.1] vs. 3.2 [1.2], *P* value <0.001; healthy, mean [SD]: 3.8 [1.0] vs. 3.2 [1.1], *P* value <0.001 and ketamine infusion project: mental illness and/or substance use disorder, mean [SD]: 3.2 [1.1] vs. 2.5 [1.1], *P* value <0.001; physical illness, mean [SD]: 3.3 [1.2] vs. 2.7 [1.2], *P* value <0.001; healthy, mean [SD]: 3.8 [1.1] vs. 3.0 [1.2], *P* value <0.001).

Likelihood of Participating in the Presence of Potential Influences by Project Types (Table A.11)

Among the entire sample, respondents were more likely to participate in the wearable device project than in the ketamine infusion project in the presence of each of the four potential influences ("you were offered $100," mean [SD]: 3.5 [1.3] vs. 2.3 [1.3], *P* value <0.001; "someone important in your life wanted you to," mean [SD]: 3.7 [1.3] vs. 3.0 [1.4], *P* value <0.001; "you had an illness being studied in this

Table A.11 Likelihood of participation in wearable device and ketamine research projects in the presence of potential influences

–	Wearable device ($n = 151$)	Ketamine infusion ($n = 151$)	*P*
Influences	Mean (SD)	Mean (SD)	value[a]
If you were offered $100 one time at the beginning of the study, how much more likely would you be to participate?	3.5 (1.3)	2.3 (1.3)	**<0.001**
If someone important in your life wanted you to, how much more likely would you be to participate?	3.7 (1.3)	3.0 (1.4)	**<0.001**
If you had an illness being studied in this research project, how much more likely would you be to participate?	3.8 (1.2)	3.3 (1.3)	**<0.001**
If you were offered $500 one time at the beginning of the study, how much more likely would you be to participate?	4.2 (1.1)	3.0 (1.4)	**<0.001**
P value[b]	**<0.001**	**<0.001**	

Scale: 1 = Not at all more likely, 2 = A little bit more likely, 3 = Somewhat more likely, 4 = Quite a bit more likely, 5 = Very much more likely

[a] *P* values correspond to paired t-test

[b] *P* values correspond to repeated measures ANOVA test

research project," mean [SD]: 3.8 [1.2] vs. 3.3 [1.3], *P* value <0.001; "you were offered $500," mean [SD]: 4.2 [1.1] vs. 3.0 [1.4], *P* value <0.001).

The likelihood of participation in the wearable device project significantly ranged from "somewhat more likely" to "quite a bit more likely" (range of means [SD] = 3.5 [1.3] to 4.2 [1.1]). For the wearable device project, participation likelihood was significantly higher if the respondent was "offered $500" compared to if the respondent was "offered $100" (mean [SD]: 4.2 [1.1] vs. 3.5 [1.3], *P* value <0.001).

The likelihood of participation in the ketamine infusion project significantly ranged from "a little bit more likely" to "somewhat more likely" (range of means [SD] = 2.3 [1.3] to 3.3 [1.3]). Participation likelihood was significantly higher if the respondent "had an illness being studied in the project" compared to if the respondent was "offered $100" (3.3 [1.3] vs. 2.3 [1.3], *P* value <0.001).

Likelihood of Participation in the Presence of Potential Influences by Health Status (Table A.12)

Likelihood of participation under each of the four influences, in the ketamine infusion project, did not differ by health status. The only exception was seen in the wearable device project, in which respondents with a physical illness and in good health were significantly more likely to participate if "someone important in your life wanted you to" in comparison with respondents with mental illness and/or substance use disorder (mean [SD]: 4.0 [1.1] vs. 3.9 [1.3] vs. 3.3 [1.3], *P* value <0.027).

Table A.12 Likelihood of participation in wearable device and ketamine infusion research projects in the presence of potential influences by health status group

	Wearable device				Ketamine infusion			
	Mental illness and/or substance use disorder ($n = 50$)	Physical illness ($n = 51$)	Healthy ($n = 50$)		Mental illness and/or substance use disorder ($n = 50$)	Physical illness ($n = 51$)	Healthy ($n = 50$)	
–	Mean (SD)	Mean (SD)	Mean (SD)	P value[a]	Mean (SD)	Mean (SD)	Mean (SD)	P value[a]
If you were offered $100 one time at the beginning of the study, how much more likely would you be to participate?	3.5 (1.3)	3.3 (1.4)	3.7 (1.2)	0.225	2.4 (1.5)	2.1 (1.2)	2.5 (1.3)	0.288
If someone important in your life wanted you to, how much more likely would you be to participate?	[xy]3.3 (1.3)	[x]3.9 (1.3)	[y]4.0 (1.1)	**0.027**	2.7 (1.2)	3.0 (1.5)	3.2 (1.3)	0.267
If you had an illness being studied in this research project, how much more likely would you be to participate?	3.8 (1.3)	3.7 (1.2)	4.0 (1.2)	0.367	3.3 (1.3)	3.0 (1.4)	3.5 (1.2)	0.161
If you were offered $500 one time at the beginning of the study, how much more likely would you be to participate?	4.2 (1.1)	4.0 (1.3)	4.4 (0.9)	0.143	3.1 (1.5)	2.7 (1.4)	3.1 (1.3)	0.157
P value [b]	**<0.001**	**<0.001**	**<0.001**		**<0.001**	**<0.001**	**<0.001**	

Scale: 1 = Not all more likely, 2 = A little bit more likely, 3 = Somewhat more likely, 4 = Quite a bit more likely, 5 = Very much more likely

x & x, y & y: P values less than 0.05 and corresponds to Fisher's Least Significant Difference (LSD) post-hoc test

[a] P values correspond to ANOVA t-test

[b] P values correspond to repeated measures ANOVA test

For both the wearable device and ketamine infusion projects, respondents varied in their ratings of the four potential influences irrespective of health status (wearable device project: mental illness and/or substance use disorder, physical illness, and healthy groups *P* value <0.001; ketamine infusion project: mental illness and/or substance use disorder, physical illness, and healthy groups *P* value <0.001).

GEE Results: Likelihood of Participation Associated with Perceived Risk for all Influences

More Likely to Participate if One Had an Illness Being Studied in a Research Study

Respondents with physical illness expressed lower mean participation likelihood than healthy respondents, controlling for all other variables (regression coefficient = −0.393, 95% CI = [−0.765, −0.021], *P* value = 0.039). Respondents with strong optimistic views expressed greater mean likelihood to participate than respondents with weak to moderate optimistic views, controlling for all other factors (regression coefficient = 0.494, 95% CI = [0.169, 0.819], *P* value = 0.003). Risk was negatively associated with participation likelihood, controlling for confounders (regression coefficient = −0.416, 95% CI = [−0.504, −0.329], *P* value <0.001). There were no other significant effects on mean participation likelihood.

More Likely to Participate if Someone in Life Wanted

Respondents with strong optimistic views expressed greater mean likelihood to participate than respondents with weak to moderate optimistic views, controlling for all other variables (regression coefficient = 0.429, 95% CI = [0.084, 0.773], *P* value = 0.015). As the level of risk increased, mean participation likelihood decreased for both research projects, controlling for all other factors (regression coefficient = −0.435, 95% CI = [−0.525, −0.346], *P* value <0.001). Gender, health status, race, trust in medical researchers' honesty, and education did not have significant effects on mean participation likelihood.

More Likely to Participate if One Was Offered $100 One Time at the Beginning of the Study

Respondents with strong optimistic views expressed greater mean likelihood to participate than respondents with weak to moderate optimistic views, controlling for all other variables (regression coefficient = 0.427, 95% CI = [0.093, 0.761], *P* value = 0.012). As the level of risk increased, mean likelihood to participate in the research projects decreased, controlling for all other factors (regression coefficient = −0.548, 95% CI = [−0.639, −0.456], *P* value <0.001). There were no other significant effects on mean participation likelihood.

More Likely to Participate if One Was Offered $500 One Time at the Beginning of the Study

Respondents with physical illness expressed on average, lower levels of participation likelihood than healthy respondents, controlling for all other variables (regression coefficient = −0.400, 95% CI = [−0.021, −0.780], *P* value = 0.039). Risk was negatively associated with mean participation likelihood, controlling for confounders (regression coefficient = −0.586, 95% CI = [−0.678, −0.494], *P* value <0.001). There were no other significant effects on mean participation likelihood.

More Likely to Participate if Offered Money

Respondents expressed on average a higher likelihood to participate if offered $500 than if offered $100, controlling for all other variables (regression coefficient = 0.678, 95% CI = [0.591, 0.765], *P* value <0.001). Respondents with physical illness expressed less mean participation likelihood than healthy respondents, controlling for all other factors (regression coefficient = −0.366, 95% CI = [−0.728, −0.003], *P* value = 0.048). Respondents with strong optimistic views expressed on average, a higher likelihood to participate than respondents with weak to moderate optimistic views, controlling for all other factors (regression coefficient = 0.346, 95% CI = [0.041, 0.651], *P* value = 0.026). Risk was negatively associated with mean participation likelihood, controlling for confounders (regression coefficient = −0.509, 95% CI = [−0.360, −0.657], *P* value <0.001). There were no other significant effects on mean participation likelihood.

LASSO Results: Factors Associated with Participation Willingness

Wearable Device Research Study

LASSO had selected 14 predictors. The top selected LASSO predictors were *perceived risk*, *whether risk outweighed benefit*, and *societal importance*. Each unit increase in risk was associated with a reduction in mean participation willingness (0.30 points). Furthermore, individuals who felt benefits of the wearable device project outweighed risks were more willing to participate (regression coefficient = 0.28). If individuals felt that others might be helped by their participation in the wearable device project, they were more inclined to participate on average (regression coefficient = 0.24). Table A.13 presents the top non-penalized coefficients.

Table A.13 Results of LASSO regression with regression coefficients ≥0.05[a]

Wearable device		Ketamine infusion	
Predictor	Coefficient estimate	Predictor	Coefficient estimate
How risky is this project?	−0.30	How risky is this project?	−0.39
After weighing the risks and benefits of this research (wearable device), do you find it to be more risky or more beneficial?	0.28	Money composite: Mean of \$100 influence and \$500 influence questions: If you were offered \$[X] one time at the beginning of the study (ketamine infusion), how much more likely would you be to participate?	0.17
How important to your decision is the possibility that others might be helped by your participation in this project (wearable device)?	0.24	How important to your decision is the possibility that others might be helped by your participation in this project (ketamine infusion)?	0.12
Working in the health field: Yes vs. no	0.17	How do the risks of this project (ketamine infusion) compare with the usual risks you live with every day?	−0.12
If you had an illness being studied in this research project (wearable device), how much more likely would you be to participate?	0.10	After weighing the risks and benefits of this research (ketamine infusion), do you find it to be more risky or more beneficial?	0.08
Money composite: Mean of \$100 influence and \$500 influence questions: If you were offered \$[X] one time at the beginning of the study (wearable device), how much more likely would you be to participate?	0.06	How important to your decision is the possibility that you might personally be helped by participating in this project (ketamine infusion)?	0.07
How do the risks of this project (wearable device) compare with the usual risks you live with every day?	−0.05	Race: White vs. non-white	0.06

[a]LASSO selected 14 predictors for wearable device and 12 predictors for ketamine infusion. Only the top 7 predictors are listed here

Ketamine Infusion Research Study

LASSO had selected 12 predictors. The top selected LASSO predictors were *risk*, *risk compared to usual risk*, *money*, and *societal importance*. For every unit increase in risk, there was a 0.39-point decrease in mean participation willingness. Furthermore, individuals who felt the risk of the ketamine infusion study was higher compared to everyday risk were less willing to participate on average (regression coefficient = −0.12). Individuals who were more influenced by a financial incentive were also more generally willing to participate in the ketamine infusion study on average (regression coefficient = 0.17).

Exploratory Relationships between Medical Research Attitudes and Attitudinal/Personality Factors

Exploratory Stigma Analysis

As shown in Table A.14, stigma level ranged from a "little bit" to "quite a bit" (overall range of means = [2.2–3.5]). Highest stigma was reported for "problems with street drugs" for all groups (overall mean(sd) = 3.5 (1.8)) and lowest stigma for "problems with diabetes" (overall mean(sd) = 2.2(1.1)). Within each health group, a similar trend was found; stigma level across health problems significantly differed.

Table A.14 Stigma associated with different health problems by respondent group

–	Stigma ratings by individuals with mental illness and/or substance use disorder	Stigma ratings by individuals with physical illness	Stigma ratings by healthy individuals	Overall	
Stigma Associated with Problems with:	Mean (SD)	Mean (SD)	Mean (SD)	Mean (SD)	*P* value[a]
Street drugs	3.9 (1.6)	3.4 (1.8)	3.1 (1.9)	3.5 (1.8)	0.121
Schizophrenia	3.7 (1.6)	3.1 (1.7)	2.8 (1.8)	3.2 (1.7)	**0.028**
Bipolar disorder	3.7 (1.3)	2.8 (1.4)	2.7 (1.6)	3.1 (1.5)	**0.001**
Alcohol use	3.5 (1.3)	3.0 (1.4)	2.7 (1.5)	3.1 (1.4)	**0.011**
Depression	3.7 (1.2)	2.6 (1.2)	2.7 (1.5)	3 (1.4)	**<0.001**
Anxiety	3.3 (1.1)	2.6 (1.2)	2.4 (1.1)	2.8 (1.2)	**<0.001**
Prescription medication use	3.2 (1.4)	2.8 (1.5)	2.4 (1.3)	2.8 (1.5)	**0.044**
Marijuana use	2.8 (1.3)	2.4 (1.4)	2.2 (1.4)	2.5 (1.4)	0.099
Diabetes	2.4 (1.2)	2.3 (1.2)	1.8 (1.0)	2.2 (1.1)	**0.03**
P value[b]	**<0.001**	**<0.001**	**<0.001**	**<0.001**	–
Stigma Total (possible range: [9, 45])	30.2 (9.0)	24.7 (10.3)	24.0 (11.4)	26.0 (10.6)	**0.001**

Participants were asked: How much stigma do you believe is associated with each of the following health conditions?

Participants rated 9 items on a 5-point Likert scale (i.e., 1 = None, 2 = A little bit, 3 = Some, 4 = Quite a bit, 5 = A lot), and a total score was calculated (possible range: 9 to 45; higher scores reflect higher level of stigma)

[a] *P* value is from 1-way ANOVA

[b] *P* value is from repeated measures ANOVA

Furthermore, stigma levels differed between respondent groups for several health problems (*P* values significant for schizophrenia, bipolar disorder, alcohol use, depression, anxiety, prescription medication use, and diabetes; refer to Table A.14). Individuals with mental illness and/or substance use disorder consistently reported higher levels of stigma as compared to both individuals with physical illness (regression coefficient = −5.5, *P* value = 0.008) and healthy individuals (regression coefficient = −7.3, *P* value = 0.001).

Research Attitudes Based on Gender

In the overall sample, both men and women agreed that society needed to devote more resources to medical research (overall mean(sd) = 4.1(0.9)). They remained neutral about whether they have some responsibility to help others by volunteering for medical research (overall mean(sd) = 3.0(1.1)).

Men and women differed on two different dimensions related to hope in medical research and confidentiality: women agreed significantly more that "medical research will find cures for many major diseases during my [their] lifetime" (mean(sd) = 3.7(0.9) men; mean(sd) = 4.1(0.7) women, *P* value = 0.005) and that "if [they] volunteered for medical research, [their] personal information will be kept private and confident ial"(mean(sd) = 3.4(1.0) men, mean(sd) = 3.8(0.9) women; *P* value = 0.046). Overall medical research attitudes significantly differed by gender (mean(sd) = 25.3(4.5) men, mean(sd) = 26.9(4.2) women; *P* value = 0.033). Refer to Table A.15.

Table A.15 Research attitudes and perceptions by gender

	Men (n = 86)	Women (n = 61)	Overall (N = 147)	
–	Mean (SD)	Mean (SD)	Mean (SD)	*P* value[a]
Society needs to devote more resources to medical research	4.1 (0.9)	4.1 (0.9)	4.1 (0.9)	0.783
I have a positive view about medical research in general	3.8 (0.9)	4 (0.8)	3.9 (0.8)	0.095
Medical research will find cures for many major diseases during my lifetime	**3.7 (0.9)**	**4.1 (0.7)**	**3.9 (0.8)**	**0.005**
Medical research can be trusted to protect the interests of people who take part in their studies	3.7 (0.9)	3.9 (0.8)	3.8 (0.9)	0.103
Participating in medical research is generally safe	3.6 (0.9)	3.8 (0.6)	3.7(0.8)	0.106
If I volunteer for medical research, my personal information will be kept private and confidential	**3.4 (1.0)**	**3.8 (0.9)**	**3.6 (1.0)**	**0.046**
We all have some responsibility to help others by volunteering for medical research	3.0 (1.1)	3.1 (1.1)	3.0(1.1)	0.445
Medical research attitudes Total (range: [7, 35])	**25.3 (4.5)**	**26.9 (4.2)**	**25.9 (4.4)**	**0.033**

Scale: 1 = Strongly disagree, 2 = Disagree, 3 = Neutral, 4 = Agree, 5 = Strongly Agree
[a] *P* value corresponds to 2-sample t-test

Table A.16 Differences in attitudinal and personality factors stratified by gender

–	Men (n = 86)	Women (n = 61)	Overall (N = 147)	
Predictor	Mean (SD)	Mean (SD)	Mean (SD)	*P* value[a]
Altruism	44.92 (10.37)	50.75 (8.95)	47.45 (10.13)	**0.001**
Trust in MR	25.89 (6.95)	28.74 (7.23)	27.15 (7.17)	**0.018**
Stigma	25.25 (10.36)	27.05 (11.19)	25.95 (10.64)	0.318
MRA	25.27 (4.54)	26.85 (4.17)	26.03 (4.49)	**0.033**
TIPI-emotional stability	5.27 (1.45)	4.79 (1.45)	5.09 (1.45)	**0.049**

MR = medical research, *MRA* = medical research attitudes, *TIPI* = ten-item personality index
[a] *P* value corresponds to 2-sample t-test

Women had significantly more trust in medical research as compared to men (mean(sd) = 28.74(7.23) women vs. mean(sd) = 25.89 (6.95) men; *P* value = 0.018), and on average self-reported higher levels of altruism as compared to men (mean(sd) = 50.75(8.95) women vs. mean(sd) = 44.92(10.37) men; *P* value = 0.001). Refer to Table A.16.

Factors Associated with Medical Research Attitudes (Valence Factor)

Refer to Table A.17 for results. Univariate linear regression analysis showed that there was a positive association between *altruism* and medical research attitudes (MRA): for every 1 unit increase in the standardized altruism score, there was a 0.290 unit increase in standardized medical research attitudes score (coefficient estimate = 0.290, *P* value = <0.001). There was also a positive association between *dispositional optimism* and medical research attitudes (coefficient estimate = 0.275, *P* value = 0.001). Further, individuals with more *trust in medical research* also tended to have more positive medical research attitudes (coefficient estimate = 0.668, *P* value <0.001).

Individuals who were more *agreeable* had on average more positive research attitudes (coefficient estimate = 0.349, *P* value <0.001). Individuals who were more *emotionally stable* also had on average more positive research attitudes (coefficient estimate = 0.190, *P* value = 0.019). Furthermore, individuals who were more *open* or more *conscious* also had on average more positive research attitudes (regression coefficient = 0.204, *P* value = 0.012 openness; regression coefficient = 0.185, *P* value = 0.023 consciousness). Refer to Table A.17.

The results of backwards linear regression model revealed that medical research attitudes were positively associated with trust in medical research and negatively associated with altruism (regression coefficients = 0.655 trust in medical research, 0.246 altruism; *P* value <0.001 and, 0.003, respectively). Individuals with higher levels of trust in medical research on average had more positive attitudes towards medical research; individuals with higher altruistic values on average expressed lower levels of trust in medical research. Candidate variable models initially included in this model were altruism, optimism, trust in medical research, stigma, and personality traits (agreeableness, emotional stability, openness, consciousness, and extraversion).

Table A.17 Association between medical research attitudes

Outcome	Predictor	Coefficient estimate	*P* value[a]
Standardized medical research attitudes	Standardized altruism	0.290	**<0.001**
	Standardized LOT-R optimism, Total	0.275	**0.001**
	Standardized trust in MR	0.668	**<0.001**
	Standardized stigma	−0.121	0.142
	Standardized TIPI- agreeableness	0.349	**<0.001**
	Standardized TIPI- emotional stability	0.190	**0.019**
	Standardized TIPI- openness	0.204	**0.012**
	Standardized TIPI- consciousness	0.185	**0.023**
	Standardized TIPI- extraversion	0.063	0.444

LOT-R = life orientation test-revised, *MR* = medical research, *TIPI* = ten-item personality index
[a] *P* values correspond to univariate linear regression

Perceived Protectiveness and Participation Willingness in the Context of Five Safeguard Procedures

Methods

Variables

The survey instrument was used as described in the General Methods/Results section. In particular, this analysis uses the following set of survey items: (1) *demographics*; (2) optimism; and (3) *safeguard procedures: perceived protectiveness and influence on willingness to participate.*

For this analysis, *perceived protectiveness of safeguard procedures and influence on willingness to participate* were analyzed in regard to five common human subjects safeguard procedures: *IRB review*, *DSMB oversight*, *informed consent*, designated decision maker (*alternative decision maker*), and special code number (*confidential coding*). After a description of each of the five safeguards, respondents were asked about their perceptions of the (a) *perceived protectiveness* of the safeguards and (b) *participation willingness* under the influence of the safeguard.

Following each safeguard summary, individuals were asked one item about the *perceived protectiveness* of the safeguard: "How much does [Safeguard Procedure X] protect people who participate in research?" (rated on a 5-point Likert scale, 1 = "Does not protect at all" to 5 = "Highly protects"). Individuals were asked to rate one item about their *participation willingness* related to each safeguard: "If you knew that a project was [protected by Safeguard Procedure X], how much would it influence your willingness to participate in the project?" (rated on a 5-point Likert scale, 1 = "Does not influence my willingness to participate at all" to 5 = "I'd be very much willing to participate"). See Table A.3 for safeguard descriptions and Table A.4 for questions.

Furthermore, a different but overlapping set of variables were used for the LASSO models ("LASSO Results" in the Results section further below). These include the initial set of covariates as described above and other measures such as select PROMIS measures and the 10 Item Personality Index (TIPI). Refer to Table A.6 for a description of this full set of covariates.

Statistical Analysis

For this portion of the project, perceived protectiveness of safeguard and influence of safeguard on willingness to participate were analyzed using the 1-way analysis of variance (ANOVA) test to evaluate between-group differences.

To address the secondary analysis, generalized estimating equations (GEE) with the Gaussian link and unstructured correlation structure were used. Participation willingness was a repeated measure since each individual rated their participation willingness for each safeguard (5 ratings per individual). GEE was used to model the relationship between participation willingness and the perceived protectiveness of safeguards, adjusting for potential confounders (i.e., gender, education, and dichotomized optimism). Education was categorized as: attainment of bachelor's degree versus no attainment of bachelor's degree. Since there were five types of safeguards considered, we used four indicator variables in the model, with IRB as the reference safeguard. Furthermore, an interaction term between safeguard procedure and perceived protectiveness was also included to assess whether the relationship between protectiveness and willingness to participate differed by safeguard.

For the exploratory analysis, least absolute shrinkage and selection operator (LASSO) was used. A LASSO model was performed for each safeguard to model the relationship between participation willingness and perceived protectiveness with respect to that safeguard. For more information about the GEE and LASSO methods, refer to the Description of General Analytic Methods section above.

Descriptive Results

Perceived Protectiveness Associated with the Five Safeguards, by Health Group

As shown in Table A.18, the overall sample rated all five of the research safeguard procedures as "somewhat" to "quite a bit protective" (range of means = [3.2–4.0]). Overall, respondents viewed the IRB review and DSMB oversight safeguards as the most protective (mean(sd) = 4.0 (0.8) and 4.0(0.9) respectively), and the alternative decision maker safeguard as the least protective (mean(sd) = 3.2(1.0)). Within each health group, a similar pattern emerged.

Table A.18 Perceptions of protectiveness by participant group and safeguard procedure type

	Mental illness	Physical illness	Healthy	Overall	
–	Mean (SD)	Mean (SD)	Mean (SD)	Mean (SD)	*P* value[a]
Alternative decision maker	2.9 (1.1)	3.4 (1.1)	3.3 (0.8)	3.2 (1.0)	**0.02**
Informed consent	3.2 (1.1)	3.6 (1.3)	3.7 (1.0)	3.5 (1.1)	0.06
Confidential coding	3.5 (1.1)	3.8 (1.0)	3.9 (1.0)	3.7 (1.0)	0.14
DSMB oversight	3.8 (0.9)	4.0 (0.9)	4.1 (0.8)	4.0 (0.9)	0.29
IRB review	4.0 (0.8)	4.0 (0.9)	4.1 (0.8)	4.0 (0.8)	0.81
All safeguards	3.5 (1.0)	3.8 (1.1)	3.8 (0.9)	3.7 (1.0)	–

Participants were asked:
"How much does this [Safeguard Procedure X] protect people who participate in research?"
Scale: 1 = Does not protect at all, 2 = Only protects a little bit, 3 = Somewhat protects, 4 = Protects quite a bit, 5 = Highly protects
DSMB = data safety monitoring board, *IRB* = institutional review board
[a] *P* values correspond to 1-way ANOVA test

Respondents with physical illness and in good health endorsed higher ratings of perceived protectiveness for all safeguards as compared to respondents with mental illness and/or substance use disorder. In particular, the alternative decision maker safeguard was viewed as significantly less protective by individuals with mental illness as compared to individuals with physical illness (mean = 2.9 and 3.4 respectively, *P* value = 0.02).

Participation Willingness Associated with the Five Safeguards, by Health Group

Respondents' willingness to participate in projects that included each safeguard also varied by safeguard procedure (range of means = [2.9–3.8]). Overall, respondents expressed the greatest level of willingness to participate in projects where the IRB review safeguard was present (mean(sd) = 3.9(1.0)). They expressed lower levels of willingness to participate in projects where the alternative decision maker safeguard was present (mean(sd) = 2.9 (1.3)). Within each health group, similar patterns were found.

Across all safeguards, respondents with physical illness and in good health reported higher levels of willingness to participate as compared to respondents with mental illness. On average, respondents with physical illness expressed higher levels of participation willingness as compared to respondents with mental illness regarding research projects that made use of the alternative decision maker safeguard (means = 3.2 and 2.4 respectively, *P* value = 0.006). Furthermore, respondents with mental illness and/or substance use disorder expressed lower levels of participation willingness in research using the informed consent safeguard when compared to respondents with either physical illness or in good health (means = 3.0 (mental illness) vs. 3.5 (for both physical illness and healthy), *P* value = 0.05). Refer to Table A.19.

Table A.19 Perceptions of participation willingness by participant group and safeguard procedure type

	Mental illness	Physical illness	Healthy	Overall	
–	Mean (SD)	Mean (SD)	Mean (SD)	Mean (SD)	*P* value[a]
Alternative decision maker	2.4 (1.2)	3.2 (1.4)	2.9 (1.2)	2.9(1.3)	0.006
Informed consent	3.0 (1.2)	3.5 (1.4)	3.5 (1.1)	3.3 (1.3)	0.05
Confidential coding	3.2 (1.2)	3.7 (1.2)	3.7 (1.0)	3.5 (1.2)	0.06
DSMB oversight	3.6 (1.2)	3.9 (1.1)	4.0 (1.0)	3.8 (1.1)	0.19
IRB review	3.7 (1.1)	3.9 (1.0)	4.0 (0.9)	3.9 (1.0)	0.46
All safeguards	3.2 (1.3)	3.6 (1.2)	3.6 (1.1)	3.5 (1.2)	–

Participants were asked:
"If you knew that a project was approved by/watched over/involved this/was using a [Safeguard Procedure X], how much would it influence your willingness to participate in the project?"
Scale: 1 = Does not influence my willingness to participate at all, 2 = I'd be a little bit more willing to participate, 3 = I'd be somewhat more willing to participate, 4 = I'd be quite a bit more willing to participate, 5 = I'd be very much more willing to participate
DSMB = data safety monitoring board, *IRB* = institutional review board
[a] *P* values correspond to 1-way ANOVA test

GEE Results: Associations between Willingness to Participate and Perceptions of Protectiveness

Association between Type of Safeguard and Stated Willingness to Participate in Research with Safeguards

Controlling for confounders, there was an association between the type of safeguard used in research (i.e., confidential coding, alternative decision maker, and informed consent safeguards) and the respondents' stated willingness to participate in research that includes the safeguard.

Participation willingness in the presence of the IRB review safeguard was consistently at greater levels relative to participation willingness in the presence of all other safeguards, including confidential coding, alternative decision makers, and informed consent (regression coefficient = −1.1, *P* value <0.001; −1.6, *P* value <0.001; −0.9, *P* value = 0.004, respectively).

Respondents' stated willingness to enroll in research was positively associated with their perceptions of the protectiveness of safeguards (regression coefficient = 0.5, *P* value <0.001), adjusting for illness type, safeguard type, gender, education, and optimism level. Moreover, the positive association between respondents' willingness to participate and perceived protectiveness varied by the type of safeguard (Fig. A.1; of IRB review, regression coefficient = 0.5, *P* value <0.001; confidential coding, regression coefficient = 0.8, *P* value <0.001; alternative decision maker, regression coefficient = 0.9, *P* value <0.001; DSMB oversight, regression coefficient = 0.6, *P* value <0.001; informed consent, regression coefficient = 0.7, *P* value <0.001).

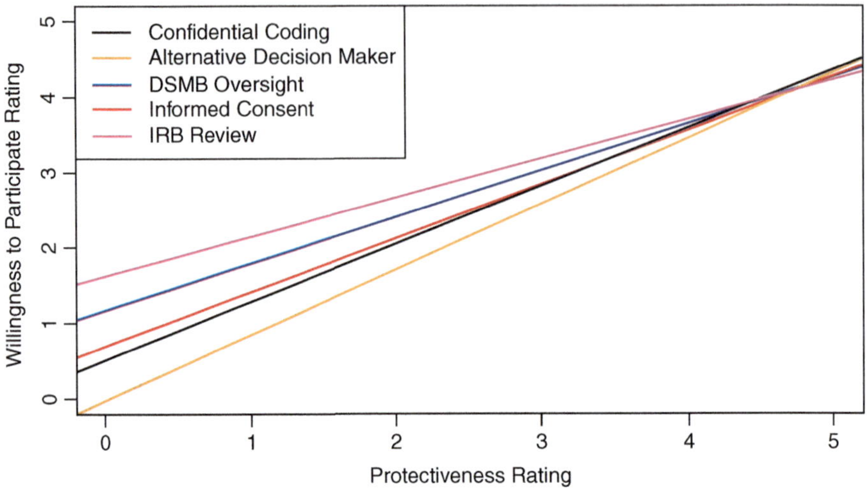

Fig. A.1 Willingness to participate positively correlates with perceived protectiveness and varies by safeguard procedure type. *DSMB* = data safety monitoring board, *IRB* = institutional review board. Protectiveness ratings ranged from 1 = "Does not protect at all" to 5 = "Highly protects"; Willingness to participate ratings ranged from 1 = "Not at all willing" to 5 = "Extremely willing"

Association Between Respondents' Stated Willingness to Participate in Research with Safeguards and Respondents' Demographic Characteristics

Our analysis indicated an association between level of optimism and stated willingness to participate. The average level expressed for participation willingness was 0.3 points higher among respondents who expressed strong levels of optimism as compared to those who endorsed weak to moderate levels of optimism (regression coefficient = 0.3, *P* value = 0.001). Education, gender, and health status were not significantly associated with participation willingness.

LASSO Results: Significant Predictors of Participation Willingness by Safeguard Type

IRB Review Safeguard

The LASSO selected the following predictors: perceived protectiveness, race (white vs. non-white), RAQ score, and Trust in Medical Research score. White respondents had a higher mean participation willingness rating as compared to non-white respondents (regression coefficient = 0.2). Positive associations were found for RAQ score, Trust in Medical Research score, and protectiveness. Table A.20 presents the estimated regression coefficients.

Table A.20 LASSO Results: predictors of participation willingness selected for each safeguard procedure type

Safeguard	Predictor	Coefficient estimate[a]
IRB	IRB perceived protectiveness	0.4
	Race: White vs. non-white	0.1
	RAQ Total	0.04
	Trust in MR Total	0.002
DSMB	DSMB perceived protectiveness	0.8
	Trust in MR Total	0.0002
Informed consent	Informed consent perceived protectiveness	0.8
	Race: White vs. non-white	0.05
	TIPI- openness to experiences	0.04
	TIPI- agreeableness	0.02
	PROMIS- cognitive	−0.006
	LOT-R optimism dichotomized	0.003
	RAQ Total	0.003
	Davis empathy scale total	0.001
	Stigma scale Total	−0.0003
Alternative decision maker	Alternative decision maker perceived protectiveness	0.9
	LOT-R optimism dichotomized	0.4
	Physical illness versus mental illness	0.1
	PROMIS depression	0.01
	RAQ Total	0.005
	Davis empathy scale total	0.003
	PROMIS self-efficacy	−0.003
	Stigma scale total	−0.001
Confidential coding	Confidential coding perceived protectiveness	0.8
	Race: White vs. non-white	0.2
	LOT-R optimism dichotomized	0.06
	RAQ total	0.009
	PROMIS depression	0.0002

DSMB = data safety monitoring board, *IRB* = institutional review board, *LOT-R* = life orientation test-revised, *MR* = medical research, *PROMIS* = patient-reported outcomes measurement information system, *RAQ* = research attitudes questionnaire, *TIPI* = ten-item personality inventory

[a]Regression coefficients are included for reference, but statistical significance is not of importance; rather because it is a variable selection technique via shrinkage, it is important to evaluate variables that were not shrunk

DSMB Oversight Safeguard

The relevant predictors selected by LASSO for mean participation willingness in the presence of the DSMB oversight safeguard were perceived protectiveness and Trust in Medical Research score.

Informed Consent Safeguard

The relevant predictors selected by LASSO for mean participation willingness in the presence of the informed consent safeguard were perceived protectiveness, race (white vs. non-white), Openness to Experience score, Agreeableness score, PROMIS- Cognitive score, LOT-R dichotomized Optimism score, RAQ score, Davis Empathy Scale score, and Stigma Scale score.

Alternative Decision Maker Safeguard

The relevant predictors selected by LASSO for mean participation willingness in the presence of the alternative decision maker safeguard were perceived protectiveness, LOT-R dichotomized optimism, health group (physical illness vs. mental illness), PROMIS- Depression score, RAQ score, Davis Empathy Scale score, PROMIS- Self-Efficacy score, and Stigma Scale score. Respondents with strong levels of optimism had on average, a 0.4 point higher mean participation willingness score, as compared to respondents with weak to moderate levels of optimism (regression coefficient = 0.4).

Confidential Coding Safeguard

The relevant predictors selected by LASSO for mean participation willingness in the presence of the confidential coding safeguard were perceived protectiveness, race (white vs. non-white), dichotomized optimism, RAQ score, and PROMIS-Depression score. White respondents, on average, had a 0.2 point higher mean participation willingness score than non-white respondents (regression coefficient = 0.2). Respondents with strong levels of optimism had on average, a 0.06 point higher participation willingness score as compared to respondents with weak to moderate levels of optimism (regression coefficient = 0.06).

Limitations

This study has several limitations. First, as recruitment was conducted online, we relied on self-reported illness status. However, for the screening survey, respondents were unaware of the inclusion criteria for the full survey, thereby mitigating the potential for biased self-reports. Second, the hypothetical nature of the questions limits the generalizability of the findings. In addition, as this was a pilot study with the primary purpose of evaluating recruitment feasibility and face validity of obtained survey responses, the sample size was modest. Further work by our team will utilize a much larger sample size to assess a range of perspectives regarding ethical aspects of psychiatric research. Despite these limitations, these findings lend support for the methodology used here, i.e., recruitment of individuals living with mental illness (in this study, self-reported mood disorders) as well as individuals in self-reported overall good health in order to evaluate a variety of ethically salient perspectives on research.

Conclusion

A number of noteworthy findings emerged from this analysis of valence factors. The examination of valence factors yielded insights that confirmed prior work [9, 10]. Other findings were unexpected and necessitate further replication in other samples. Taken together, several valence factors emerged as significant factors associated with participation willingness. Some well-studied factors, such as perceived risk, did not vary by personal attributes (e.g., self-reported illness) but depended on one's level of self-reported trust in medical research. Other understudied valence factors (e.g., perceived protectiveness) did not vary as expected by self-reported illness and predicted high willingness to participate in research. Our findings offer new insights regarding how information regarding safeguards can be used to bolster public trust in the context of research. A number of valence factors (e.g., altruism, stigma) did appear to vary across other attributes such as gender and self-reported illness. Table A.21 includes more examples of findings in support of study hypotheses regarding valence factors.

Our hope is that readers and researchers consider how decisions to enroll in research extend beyond a study's illness- and demographics-related inclusion and exclusion criteria. Decisions to enroll in research implicate a variety of valence factors, e.g., individual dispositional and attitudinal traits that may systematically differ across sociodemographic characteristics. Focusing on valence factors provides an opportunity to enrich engagement and dialogue for community-based research and public health outreach efforts.

Table A.21 Emerging evidence from the pilot study in support of study hypotheses

Hypotheses	Evidence in support of hypotheses
Measures of valence factors (e.g., perceptions of risk, stigma, and perspectives regarding medical research) will vary by study attributes (e.g., innovative project type) and/or personal attributes (e.g., self-reported illness)	Perceptions of risk • Perceptions of risk regarding the projects that involve wearable devices or ketamine infusion do not vary by self-reported illness (mental illness and/or substance use disorder, physical illness, and good health), regardless of how the questions were asked. Stigma • Perceived stigma varied by self-reported health status. Individuals with mental illness and/or substance use disorder consistently reported higher levels of stigma as compared to participants with physical illness and in good health. Perspectives regarding medical research • Men and women differed in their expressed hope in medical research and attitudes towards medical research.
Respondents will express greater willingness to participate when they express higher levels of *positive valence factors* **such as** perceived helpfulness of research, optimism, and trust in medical research	Dispositional optimism • Respondents with highly optimistic views expressed greater participation likelihood than respondents with weak to moderate optimistic views, controlling for all other factors in the presence of influences (e.g., if one had an illness being, if someone in life wanted. If one was offered $100 one time at the beginning of the study). Perceived risk • Participation willingness for a wearable device research study was negatively associated with perceived risk. Survey respondents who felt that benefits of the wearable device study outweighed risks were more willing to participate compared to those who felt risks outweighed benefits. • Participation willingness in a ketamine infusion research study was also negatively associated with risk perception. Furthermore, individuals who felt the risk of the ketamine study was higher compared to everyday risk were less willing to participate on average. Individuals who were more influenced by a financial incentive were also more generally willing to participate in the ketamine infusion study on average. Perceived helpfulness of research • Respondents were more inclined to participate in a wearable device study when they felt that others might be helped by their participation in the wearable device study. Other influencing factors • Respondents were more likely to participate in the wearable device project than in the ketamine infusion project, even in the presence of influencing factors such as being offered money (e.g., $100 or $500) or if "someone important in your life wanted you to." • Scores on research attitudes questionnaire, Trust in Medical Research, openness to experience subscale, life orientation test-revised optimism subscale, and the Davis empathy scale were among significant predictors of participation willingness. • Perceptions of safeguards were significant predictors of participation willingness.

References

1. Roberts LW, Warner TD, Brody JL, Roberts B, Lauriello J, Lyketsos C. Patient and psychiatrist ratings of hypothetical schizophrenia research protocols: assessment of harm potential and factors influencing participation decisions. Am J Psychiatry. 2002;159(4):573–84.
2. Roberts L, Green Hammond K, Warner T, Lewis R. Influence of ethical safeguards on research participation: comparison of perspectives of people with schizophrenia and psychiatrists. Am J Psychiatry. 2004;161(12):2309–11.
3. Roberts LW, Kim JP. Giving voice to study volunteers: comparing views of mentally ill, physically ill, and healthy protocol participants on ethical aspects of clinical research. J Psychiatr Res. 2014;56(1):90–7.
4. Roberts LW, Kim JP. Healthy individuals' perspectives on clinical research protocols and influences on enrollment decisions. AJOB Empir Bioeth. 2017;8(2):1–10.
5. Mainous AG, Smith DW, Geesey ME, Tilley BC. Development of a measure to assess patient Trust in Medical Researchers. Ann Fam Med. 2006;4(3):247–52.
6. Rubright JD, Cary MS, Karlawish JH, Kim SYH. Measuring how people view biomedical research: reliability and validity analysis of the research attitudes questionnaire. J Empir Res Hum Res Ethics. 2011;6(1):63–8.
7. Segerstrom SC, Evans DR, Eisenlohr-Moul TA. Optimism and pessimism dimensions in the life orientation test-revised: method and meaning. J Res Personal. 2011;45(1):126–9.
8. Tibshirani R. Regression shrinkage and selection via the lasso. J R Stat Soc Ser B Methodol. 1996;58(1):267–88.
9. Kim JP, Tsungmey T, Rostami M, Mondal S, Kasun M, Roberts LW. Factors influencing perceived helpfulness and participation in innovative research: a pilot study of individuals with and without mood symptoms. Ethics Behav. 2021;0(0):1–17.
10. Kim JP, Ryan K, Tsungmey T, Kasun M, Roberts WA, Dunn LB, et al. Perceived protectiveness of research safeguards and influences on willingness to participate in research: a novel MTurk pilot study. J Psychiatr Res. 2021;138:200–6.

Index

L. W. Roberts (ed.), *Ethics and Clinical Neuroinnovation*,
https://doi.org/10.1007/978-3-031-14339-7

GPSR Compliance

The European Union's (EU) General Product Safety Regulation (GPSR) is a set of rules that requires consumer products to be safe and our obligations to ensure this.

If you have any concerns about our products, you can contact us on ProductSafety@springernature.com

In case Publisher is established outside the EU, the EU authorized representative is:

Springer Nature Customer Service Center GmbH
Europaplatz 3
69115 Heidelberg, Germany

Batch number: 10732076

Printed by Printforce, the Netherlands